W9-CCK-582

Cell Interactions In Atherosclerosis

Edited by

Horst Robenek, Ph.D.
Professor
Institute for Arteriosclerosis Research
Department of Cell Biology
and Ultrastructure Research
University of Münster
Münster, Germany

and

Nicholas J. Severs, Ph.D.
Reader
Department of Cardiac Medicine
National Heart and Lung Institute
London, England

CRC Press
Boca Raton Ann Arbor London Tokyo

Library of Congress Cataloging-in-Publication Data

Cell interactions in atherosclerosis / editors, Horst Robenek and Nicholas John Severs.
p. cm.
Includes bibliographical references and index.
ISBN 0-8493-5505-2
1. Atherosclerosis—Pathophysiology. 2. Atherosclerotic plaque. 3. Vascular endothelium. I. Robenek, Horst. II. Severs, Nicholas John.
[DNLM: 1. Atherosclerosis—physiopathology. 2. Cell Communication—physiology. 3. Extracellular Matrix—physiology. 4. Lipoproteins—physiology. WG 550 C3925]
RC692.C42 1992
616.1'3607—dc20
DNLM/DLC
for Library of Congress 92-20114
CIP

Direct all inquiries to CRC Press, Inc., 2000 Corporate Blvd., N. W., Boca Raton, Florida, 33431.

International Standard Book Number 0-8493-5505-2

Library of Congress Card Number 92-20114
Printed in the United States 1 2 3 4 5 6 7 8 9 0
Printed on acid-free paper

PREFACE

Atherosclerosis, the pathological process that obstructs arteries, causing coronary heart disease and stroke, is the leading cause of death and disability in most industrialized countries of the developed world. The disease is not new — examination of the arterial remains of Egyptian and Grecian mummies suggests its presence in human populations dating back to 1000 years B.C. and earlier. However, it is only during the present century that the prevalence of the disease has, in many parts of the world, reached epidemic proportions. The reasons for this increase are not entirely clear, and despite advances in understanding the pathogenesis of the disease, the recognition of risk factors, and improvements in medical treatment, atherosclerosis remains a major health problem to the individual and an intractable socioeconomic burden to the community.

Atherosclerosis is often viewed as a product of environmental and genetic interaction. The effects of this interaction are translated into aberrant behavior in arterial cells, and this in turn leads to the formation of raised lesions on the internal surface of the artery. An understanding of the nature of atherosclerosis thus depends on a detailed knowledge of the nature of the cells that make up the arterial wall, how their coordinated activities control normal function of the healthy artery, and how their interactions in unfavorable conditions can culminate in symptomatic disease. Fundamental research in the field of the cell biology of atherosclerosis provides an essential part of this understanding, and thus contributes to the body of knowledge from which rational strategies for reducing the incidence and effects of the disease may ultimately be developed. Apart from such potential practical applications, the cell biology of atherosclerosis also poses problems that to many are fascinating scientific challenges in their own right.

The development of new techniques in cell biology has been an important stimulus to progress in atherosclerosis research. The scope of these techniques has expanded significantly in recent years, particularly in the ways cells can be manipulated in culture to dissect out different aspects of their functions and interactions. At the same time, morphological methods for detecting and visualizing structural components of functional importance and identifying their chemical makeup have undergone a revolution. And with the advent of recombinant DNA technology, the localization of specific messenger RNAs within cytochemically identified cells in tissue sections has now become possible. Through these and other developments, new opportunities for studying the fundamental cell processes involved in atherosclerosis have been opened up. To bring the results of these advances to the attention of the wider scientific and medical audience, we have assembled in this book a collection of topics of current interest on the theme of cell interactions in atherosclerosis. To make this material as accessible as possible to those entering the field from all disciplines, the first chapter provides a general review of the structure and function of the cells of the arterial wall. The three principal types of arterial cell — the endothelial cell, the smooth muscle cell, and the macrophage — are introduced, and their hypothesized roles in the pathogenesis of atherosclerosis discussed. This is intended to provide the general background from which the informed but nonspecialist reader can move directly to any of the specialized topics elsewhere in the book that may take their interest.

The chapters are loosely grouped into general themes. Chapters 2 and 3 continue with the morphological slant of Chapter 1. A spectacular revelation of the three-dimensional structure of the inner layer of the artery, the site at which lipid is deposited when atherosclerotic lesions develop, is given in Chapter 2. Chapter 3 highlights the parallels between atherosclerosis and the cellular changes that occur in the arteries of transplanted organs, which are prone to a form of accelerated atherosclerosis.

Chapters 4 to 7 are centered on the extracellular matrix and growth factors. The extracellular matrix is the material that cells secrete around themselves; components of the matrix determine physical properties such as tensile strength and elasticity in the normal vessel wall, and enhanced production of these components accounts for bulk growth of the atherosclerotic lesion. Chapter 4 includes a detailed review on collagens — the most abundant component of the lesion — and presents evidence that macrophages play a role in activating the synthesis of collagen by smooth muscle cells. Matrix components have profound effects on the behavior and function of the cells they surround, and are now believed to influence many of the altered cellular properties involved in atherosclerosis. Elegant experiments, detailing how the presence of a collagen lattice alters the proliferative and synthetic activities of cultured smooth muscle cells and their responses to growth factors, are presented in Chapter 5. This theme is continued in Chapter 6, with a succinct but comprehensive discussion on how a complex interplay between matrix components (e.g., collagens and fibronectin), growth factors, and other agents in the blood and interstitial fluid modulates migratory activity in endothelial cells and smooth muscle cells. These processes underlie the responses of vascular cells to arterial injury of the type sustained when invasive medical procedures are carried out to relieve obstructive atherosclerosis; all too often, the cellular responses provoked by such procedures lead to reobstruction of the artery. Among other components of the matrix that influence properties of the vascular wall are proteoglycans and thrombospondin, and new approaches for detecting and mapping these components in the arterial wall are presented in Chapter 7.

Chapters 8 to 10 are centered on lipoproteins, the blood-borne particles that transport cholesterol to and from cells. The way the different cell types of the arterial wall interact with lipoproteins by means of their specific cell surface receptors is the basis for understanding the link between cholesterol and atherosclerosis. Chapter 8 gives a detailed exposition of oxidative modification of low-density lipoprotein, a critical process which alters the lipoprotein's properties in such a way that macrophages avidly accumulate massive quantities of cholesterol in the arterial wall. The discovery of this process and its effects on vascular cells is pivotal to the lipid hypothesis of atherogenesis in its modern form. Chapter 9 documents how cells that have been isolated from early lesions respond to "atherogenic" lipoproteins when cultured; this approach allows hypotheses derived from studies on cells cultured from healthy arteries and blood to be related convincingly to the cells of the atherosclerotic lesion. Through advances in cytochemical techniques, it has become possible to tag lipoproteins, so that the properties of different lipoprotein receptors can be examined at the ultrastructural level, and the pathways of internalized lipoproteins traced with precision through the cell. Such visual demonstrations of lipoprotein receptors in action, set in the general context of how cellular activities determine cholesterol levels in the artery wall, is the theme of the final chapter.

We wish to thank all our contributors for their collaboration in helping us put together this volume, and the staff at CRC Press for their cooperation in preparing it for publication.

We are particularly indebted to Monika Rohe whose organizational and secretarial skills have been of such great help throughout editorial preparation.

Horst Robenek
Nicholas J. Severs

THE EDITORS

Horst Robenek, Ph.D., is a University Professor of Cell Biology and Head of the Department of Cell Biology and Ultrastructure Research at the Institute for Arteriosclerosis Research, University of Münster, Münster, Germany.

Dr. Robenek obtained his doctorate in 1976 from the Biology Faculty at the University of Münster. He then served as a Scientific Assistant at the Institute of Medical Cytobiology at the Medical Faculty of the University of Münster. In 1984 he received the degree of Privatdozent and the Venia Legendi for Medical Cell Biology. In 1987 he was awarded the German National Heisenberg Fellowship in Science. In 1989 he was appointed to a Professorship at the University of Münster and to the position of Head of the Department of Cell Biology and Ultrastructure Research at the Institute for Arteriosclerosis Research.

Dr. Robenek is a member of the American Society for Cell Biology, the German Society for Cell Biology, the German Society for Electron Microscopy, and the German Arteriosclerosis Society.

Dr. Robenek has been the recipient of research grants from the German National Research Council and industry. He is the Vice Chairman of the Special Collaborative Program entitled "Pathomechanisms of Cellular Interactions" funded by the German National Research Council. He serves on the editorial boards of the *European Journal of Cell Biology* and the *Journal of Nutrition, Metabolism and Cardiovascular Diseases.* He has organized a number of national and international workshops and scientific conferences in the area of cytochemistry and cell biology.

Dr. Robenek has published more than 150 research papers in leading journals of cell biology, cytochemistry, and electron microscopy, and has presented over 50 invited lectures at international and national meetings. He is also the author of numerous invited review articles. His current research interests include lipoprotein metabolism and cellular cholesterol homeostasis.

Nicholas J. Severs, Ph.D., is Reader in Cell Biology in the Department of Cardiac Medicine, National Heart and Lung Institute (formerly, Cardiothoracic Institute), University of London, London, England.

Dr. Severs received his B.Sc. and Ph.D. degrees from the University of London in 1972 and 1976, respectively. After doing post-doctoral work as a Temporary Lecturer in the Cell Pathology Unit, School of Pathology, Middlesex Hospital Medical School, he was appointed as Lecturer in the Department of Cardiac Medicine at the Cardiothoracic Institute in 1979. He became a Senior Lecturer in 1986, and Reader in Cell Biology in 1992.

Dr. Severs is a member of the British Society for Cell Biology, the Royal Microscopical Society, the British Society for Cardiovascular Research and the

International Society for Heart Research. He has served on the editorial boards of the *Journal of Cell Science* and *Cardioscience*.

Dr. Severs has been the recipient of research grants from the British Heart Foundation, Medical Research Council and Wellcome Trust. He has published more than 90 research papers and review articles, and was awarded the Robert Feulgen Prize 1992 of the International Association of Histochemists in recognition of his work using immunocytochemical techniques in the study of heart disease. His current research interests center on membrane interactions in cardiovascular cell function.

CONTRIBUTORS

Leonard Bell
Section of Cardiovascular Medicine
Pathology Department
Yale University, School of Medicine
New Haven, CT

Joy S. Frank
Department of Medicine
and Physiology
Cardiovascular Research Laboratories
UCLA School of Medicine
Los Angeles, CA

Olli Jaakkola
Department of Biomedical Sciences
University of Tampere
Tampere, Finland

Elisabeth Jaeger
Institute for Arteriosclerosis
Research
Department of Biochemistry
University of Münster
Münster, Germany

Joseph A. Madri
Section of Cardiovascular Medicine
Pathology Department
Yale University, School of Medicine
New Haven, CT

Patricia F.E.M. Nievelstein-Post
Department of Medicine
and Physiology
Cardiovascular Research Laboratories
UCLA School of Medicine
Los Angeles, CA

Wulf Palinski
Department of Medicine
University of California, San Diego
UCSD School of Medicine
La Jolla, CA

Jürgen Rauterberg
Institute for Arteriosclerosis Research
Department of Biochemistry
University of Münster
Münster, Germany

Horst Robenek
Institute for Arterioslerosis Research
Department of Cell Biology and
Ultrastructure Research
University of Münster
Münster, Germany

Albert Roessner
Gerhard Domagk Institute for
Pathology
University of Münster
Münster, Germany

Nicholas J. Severs
Department of Cardiac Medicine
National Heart and Lung Institute
London, England

Michael Thie
Institute for Arteriosclerosis
Research
Department of Cell Biology and
Ultrastructure Research
University of Münster
Münster, Germany

Wolfgang Völker
Institute for Arteriosclerosis Research
Department of Cell Biology and
Ultrastructure Research
University of Münster
Münster, Germany

Ekkehard Vollmer
Gerhard Domagk Institute for
Pathology
University of Münster
Münster, Germany

TABLE OF CONTENTS

Chapter 1

CONSTITUENTS OF THE ARTERIAL WALL AND ATHEROSCLEROTIC PLAQUE: AN INTRODUCTION TO ATHEROSCLEROSIS

Nicholas J. Severs and Horst Robenek

TABLE OF CONTENTS

ISBN 0-8493-5505-2

I. BACKGROUND AND INTRODUCTION

Coronary heart disease is the leading cause of mortality and morbidity in most industrialized countries.[1-3] The conditions known as angina, ischemic heart disease, myocardial infarction, and sudden (ischemic) cardiac death are all late and overlapping manifestations of the same underlying disease process — atherosclerosis of the coronary arteries. Put simply, atherosclerosis is the process by which portions of the inner layer of an artery become thickened with fibromuscular material and lipid, ultimately causing narrowing or occlusion of the vessel lumen.[4-6] The earlier stages of this disease process are silent, clinical symptoms usually only becoming apparent after decades of lesion progression. When advanced, atherosclerotic lesions (plaques) in the coronary artery restrict blood supply to the portion of heart muscle served by the diseased vessel, causing transient ischemic episodes that are commonly experienced as angina. Though this condition is associated with increased risk of myocardial infarction, it may, if the lesions involved are predominantly fibromuscular, persist in stable form for many years. Atherosclerotic plaques are, however, prone to fissure or rupture, especially when a soft deformable core of lipid is present, and this can precipitate thrombosis in the vessel lumen — an acute, life-threatening event that can result in unstable angina, myocardial infarction, and ventricular fibrillation.[7-9] Although atherosclerosis is particularly prevalent in coronary arteries, it also commonly affects the aorta (especially the abdominal segment) and the carotid, cerebral, iliac, and femoral arteries, forming the underlying pathology for aneurysm and dissection of the aorta, stroke, and intermittent claudication. However, not all arteries are equally prone to disease; the renal, brachial, and pulmonary arteries, for example, are seldom seriously affected.

Because of the epidemic proportions of the coronary heart disease problem, the research effort directed at understanding, diagnosing, and treating the disease is vast, spanning virtually all scientific disciplines. As a result, much has been learned about the nature of atherosclerosis, its complications, and their role in symptomatic disease.[10] The impact of this information explosion on the coronary heart disease problem has, however, been disappointingly small. A marked reduction in disease incidence has been recorded in a number of countries (notably the U.S. and New Zealand),[3,11] but the relative contributions of improvements in medical intervention, life style changes, and other factors to this reduction have been difficult to assess,[12] and overall incidence still remains unacceptably high. Meanwhile, the incidence of the disease in a number of other countries (e.g., the U.K. and Germany) has shown only minor decline or little change, and in those of Eastern Europe it has continued to climb.[3,13,14] This state of affairs stems, in part, from the difficulty in implementing, on a sufficiently large scale, the full benefits of existing knowledge, and, in part, from the realization that despite the intensity of the research effort, the precise mechanisms of the disease remain imperfectly understood. From the current standpoint, the pathogenesis of atherosclerosis is best viewed as a multifactorial process in which numerous factors may interact in different ways and assume different degrees of importance in different individuals. Consequently, a number of different, but overlapping, pathways of pathogenesis may be involved.

The fact that identified "risk factors" fail to account in full for the observed incidence of disease highlights awareness that many gaps in our understanding of the underlying causes still remain. And as our knowledge has expanded, further new questions and problems have constantly emerged. This is as true at the practical level of therapy as it is at the fundamental scientific level. For example, the newer invasive interventions (e.g., atherectomy, stenting, laser and thermal ablation) have met with the same problem of a high incidence of short-term restenosis[15,16] that limits the success of the established technique of balloon angioplasty,[17-21] and accelerated atherosclerosis in the longer term is a major factor limiting survival following bypass grafting and cardiac transplantation.[22-28] Another topical issue, brought to the fore by recent serial angiography studies,[29,30] is the possibility of lesion regression; here, a rational basis for developing therapeutic strategies would be assisted by knowing which plaque constituents at which stage of lesion development can be made to regress, and how. These are but two of the practical issues in which answers may come from understanding the interactions and responses of the cells involved.

The cell biology of atherosclerosis is, however, concerned with more than answers to specific questions like these. A coherent framework for understanding atherosclerosis must embrace all relevant fields, from epidemiology to molecular genetics, and an understanding of what is happening at the cellular level underpins information from all these sources. Risk factors such as abnormal serum lipids, hypertension, and cigarette smoking[10,31-33] exert their effects as "extrinsic" factors that act upon and induce functional alterations in the cells of the vessel wall, pharmacological effects are mediated by the cell's membranes, and the effects of inherited deficiencies in lipoprotein metabolism[33-35] manifest themselves as abnormalities of cell behavior and function. The scope of techniques in cell biology has expanded significantly in recent years, particularly in terms of the ability to manipulate cells and explore their interactions, in the sensitivity and power of detection of structural detail, and in identifying the chemical makeup of cellular constituents. As a result, new opportunities for studying the fundamental cell processes involved in atherosclerosis have been opened up. It is the purpose of this book to focus in detail on some of these new developments and bring them to the attention of a wider audience. To this end we have assembled together a range of topics on the theme of ultrastructure, membranes, and cell interactions in atherosclerosis, selected for their current interest and novelty. To make this material as widely accessible as possible, and recognizing that not all readers interested in the subject will necessarily be familiar with all the relevant background, this opening chapter provides a general introductory perspective and reference point for the chapters that follow. Of necessity, our treatment of individual topics is selective, and citations to published work, rather than being comprehensive, are illustrative, emphasizing, in particular, recent studies and reviews that offer openings to the wider literature. Further detailed background is available in the numerous books previously published on the subject of atherosclerosis and its consequences; useful sources include an earlier CRC monograph edited by White,[36] and a recent two-volume text by Davies and Woolf[37] and Poole-Wilson and Sheridan.[38]

The structure of the normal artery wall provides our starting point for examining the nature of atherosclerosis and the atherosclerotic lesion.

II. STRUCTURE OF THE ARTERIAL WALL

A. Basic Architecture and Structural Variation in the Arterial Tree

Arteries are essentially thick, muscular, elastic tubes. The arterial wall consists of three circumferential layers or tunicae: the tunica intima (innermost layer), the tunica media (middle layer), and tunica adventitia (outermost layer). Figure 1 illustrates these layers and their makeup, taking the rabbit coronary artery as an example. Principal constituents of the arterial wall are endothelial cells and smooth muscle cells together with their secreted products of the extracellular matrix, elastin, collagen, and proteoglycans. Other major cell types present include macrophages, fibroblasts, and nerve cells. The arrangement of these constituents in the different layers of the arterial wall follows a general plan that is common to all types of artery.

The intima consists of an inner lining of endothelial cells that rests on an elastic lamina, the "internal elastic lamina". A layer of connective tissue that varies in thickness in different vessels is interposed between the lamina and

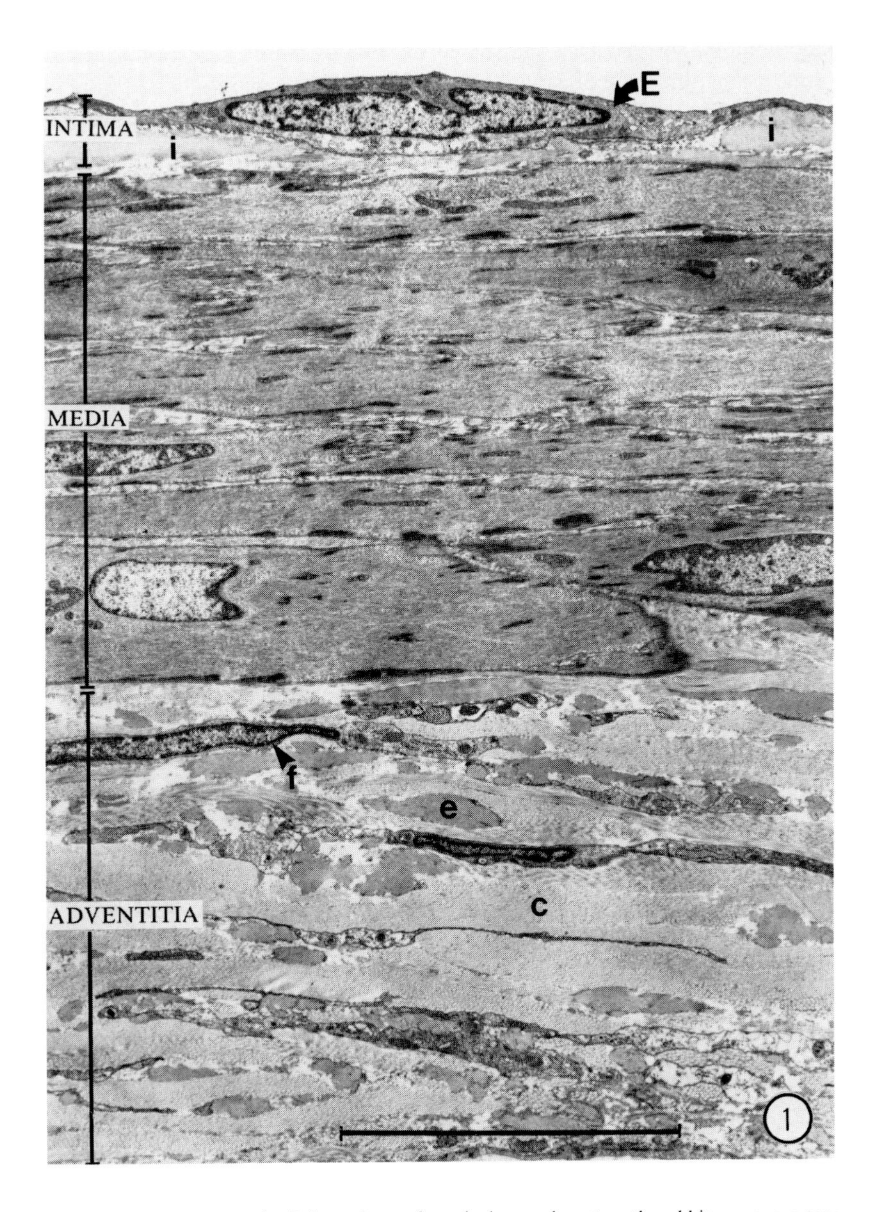

FIGURE 1. Structure and cellular makeup of a typical muscular artery, the rabbit coronary artery, as seen by thin-section electron microscopy. The arterial wall consists of three layers: the intima, media, and adventitia. The intima comprises a single layer of flattened endothelial cells (E) seated on the internal elastic lamina (i). The media contains smooth muscle cells embedded in only a small quantity of extracellular matrix. The adventitia consists mainly of extracellular matrix components, notably collagen (c) and elastin (e), with scattered fibroblasts (f). (Bar: 5 μm).

endothelium. The media is composed of smooth muscle cells embedded in variable quantities of extracellular elastin and collagen and is divided from the adventitia by an external elastic lamina. The adventitia comprises a sleeve of fibrous connective tissue, which in larger vessels is penetrated by small blood vessels, the vasa vasorum, supplying the outermost regions of the artery wall.

Within this basic plan a number of structural variations occur according to position in the arterial tree.[39] The arterial system is generally classified into three categories of vessel: elastic arteries, muscular arteries, and arterioles. Elastic arteries are the largest arteries, responsible for conducting blood from the heart to major sites within the body. The aorta, carotid, iliac, and pulmonary arteries belong to this group. Muscular arteries are the medium caliber vessels that branch from elastic arteries to distribute blood to specific organs. Examples are the coronary and femoral arteries. Arterioles are smaller still and represent the ultimate branches of the arterial system before it joins to the capillary bed.

The hallmark of the elastic artery is found in the structure of its media (Figure 2). This consists of a series of thick, fenestrated elastic layers alternating with essentially single layers of smooth muscle cells, the orientation of which is uniform in a given layer, but may vary in different layers. In muscular arteries, by contrast, the media consists predominantly of smooth muscle cells arranged in circular or helical configuration, with a much smaller proportion of elastin and collagen (Figure 1). The connective tissue components may increase in quantity as the muscular artery approaches the elastic artery, and close to the point at which it joins, series of elastic laminae, similar to, but less distinct than, those of elastic arteries, may be present.

The details of intimal structure also differ between elastic and muscular arteries. In elastic arteries, the intima is thicker than in muscular arteries, containing a layer of connective tissue (elastin, collagen, and elastic fibrils) in which scattered smooth muscle cells and occasional macrophages are observed. Although a somewhat similar layer may be found in the larger muscular arteries, in most species this gradually diminishes so that in the smaller muscular arteries, the endothelium generally sits directly on the internal elastic lamina, as illustrated in Figure 1. Muscular arteries have a prominent internal elastic lamina with periodic perforations through which endothelial cells and smooth muscle cells can interact; elastic arteries, by contrast, have a less obvious internal elastic lamina formed from the outermost of the series of elastic layers.

The adventitia in elastic arteries tends to be thin in relation to the thickness of the media, whereas a more substantial adventitia, often comparable to the thickness of the media, is present in muscular arteries. The major constituents of the adventitia are fibroblasts, collagen, elastin, occasional smooth muscle cells, and, in addition to the vasa vasorum, nerve cells and lymphatic capillaries.

As muscular arteries branch and become smaller, they eventually become arterioles. Arterioles are defined by the presence of just one or two layers of circumferentially arranged medial smooth muscle cells. In larger arterioles the

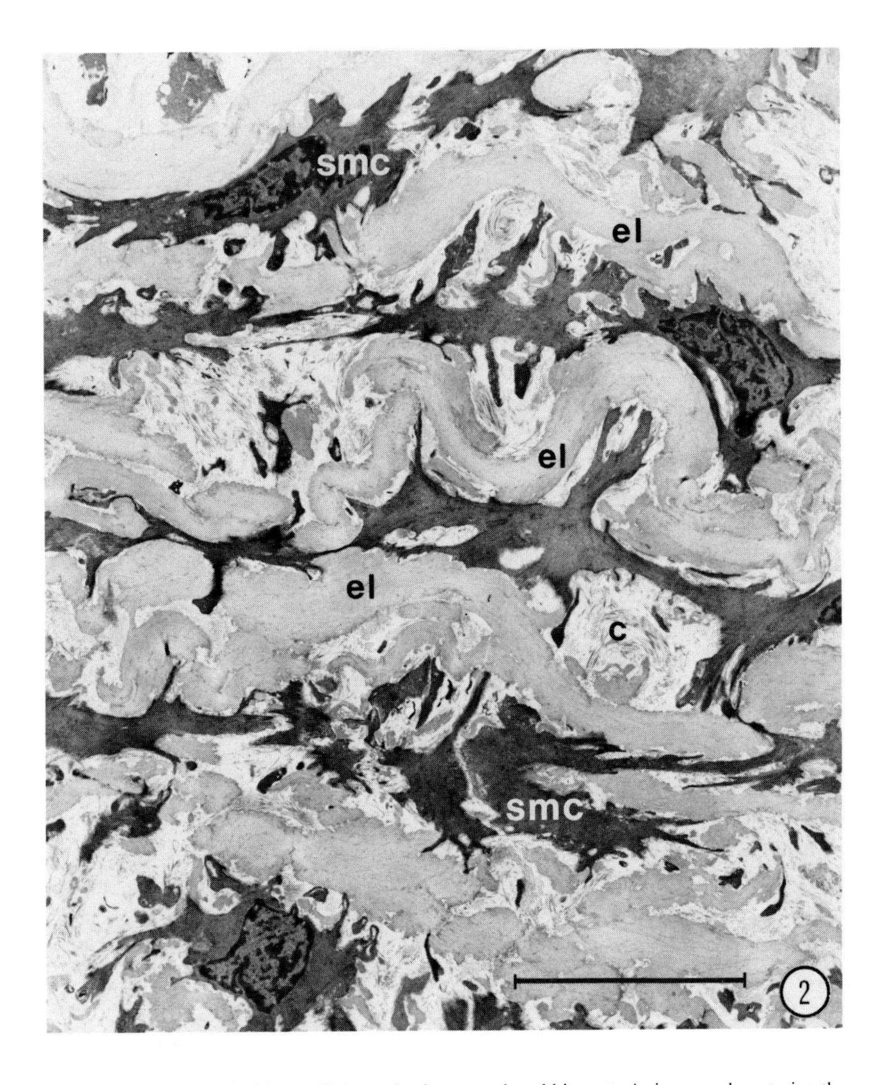

FIGURE 2. Structure of the media in an elastic artery, the rabbit aorta. As in muscular arteries, the sole cell type of the media is the smooth muscle cell (smc). However, in contrast to muscular arteries, extracellular matrix is abundant. Series of thick elastic laminae (el) alternate between the layers of individual smooth muscle cells. (c, collagen). (Bar: 10 μm).

endothelium sits on a thin internal elastic lamina, and an external lamina separating the media from the adventitia may be discernible. These laminae disappear, however, as the diameter of the vessel decreases. The termination of arterioles is marked by the presence of "precapillary sphincters", 5 to 50-μm-long zones containing one or two layers of smooth muscle cells that, by contraction or relaxation, control the flow of blood to the capillary beds.

B. The Human Coronary Artery and Atherosclerosis

Atherosclerosis may occur in both elastic and muscular arteries, but does not develop in arterioles. In the coronary artery the major extramural vessels are most prone to disease, especially at sites of branching and turbulent flow. Extension of disease into the epicardial branches may occur, but the smaller intramural branches are seldom affected; susceptibility to disease thus declines as the size of the vessel decreases. Abnormalities in function of arterioles and smaller branches of the coronary arteries may nevertheless account for some forms of angina in which significant atherosclerosis does not appear to be involved, a heterogeneous disease entity designated syndrome X.[40]

A feature distinguishing human coronary arteries from most other muscular arteries is the presence of a thickened intima.[41,42] The endothelium in the human coronary artery does not rest directly on the internal elastic lamina, but overlies a deeper intimal zone consisting of smooth muscle cells and their extracellular matrix products (Figure 3a–c). At branch points, more pronounced eccentric fibrous thickenings ("intimal pads" or "focal cushions") are superimposed upon this diffuse thickening. The intimal smooth muscle cells toward the luminal side tend to lose their contractile features; those below retain a contractile form and adopt a predominantly longitudinal orientation. Apart from smooth muscle cells, isolated macrophages are present, and the internal elastic lamina undergoes splitting, breakage, and duplication. Diffuse intimal thickening and "intimal pads" develop in the coronary arteries from infancy or earlier. As similar structures are found in some other types of artery that are not prone to atherosclerosis (e.g., the renal artery and the helicine arteries of the penis),[43,44] their appearance is regarded as a normal developmental event rather than a pathological feature. However, the occurrence of eccentric intimal thickenings in regions that are commonly afflicted with disease in later life (e.g., the branch points of extramural coronary arteries) and their absence from the coronary arteries of nonhuman primates that are not susceptible to the disease suggest that the presence of these structures predisposes certain sites to further change that may ultimately lead to overt atherosclerosis.[42]

III. THE ATHEROSCLEROTIC LESION

A. The Fibrous Plaque and Complicated Lesion

The characteristic lesion of atherosclerosis is termed the fibrous plaque (also known as the fibrolipid plaque, fibrofatty plaque, atherosclerotic plaque, atheromatous plaque, or simply atheroma[4-6,42,45-49]). The plaque is a raised region of the intima, consisting of increased numbers of smooth muscle cells with massive accumulations of their connective tissue products, collagen, elastin, and proteoglycans (Figure 3d). As sketched in Figure 4, a core or pool of lipid and necrotic debris is typically present at the base of the lesion, a "cellular" region containing smooth muscle cells and macrophages lies above, and a denser fibrous layer of connective tissue containing occasional attenuated smooth muscle cells forms a cap.

The macrophages and smooth muscle cells of the "cellular" region frequently show intracellular accumulation of cholesterol and cholesterol esters. A characteristic cell morphology observed in these regions is the "foam cell" (Figure 3e), formed predominantly from macrophages,[50] but also sometimes from smooth muscle cells[51] that have taken up large quantities of lipid. Death and disintegration of "foam cells" lead to accumulation of extracellular lipid and necrotic debris in the "core". Crystallization of cholesterol (Figure 3f) and calcification may occur. Apart from smooth muscle cells, macrophages, and "foam cells", lymphocytes are also found in some lesions.

The fibrous cap may have a complete covering of endothelium, though some degree of endothelial loss is common.[52] Thrombi may form over areas denuded of endothelium or when plaques develop fissures, and these may, with further growth of the plaque, subsequently become incorporated as a constituent. Individual atherosclerotic lesions exhibit considerable variability in overall composition and in the relative abundance, organization, and distribution of the various constituents. In some, for example, the fibromuscular thickening is of fairly uniform structure, without clear distinction into "cellular" vs. dense fibrous zones. Whereas some plaques contain a massive lipid core with just a thin fibrous cap, others appear as extensive fibromuscular thickenings with little lipid. A distinction is sometimes drawn between earlier "atheroma", in which the lipid core has become discernible, and more advanced "fibroatheroma", in which the "collagenous cap" has also developed.[42] Where an advanced lesion shows calcification, hemorrhage, or thrombus, it is often referred to as a "complicated lesion".

Although the atherosclerotic plaque is an intimal lesion, its presence is frequently accompanied by changes in the media and adventitia. Notable among these are a reduction of medial smooth muscle below the plaque, the presence of lipid in the media, and an upward growth of the vasa vasorum into the plaque. Lymphocytic infiltration of the adventitia may be found in association with advanced plaques, especially after thrombosis.[6]

B. The Fatty Streak

A second type of lesion, regarded as a precursor to the overt atherosclerotic plaque, is the fatty streak.[4-6,42] Fatty streaks are macroscopically flat or only slightly elevated lesions and tend initially to be round in shape, later becoming elongated in line with the direction of blood flow. The major constituent of the fatty streak comprises subendothelial accumulations of lipid, predominantly within "foam cells" (Figure 3g). Scattered extracellular deposits of lipid, and lipid accumulation in smooth muscle cells of the media and intima, are associated features. Lymphocytes occur in some lesions. Fatty streaks are found in the aorta as early as infancy and in the coronary arteries from early puberty. Their presence

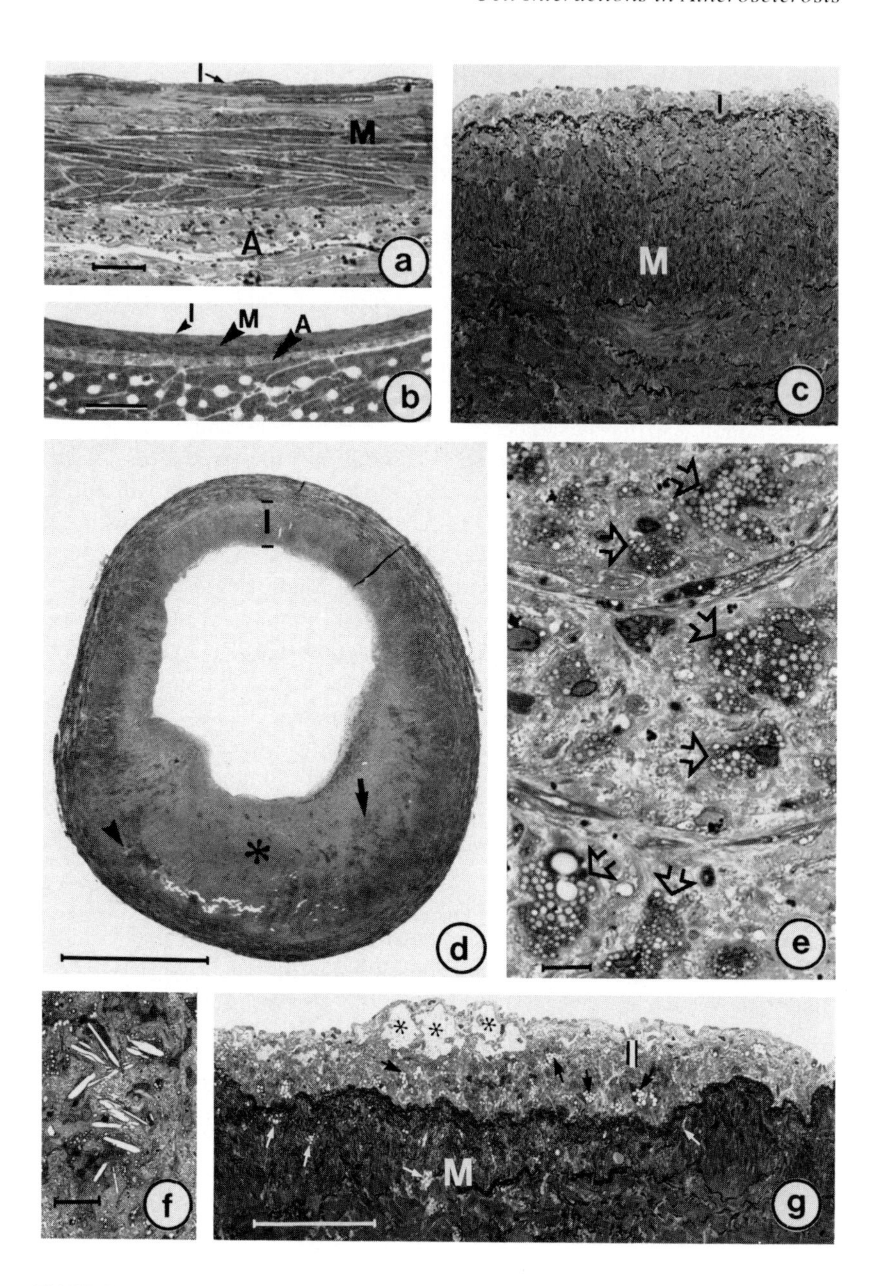

FIGURE 3. Light microscopical features of normal and diseased coronary artery. The structure of the normal coronary artery of rabbit and man is compared in (a)–(c). (a) Shows rabbit coronary artery by high magnification light microscopy, allowing comparison with the electron microscopical survey view shown in Figure 1. The intima (I) consists of a single layer of endothelial cells, just discernible where their nuclei cause a bulging of the cell. (M, media; A, adventitia). (Bar: 10 μm). (b) and (c) allow comparison of the structure of rabbit and human coronary artery at the same magnification. At this lower magnification the intima (I) in the rabbit artery (b) is so thin that it cannot

in children of all populations, irrespective of the incidence of clinical manifestations of atherosclerosis in the corresponding adult populations, indicates that fatty streaks do not inexorably progress to advanced atherosclerotic lesions. Though some have contested the view, the weight of evidence nevertheless indicates that most, if not all, advanced atherosclerotic lesions start as fatty streaks. The evidence for this conclusion includes the sequential changes observed in lesion structure in animal models of atherosclerosis,[53,54] and a general, though somewhat approximate, concordance in the preferred anatomical sites of advanced lesions and fatty streaks. Most compelling of all, however, has been the demonstration, from a detailed autopsy series of human subjects, that lesions of intermediate structure ("preatheroma") occur as part of the progression of changes over the first 3 decades of life.[42]

The processes by which the constituents of the normal artery participate in the evolution of atherosclerotic plaques are complex. Each of the principal cell types of the artery wall — the endothelial cell and the smooth muscle cell — is intimately involved. So too are two cell types originating from the blood: monocyte/macrophages and platelets. Increasingly, atherosclerosis is being understood in terms of a multiplicity of interactions between these cell types and their secreted products.[5,48,55] A further critical factor is the interaction of arterial cells with, and their handling of, blood-borne lipoprotein particles. It is also suspected that lymphocytes and fibroblasts may participate in the web of cellular interactions, though their involvement is less clearly established. In all these cellular interactions, membranes have a central part to play. Apart from fulfilling a barrier function, they are the seat of the specific recognition, signal transduction, transport, communication, adhesive, and secretory mechanisms that permit cells to interact dynamically with one another and with the external milieu. In the following sections, therefore, we will take a more detailed look at each of the principal cell types of the arterial wall, examining their normal structure and functions, their membranes, and how they interact, with particular emphasis on those activities relevant to atherosclerosis.

---

easily be distinguished, whereas in man (c) a much thicker intima (I) is apparent. Discontinuous elastic layers are visible as wavy dark lines throughout the media (M). (Bar: 50 μm). (d) Shows a transverse section through an atheromatous coronary artery from a 59-year-old male with ischemic heart disease. The plaque comprises an eccentric thickening (*) of the intima, with diffuse intimal thickening (I) apparent over the full circumference of the vessel. (Bar: 1 mm). (e) Shows a group of foam cells (arrows) (lipid-filled macrophages and smooth muscle cells) from the region indicated with an arrow in (d). (Bar: 10 μm). (f) Cholesterol crystals from the area indicated with an arrowhead in (d). (Bar: 20 μm). (g) Illustrates a fatty streak from the coronary artery of a 28-year-old male. Lipid-filled foam cells (black arrows) are abundant in the intima (I), and pools of lipid (*) have started to form beneath the still-intact endothelium. Lipid accumulation is also apparent within medial smooth muscle cells (white arrows). (Bar: 100 μm). (Toluidine-blue stained epoxy-resin sections. Human coronary arteries are from explanted hearts of patients receiving cardiac transplants; (c) and (g) are from the same patient receiving a transplant for a cardiac condition unrelated to atherosclerosis).

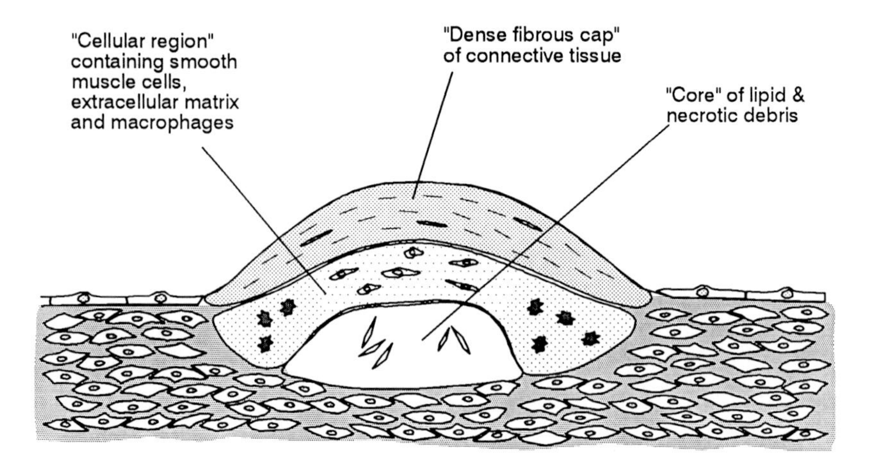

FIGURE 4. Diagrammatic representation of the atherosclerotic plaque. The plaque consists of a raised lesion of the intima. Common features include a "core" of lipid and necrotic debris, a cellular region rich in extracellular matrix products, containing smooth muscle cells and macrophage/foam cells and a dense fibrous "cap". This represents a highly idealized structure, and in practice, different plaques show considerable variation on this general theme.

IV. THE ENDOTHELIAL CELL

Endothelial cells are thin, "scale-like" cells (Figure 5a) that are joined together to form a continuous monolayer lining the inside of all blood vessels and the chambers of the heart. A primary function of the endothelium is to form a selective permeability barrier between the blood and underlying tissue fluid compartments.[56,57] Normal intact endothelium provides a nonthrombogenic inner surface to blood vessels, preventing platelet adherence and the activation of coagulation systems. Once considered a largely inert lining, it is now known that the endothelium is highly active metabolically and has a multiplicity of regulatory functions, many of which are relevant to the pathogenesis of atherosclerosis.[5,55] Apart from the transport of lipoproteins and the regulation of coagulation, for example, endothelial cells manufacture and release a range of substances that control growth, differentiation, and function of smooth muscle cells and macrophages.

Endothelial cells have an unmistakable shape and ultrastructural appearance (Figures 5a,b). They are so thin and flat that the nucleus often causes a prominent luminal bulge. Most of the organelles (the mitochondria, endoplasmic reticulum, ribosomes, Golgi apparatus and vesicles) are distributed in a band-like zone around the nucleus (Figure 5b). Dense, ovoid membrane-bound bodies with a tubular internal structure, originally termed Weibel-Palade bodies, are a characteristic component; these form storage sites for von Willebrand factor (factor-VIII-related antigen),[58,59] an adhesive glycoprotein that is secreted upon vascular damage, mediating attachment of platelets to subendothelial connective

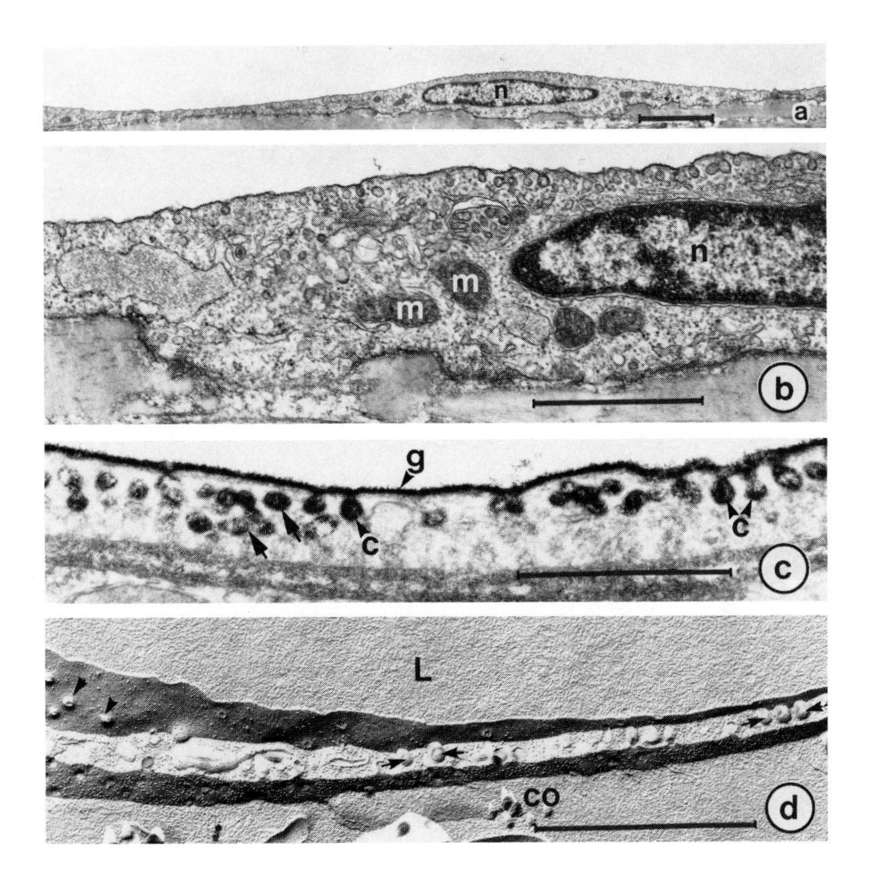

FIGURE 5. The vascular endothelial cell. (a) Endothelial cells are characteristically thin and flat, bulging in the center to accommodate the flattened nucleus (n), as illustrated in this survey view. (Bar: 2 μm). (b) Most of the organelles (e.g., mitochondria, m) are situated in a zone adjacent to the nucleus (n). (Bar: 1 μm). (c) The luminal surface has a prominent glycocalyx (g), revealed here by ruthenium red staining. Apart from staining anionic sites, ruthenium red acts as an extracellular marker, filling endothelial vesicles that open out at the luminal surface (caveolae, c). Ruthenium-red-positive vesicles seen individually and as fused chains that appear to lie free in the cytoplasm (arrows) are in reality connected to the luminal membrane. (Bar: 0.5 μm). (d) Freeze-fracture replica view of a cross-fractured endothelial cell, showing details of endothelial vesicles (small arrows) in the cytoplasm, and vesicles attached to the luminal and abluminal membranes (arrowheads). (L, lumen; co, collagen fibrils in the subendothelial space.) (Bar: 1 μm). (a), (b), and (d) are from rabbit coronary artery; (c) from rat.

tissue. Endothelial cells are coated with a carbohydrate-rich glycocalyx on their luminal surface and rest on a basal lamina formed from laminin, type IV collagen, and other matrix components that are secreted via the abluminal membrane (Figure 5c).

One of the most conspicuous features of the vascular endothelial cell is an abundance of 70-nm-diameter smooth-surfaced vesicles (Figures 5c,d and 6a).

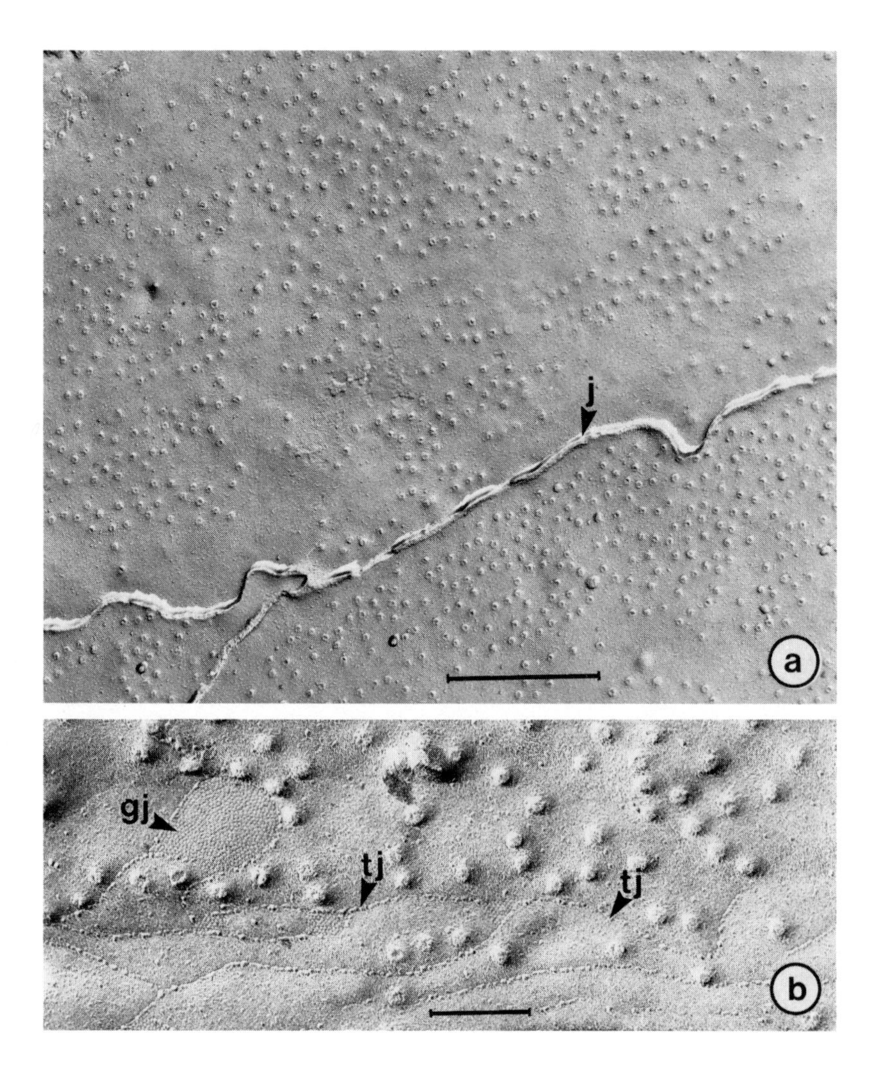

FIGURE 6. Membrane features of coronary endothelial cells revealed by freeze-fracture electron microscopy. (a) En face aspect (E-face) of the luminal plasma membranes of endothelial cells from rabbit coronary artery. Note the abundance of caveolae (attached vesicles), fractured across their necks to reveal "crater-like" structures in the membrane. Portions of three cells are present in this field; (j) indicates the junctional boundary between the neighboring cells. (Bar: 1 μm). (b) Shows a view in which the fracture plane has travelled down into a boundary of the type indicated in (a). The plasma membranes of the adjacent cells in these regions overlap one another and establish intercellular junctions. Arterial endothelial cells typically have junctional complexes in which occludens junctions (tj) and gap junctions (gj) are closely associated. Occludens junctions (tight junctions) contribute to the permeability properties of the paracellular pathway; they consist of cylindrical-like fusions between the adjacent membranes, seen here as rows of particles. Gap junctions are aggregates of protein channels spanning the two membranes, permitting communication from one cytoplasmic compartment to the next. In this example from a small branch of a human coronary artery, the protein channels have been fractured out of the membrane to leave pit-like

These occur both as free-floating structures in the cytoplasm, and in the form of vesicles attached to the plasma membrane, referred to as caveolae.[60] Single vesicles, chains comprising two or more fused vesicles, and more complex branching chains (racemose structures) are also observed. Distinct from this vesicular system are clathrin-coated vesicles and pits, which, though regularly observed, are far less common.

The bidirectional transport of a multitude of substances (nutrients, waste products, signaling molecules, and respiratory gases) between the plasma and underlying cells of the tissue or vessel wall is conducted through the endothelium. The two principal pathways proposed for this traffic are the plasmalemmal vesicle system[56,57,61] and the junctional boundaries between adjacent cells.[62] An additional specialization, the endothelial fenestration,[63] facilitates transendothelial transport in some types of capillary, but this structure is not present in elastic or muscular arteries.

A. Transendothelial Transport and Lipoproteins

A transport role for plasmalemmal vesicles has been extensively discussed in both microvascular and arterial endothelium.[56,57,61,64] A widely invoked hypothesis of vesicular transport involves the coupling of endocytosis to exocytosis, a process referred to as transcytosis.[56,65] This is suggested to involve a population of free-floating vesicles that shuttles back and forth across the cell, repeatedly taking up and discharging its cargo by transient connection with the plasma membrane on either side. The transient fusions may occur at the general plasma membrane or at caveolae, which for the most part appear to represent semipermanent invaginations, rather than fusing vesicles.[60] The composition of the vesicle membranes, which differs from that of the plasma membrane, is apparently preserved despite the repeated fusion and fission events. On occasion complete transendothelial channels may be formed where a string of fused vesicles or cavaeolar systems temporarily extends from the luminal to the abluminal surface. Serial section studies of microvascular endothelium indicate that individual free vesicles are in reality far less common than single sectional views originally suggested. Chains of vesicles are found to extend from the luminal and abluminal fronts; transport may therefore take place primarily by diffusion along such preformed channels, rather than via discrete vesicles.[61]

The plasmalemmal vesicle system has been suggested to be the major route for the transendothelial transport of low-density lipoprotein (LDL)[66] — the principal carrier of cholesterol in the plasma. Evidence for the involvement of

---

imprints (E-face view). (Bar: 0.2 μm). (E-face = en face interior aspect of half-membrane sheet left attached to the extracellular space after splitting the membrane by freeze-fracture; P-face = en face interior aspect of half-membrane sheet left attached to the protoplasm of the cell after splitting the membrane by freeze-fracture. For further details of freeze-fracture nomenclature, see Branton et al.[237]).

endothelial vesicles in the transport of LDL has been documented in coronary artery and aorta.[66] LDL transport across the endothelium via this route is a low-affinity, nonsaturable process. After delivery to the abluminal surface, the lipoprotein is taken up by other cells of the vessel wall, e.g., smooth muscle cells and fibroblasts. Uptake of native LDL by these cells is accomplished by receptor-mediated endocytosis, a high-affinity process involving the binding of LDL (or other proteins) to specific cell-surface receptors that work in concert with a protein, clathrin, that coats the cytoplasmic surface of receptor-bearing membrane domains.[67,68] Clathrin is initially organized as flat polygonal lattices beneath the plasma membrane, interacting with adaptor or assembly proteins that, in turn, interact with the cytoplasmically exposed segments of the receptor molecules.[69] When LDL binds, the clathrin shapes the membrane into a deep invagination and generates forces that pinch it off as a clathrin-coated vesicle. The adaptor proteins are thought to be involved in targeting the vesicle to the correct destination, which in the case of LDL is the endosome. Disassembly of the clathrin coat allows the vesicle to fuse with the endosome, a compartment characterized by a low pH. In this environment the LDL and the receptor are uncoupled from one another; the receptor is then recycled back to the cell surface, while the LDL is directed to the lysosomes. Hydrolysis by lysosomal enzymes releases cholesterol, which under normal circumstances is used for incorporation into the cell's membranes. Endothelial cells are themselves equipped with LDL receptors, and LDL internalized by this route is not transmitted to the abluminal surface, but retained for use within the cell.

LDL may become oxidatively modified during its transit through the endothelial cell or after its arrival in the subendothelial space, destroying its ability to bind to the LDL receptor (see Section VII B). Endothelial cells have receptors for lipoproteins other than LDL and play important roles in many other facets of lipoprotein metabolism. For example, the enzyme lipoprotein lipase, located at the endothelial luminal surface, is responsible for converting very-low-density lipoprotein (VLDL) to intermediate-density lipoprotein and LDL and for converting chylomicrons to chylomicron remnants (for further explanation of these lipoproteins, see Section VII B). It is also well established that endothelial cells exercise important influences on the expression of LDL receptors and on the activities of cholesterol-metabolizing enzymes in smooth muscle cells.[70,71]

B. Intercellular Junctions and Endothelial Cell Interactions

A second potential pathway for endothelial transport lies between adjoining endothelial cells. Where neighboring cells meet, their margins become highly attenuated and overlap. The adjacent plasma membranes at these sites are linked by intercellular junctions (Figures 6a,b) that vary in type and form in different segments of the vasculature.[72,73] Arterial endothelial cells typically have prominent gap junctions, but rather rudimentary occludens junctions (Figure 6b). Occludens junctions consist of thread-like fusions between adjoining plasma membranes that create a barrier between the overlapping cell margins, influencing permeability properties across the cell layer.[74] The sites of fusion, comprising pairs of offset cylindrical inverted lipid micelles,[75] apparently stabilized by associated cytoplasmic proteins,[76,77] have a discontinuous structure in arterial endothelial cells. This suggests a partial, rather than a tight, seal in which tortuous permeability pathways between the cells may permit passage of macromolecules up to several nanometers in diameter. Some investigators suggest that these pathways, rather than endothelial vesicle systems, form the principal route for transendothelial transport processes.[62]

Gap junctions are quite distinct from occludens junctions. They are the junctions responsible for intercellular communication, a function mediated by direct linkage of adjacent cytoplasmic compartments via protein channels in the apposing plasma membranes.[78,79] The permeability of endothelial gap junctions to small molecular tracers has been demonstrated *in vivo* and *in vitro*,[80,81] and intercellular communication via this route may play an important part in the migratory responses of endothelial cells to injury and in contact inhibition of growth, the properties that govern maintenance of the characteristic monolayer organization.[82]

In addition to gap junctions and tight junctions, at least two types of adherentes junctions have been defined in endothelial cells — a form of zonula adherens that anchors neighboring cells together, and assymetric focal adhesion structures that attach the cells to the basal lamina.[83] The focal adhesions contain talin and receptors for laminin and type IV collagen (i.e., integrins); the zonula adherentes junctions contain plakoglobin on their cytoplasmic surface and a calcium-dependent cell adhesion molecule (cadherin)[84,85] in the intermembrane space.[86] Integrins are heterodimeric membrane glycoproteins that bind specific cytoskeletal and extracellular matrix proteins.[87] Specificity of binding is governed by the precise makeup of individual integrins; typically, a single β subunit (of which there are several types, denoted β 1, β 2, etc.) associate with different sets of α subunits (see Chapter 6 by Joseph Madri and Leonard Bell for a comprehensive discussion). Apart from their role in cell-substratum interactions, these molecules have also been reported at the interacting borders of some types of endothelial cells[88,89] and play a role in endothelial-cell-leukocyte adherence.[90] Another recently identified molecule at the endothelial border is a calcium-independent cell adhesion protein, PECAM-1 (endoCAM/CD 3D), which belongs to the immunoglobulin gene superfamily.[91-93] This molecule is also present in the plasma membranes of some leukocytes and platelets, and it, too, is thought to contribute to the interaction of these cells with endothelial cells.

On their cytoplasmic side, the adherentes junctions are intimately linked, via vinculin and α-actinin, to a system of actin microfilaments that may be disposed in a circumferential band and/or as superficial stress fiber bundles. These actin microfilaments are associated with myosin and, together with microtubules and intermediate filaments of vimentin, make up the cytoskeleton.[94] The cytoskeleton, through the attachment of the actin microfilament system to the adherentes junctions, forms part of the essential apparatus for cell-cell and cell-substratum adhesion. Temporary parting of endothelial cells, required for example to admit monocytes to the subintima, is mediated by regional microfilament-induced cellular contraction, involving separation of the zonula adherens and removal of occludens junctional elements, which are known to be highly labile structures.[95]

Apart from the maintenance and control of endothelial integrity, the cytoskeleton plays a major role in injury and repair processes.[94] Disturbance or damage to the endothelium leads to extensive remodeling of the cytoskeleton, which may result in a stable adaptation or may further promote or be associated with cellular dysfunction. Injury in the form of cell death results in cytoskeleton-mediated extrusion of lamellipodia from neighboring survivors; these undermine still present dying cells until they slough off, thus ensuring that, by cell spreading, continuity of the barrier is maintained. Where frank denudation of endothelium occurs, surviving cells at the margin actively crawl into the damaged zone. Migration is accompanied by mitotic division until the monolayer is restored, whereupon further cell division is abruptly halted.

Normal undamaged endothelium in the adult is characterized by an exceptionally low turnover rate, estimated to be in the order of a few cells per 1,000 to 10,000 per day. The distribution of replicating cells is not uniform, however; major areas are quiescent, while focal spots of higher turnover appear to represent areas subject to repeated stress or injury.

C. Endothelial Control of Hemostasis and Smooth Muscle Cell Functions

The endothelium lies at the hub of hemostatic regulatory mechanisms. As well as controlling coagulation, it detects and transmits signals for smooth muscle contraction and relaxation. Anticoagulant properties are maintained by the synthesis and release of prostacyclin (which inhibits platelet aggregation)[96] and plasminogen activators (which control fibrinolysis), and by the presence of heparin-like molecules,[97] activated protein C,[98] and plasminogen bound to the luminal surface. Opposing properties are triggered by perturbation of the endothelium, e.g., in response to inflammatory stimuli such as interleukin-1; under these conditions tissue factor (the major initiator of coagulation) is expressed, a decrease of plasminogen and protein C binding sites occurs, inhibitors of plasminogen activator appear on the luminal surface, and von Willebrand protein is released.[99-101] The endothelial plasma membrane houses a multitude of receptors for blood-borne vasoactive agents, including catecholamines, acetylcholine, histamine, serotonin, and angiotensins. Angiotensin-converting enzyme (which converts angiotensin I to the vasoconstrictor angiotensin II), for example, has been localized cytochemically to the luminal plasma membrane. The same enzyme also contributes to the degradation of bradykinin, another vasoactive peptide. Among the vasoactive substances degraded by endothelial cells are epinephrine and 5-hydroxytryptamine; these are actively taken up from the blood stream. Prostaglandins are taken up similarly and converted to prostacyclin, which apart from inhibiting platelet aggregation, act to relax smooth muscle cells.

The discovery of endothelial-derived relaxant factors (EDRF)[102] and endothelins[103] has emphasized yet further the fundamental role of the endothelium in control of smooth muscle cell relaxation, contraction, and differentiation. In brief, the vasodilative action of acetylcholine, bradykinin, and substance P in coronary (and other) arteries is mediated by release of EDRF from the basal side of the endothelium.[102] EDRF has been identified as nitric oxide[104] synthesized from L-arginine,[105] though discussion on the existence of other forms of EDRF continues.[106] Apart from causing smooth muscle cell relaxation, nitric oxide and prostacyclin inhibit smooth muscle cell proliferation.[107] Acting in opposition are the endothelins — 21-amino-acid peptides that have potent vasoconstrictive effects and are mitogenic for smooth muscle cells.[108] Some other vasoconstrictors, e.g., angiotensin II, also appear to have a similar dual effect.

These are far from the only endothelial cell products that influence the proliferative behavior of smooth muscle cells. One of the endothelial cell products most widely discussed in the context of atherosclerosis is platelet-derived growth factor (PDGF).[109-112] PDGF owes its name to the cell type from which it was first isolated, but apart from being found in platelets, it is produced by a number of other cell types, including endothelial cells.[113] The platelet PDGF molecule comprises two polypeptide chains (designated A and B); cells other than platelets usually produce only one of the two chains (e.g., endothelial cells synthesize the B chain) and secrete homodimeric forms. PDGF acts as a potent mitogen for smooth muscle cells, an effect that is antagonized by heparin, yet another endothelial product.[114] Apart from its anticoagulant properties, heparin interferes with the binding of PDGF to its receptor on the smooth muscle cell, thereby inhibiting smooth muscle cell proliferation.

Endothelium is thus strategically placed to exercise multiple regulatory roles over smooth muscle. It is not surprising, therefore, that endothelial dysfunction is widely considered to be a prime mediator of altered smooth muscle cell behavior in atherosclerosis.[5,55,94]

V. THE SMOOTH MUSCLE CELL

Vascular smooth muscle cells — the sole cell type of the arterial media — are elongated, flattened contractile cells. In muscular arteries they typically have smoothly tapering ends (Figure 1); in elastic arteries their shape is more irregular (Figure 2). Through periodic contacts, they are organized together with their long axes in parallel, forming layers ensheathed in connective tissue. The arrangement of these layers varies in different vessels. In the coronary artery, the cells mostly describe a low-pitch helical course around the vessel.

A. The Vascular Smooth Muscle Cell as a Specialized Contractile Cell

The bulk of the smooth muscle cell cytoplasm is packed with bundles of myofilaments and intermediate filaments (Figure 7a). The nucleus is an irregular ovoid shape, and a filament-free cone of cytoplasm on either side accommodates most of the mitochondria, the Golgi apparatus, lysosomes, endoplasmic reticulum, and other organelles. Elements of the sarcoplasmic reticulum, and some mitochondria, are distributed widely throughout the cell.

As in other muscle cells, the myofilament bundles consist of actin (thin) filaments and myosin (thick) filaments.[115,116] Other key proteins of the contractile machinery are caldesmon, tropomyosin, and calmodulin. In contrast to striated muscle, the thin and thick filaments are not organized in a regular repeating manner, so banding patterns are not apparent. The myosin is less abundant than the actin and, owing to its lability, requires particular care in electron microscopical preparative technique for successful visualization. Conspicuous densely-staining structures, the dense bodies (Figures 7a,d), act as insertion sites for the actin filaments, allowing transmission of tension from one contractile bundle to the next. Most of the dense bodies are wedge-shaped and attached to the plasma membrane, but some occur as free-floating structures in the cell interior. Intermediate filaments, consisting of desmin and vimentin,[117] form an interweaving longitudinally organized network, which also links to the dense bodies, so that the cytoskeleton and contractile systems are mechanically coupled.[118,119] Filamin is present within the intermediate filament system bundles, apparently associated with actin that is separate from that of the contractile system.[120] Membrane-associated dense bodies are rich in α-actinin, talin, and filamin and are fastened to the plasma membrane by longitudinally arranged surface patches of vinculin.[121]

The plasma membrane between the membrane-associated dense bodies contains abundant caveolae strewn in rows and bands parallel with the long axis of the cell (Figure 7b). Elements of the sarcoplasmic reticulum run in close association with the caveolae, establishing couplings both with them and with the general plasma membrane.[122] Although smooth muscle cells have no transverse tubules, the caveola-junctional sarcoplasmic reticulum associations may form the counterpart of the "interior couplings" typical of striated muscle. The plasma membrane of the smooth muscle cell has a range of receptors, notably for growth factors, such as PDGF, and for LDL, which are internalized by classical receptor-mediated endocytosis. A surface coat and external lamina* cover the plasma membrane, to which matrix components attach. Further anchorage of the smooth muscle cell to the extracellular matrix may occur at the dense bodies, which are thought to form transmembrane junctions akin to the focal adhesions of cultured cells.[123] Smooth muscle cells are anchored to one another by adherens junctions and are electrically coupled by gap junctions[124]

---

* The term "external lamina" as used here for the smooth muscle cell is synonymous with "basal lamina" used to describe the corresponding structure of the endothelial cell. The former term is strictly more accurate for the smooth muscle cell because, in contrast to endothelial and epithelial cells, the lamina is not basal in location — it completely surrounds the cell. As will be seen elsewhere in this book, the earlier term "basement membrane" is still preferred by some authors, even for smooth muscle cells where the structure neither forms a "basement" nor is a "membrane" in the current sense of the word.

(Figure 7c), through which coordinated contractile responses are mediated. In muscular arteries and arterioles, contacts between the superficial layer of medial smooth muscle cells with endothelial cells are established by processes from either cell type interacting through discontinuities in the internal elastic lamina (Figure 7d). A type of adherens junction is often present at this site, and gap junctions can also be formed, at least in some vessels of some species.[125] These interactions may facilitate intercellular communication between the media and intima, by chemical signaling both across the greatly reduced extracellular space and, where gap junctions are present, directly between the cytoplasmic compartments of smooth muscle and endothelial cells. Myoendothelial contacts, though present in coronary arteries, become particularly abundant as coronary arterioles approach capillaries.[39]

Smooth muscle cells have the capacity to undergo spectacular changes in length. They contract relatively slowly and are capable of maintaining tension for prolonged periods with very low energy consumption, properties ideally suited to the control of blood vessel caliber and regulation of blood distribution and flow. As in striated muscle, contraction is activated by a rise in intracellular calcium, brought about by calcium influx through the plasma membrane and by release of stored calcium from the sarcoplasmic reticulum. However, the mechanisms of excitation-contraction coupling and of contraction in smooth muscle differ in detail from those of other types of muscle. Release of inositol 1,4,5-triphosphate, formed when an agonist interacts with a plasma membrane receptor, is thought to be an important pathway for triggering intracellular calcium release.[126] Interaction of thick and thin filaments is governed by calcium-regulated phosphorylation of myosin via a phosphorylating enzyme (kinase), and the regulatory protein caldesmon confers calcium sensitivity on actin (to which it is bound) through an ability to interact with calmodulin (calcium-binding protein), tropomyosin, and myosin.[127,128] These regulatory mechanisms, together with a high ratio of actin-to-myosin length[129] and an ability of caldesmon to crosslink thick and thin filaments, go some way in explaining the unique contractile properties of smooth muscle.[127] Smooth muscle contractile state is governed externally by a complex interplay between the autonomic nervous system and the endothelium, which, as we saw in the previous section, presides over the action of a multitude of vasoactive agents carried in the blood. The smooth muscle cell plasma membrane is accordingly equipped with the signal detection and transduction machinery to set the appropriate responses to all these stimuli into action.

B. The Vascular Smooth Muscle Cell as a Synthetic Cell

In the normal vessel of the neonate and adult, smooth muscle cells are, first and foremost, contractile cells, but they nevertheless have an important synthetic capacity, producing the range of diverse components that make up the extracellular matrix of the media. The three main classes of matrix component synthesized by smooth muscle cells are collagen, elastin, and proteoglycans; other

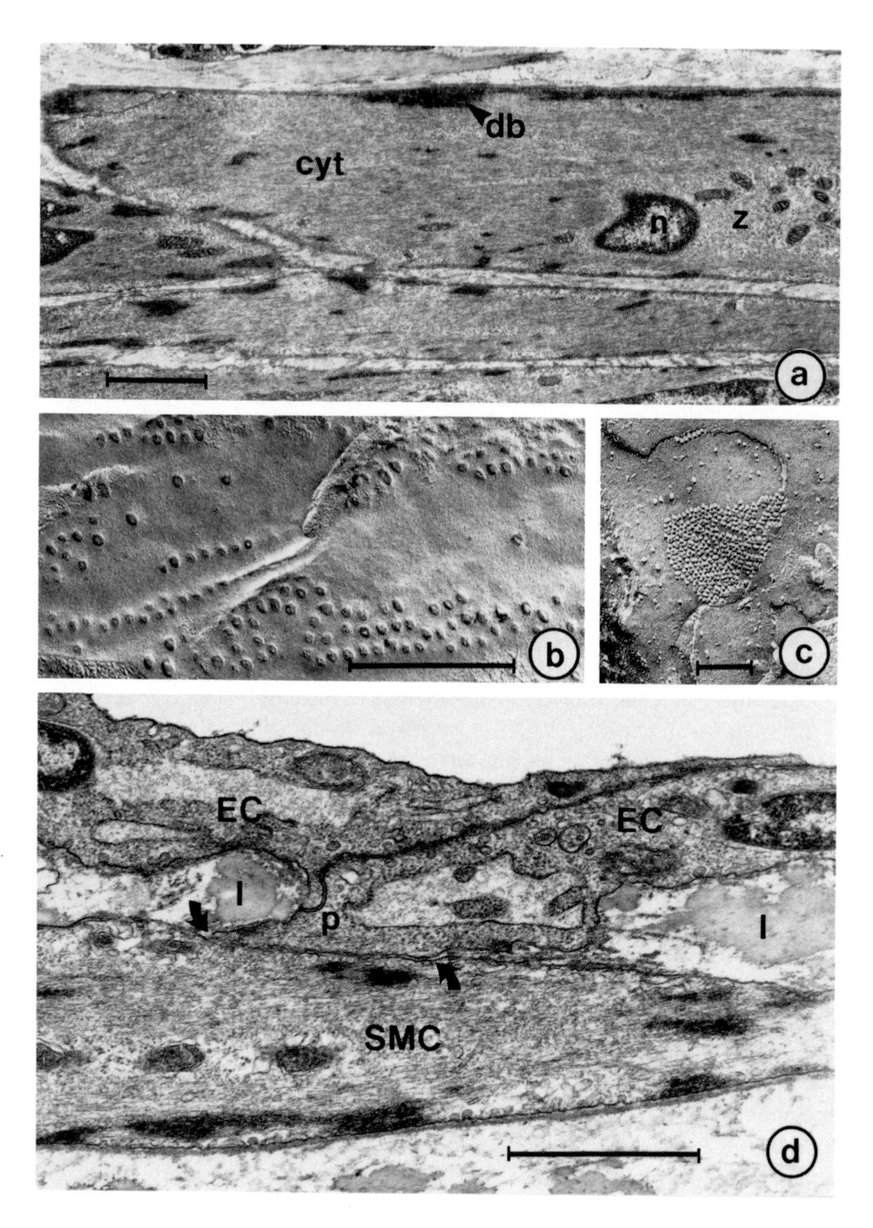

FIGURE 7. The smooth muscle cell. Typical features of the contractile phenotype. (a) The cytoplasm (cyt) of contractile smooth muscle cells is packed full of contractile filaments anchored to dense bodies (db). A filament-free zone (z) at the tips of the nucleus (n) contains most of the other organelles. (Bar: 2 μm). (b) En face freeze-fracture views of the plasma membrane reveal caveolae, organized in distinctive rows and bands between the membrane areas at which dense bodies are attached to the plasma membrane. (Bar: 1 μm). (c) Gap junctions ensure electrical coupling between adjacent smooth muscle cells; in this example (a P-face view), the junction is seen as an aggregate of particles that represent its protein channels (connexons). (Bar: 100 nm). (d) Smooth muscle cells

constituents of the matrix include the glycoproteins fibronectin, laminin, thrombospondin, and tenascin.

Collagens are a group of fibrous proteins that occur in abundance throughout the animal kingdom. The collagen molecule consists of three polypeptide chains (α chains) that are wound around each other to give a characteristic triple-stranded helical structure. Fifteen different types of collagen, comprised of various combinations of α chains (designated $\alpha 1$, $\alpha 2$ with Roman numeral suffixes), have been identified to date. Collagen types I, II, and III are crosslinked to give large polymers, forming the familiar banded collagen fibrils that give the tissue tensile strength. Type IV collagen does not assemble into a fibril; together with laminin, it forms the external lamina. (For a more detailed treatment of vascular collagen, the reader is referred to Chapter 4 by Rauterberg and Jaeger.)

Like collagen, elastin molecules are extensively crosslinked to one another, but they form sheets and filaments with a very different property. The elastin molecule is characterized by an unusual "unfolded" molecular structure, allowing variable coil conformations that impart an elastic property to the tissue. The third main class of matrix components, proteoglycans, consist of numerous repeating glycosaminoglycan chains (e.g., of chondroitin, dermatan, or heparan sulfate) covalently bound to long protein cores, and form a hydrated gel in which the other matrix components are embedded (see Chapter 7 by Völker).

It has become increasingly clear that the various classes of extracellular matrix components are not only responsible for the overall structural integrity of the vessel wall, but can also have profound influences on cell adhesion, motility, and differentiation.[130,131] For this reason they play an important role in the cellular events underlying atherosclerosis and the response of the vascular wall to injury: themes that are explored in detail in Chapter 5 by Thie and Chapter 6 by Madri and Bell.

C. Reversible Modulation of Contractile and Synthetic Phenotypes

An important property of the smooth muscle cell is its ability, under the influence of chemoattractants and mitogens *in vivo* or of culture conditions *in vitro*, to undergo profound functional alteration.[132,133] With appropriate stimulation, smooth muscle cells can be induced to migrate, proliferate, and become highly active synthetically, losing their contractile function as they do so. In the synthetic phenotype, the contractile apparatus of the cell becomes rudimentary, the Golgi apparatus and rough endoplasmic reticulum become prominent, and greatly augmented production of external lamina and extracellular matrix components occurs (Figure 8). These changes represent a reversion to the

---

(SMC) that lie immediately beneath the internal elastic lamina (I) often interact, via interruptions in the lamina, with the endothelial cells (EC) that lie above. In this example a process from an endothelial cell (p) comes into close contact (arrows) with a smooth muscle cell. (Bar: 1 μm.)

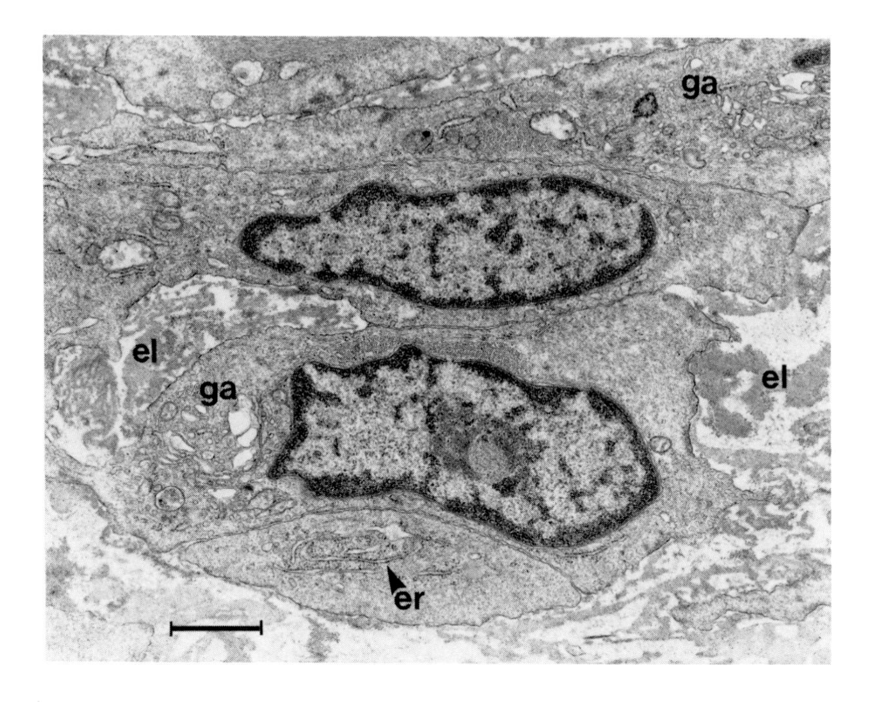

FIGURE 8. The synthetic-state smooth muscle cell. Typical features of smooth muscle cells in the synthetic phenotype include enlarged and conspicuous Golgi apparatus (ga) and rough endoplasmic reticulum (er), greatly reduced complement of contractile filaments, and enhanced production of external lamina components (el). From intima of fatty streak lesion shown in Figure 3g. (Bar: 1 μm.)

phenotype expressed during development and occur as part of the natural repair response to arterial injury. Such modulation of differentiation to the synthetic state is an indispensable step for participation of smooth muscle cells in the formation of the atherosclerotic lesion.

The changes occurring upon phenotypic alteration to the synthetic state are both quantitative and qualitative in nature. Among these changes are a switch from production of the α-isoform of actin to the β-isoform, of desmin to vimentin, and a reduction in tropomyosin, caldesmon, and vinculin expression.[134-137] The ratio of collagen type I to collagen type III increases,[138] heparan sulfate proteoglycan is decreased, while chondroitin sulfate and dermatan sulfate proteoglycans are increased.[139] The altered gene expression underlying these changes is modulated by the local balance of growth factors (e.g., PDGF and transforming growth factor-β) and inhibitors (e.g., EDRF, heparins, and cytokines such as interleukin-1).[140] A particularly important discovery is that synthetic state smooth muscle cells, including those isolated from human atherosclerotic lesions, themselves secrete PDGF,[141-143] thus creating a self-stimulating ("autocrine") cycle of proliferation and production of extracellular matrix components. This process may become augmented by insulin-like growth factor or inhibited by prostaglandins, both of which are also synthetic products of the smooth muscle cell.

Not all contractile smooth muscle cells appear to be equally sensitive to the effects of chemoattractants and mitogens. Following balloon-catheter injury in experimental animals, the medial smooth muscle cells that migrate into the intima include some that first undergo replication and some that do not, and only a proportion of the population arriving in the intima appears to go on to further division.[5,144,145] Variable expression of growth factor receptors appears to be one mechanism by which the responses of different subpopulations of smooth muscle cells may be modulated. Thus, the description of smooth muscle cells as being of either a contractile or a synthetic phenotype may be oversimplistic, as variable expression of properties over a continuum between these forms may occur. Nevertheless, the fact that smooth muscle cells can be made so readily to abandon their "proper" contractile function to migrate, divide, and/or synthesize massive quantities of extracellular matrix is identified in all hypotheses of atherosclerosis as the basis for the bulk growth of fibromuscular material in the plaque. Where individual hypotheses differ is in emphasis of precisely which factors lead up to these events.

VI. THE MACROPHAGE

Macrophages are mobile, phagocytotic cells specialized in mopping up, by various forms of endocytosis, unwanted cellular and extracellular debris, invading microorganisms, and other foreign matter. They thus play a major role in host defense and repair of damaged tissue. Macrophages are widely distributed in the loose connective tissue of most organs and are recruited in abundance to localized sites in inflammatory responses. They occur as normal, isolated inhabitants[42,146] of the intima and adventitia of arteries, but it is as major constituents of the intima in developing and progressing atherosclerotic lesions (Figure 9) that macrophages have increasingly attracted interest as crucial players in the atherosclerotic process.[50,147-150]

A. Plasticity of Macrophage Function

Macrophages originate from monocytes, which gain access to the tissue by adhering to the endothelial cell surface and then burrowing through temporary openings formed at the junctional boundaries between adjacent endothelial cells.[148] The newly arrived monocytes then undergo transformation into macrophages, enlarging, becoming active in endocytosis, and producing lysosomes. The mobility and activity of macrophages and their monocyte precursors is governed by chemotactic and stimulatory agents released by other cells and by components of the extracellular matrix. Apart from developing from monocytes, fully formed macrophages may themselves divide, on occasion, to give rise to new macrophages.

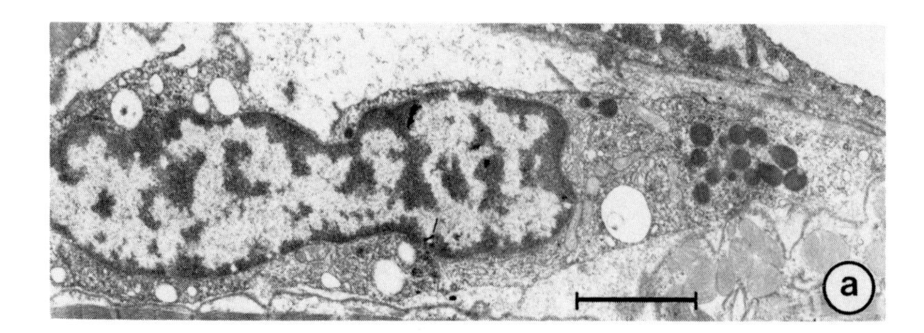

FIGURE 9a. The macrophage. (a) shows a resident macrophage beneath the endothelium in a rat coronary capillary. (Bar: 2 μm.)

The shape and appearance of the macrophage vary according to its activity. Actively wandering ("free") cells generally have an irregular profile, with the cell surface thrown out into pseudopods, while resting ("fixed" or "resident") cells tend to have smoother, drawn-out features (Figure 9). The cytoplasm typically contains polymorphic vacuoles, lysosomes, and residual bodies, representing various stages in the digestion of engulfed products. Plasma membrane vesicles and caveolae are common, reflecting the extensive endocytotic and exocytotic activity. Macrophages have no external lamina. Their nucleus is irregular in shape, and the usual complement of organelles is present. The endoplasmic reticulum and Golgi apparatus are well developed, reflecting the cell's activity in synthesis and secretion. Among the products synthesized for secretion are lysozyme (an enzyme that digests bacterial cell walls), neutral proteases (which degrade connective tissue), interferons (antiviral proteins), and components of the complement system (which lyse cells and contribute to phagocytic recognition).

Macrophages produce many other immunoregulatory molecules and a host of chemoattractants, growth factors, and cytokines, which influence the activities of smooth muscle cells, endothelial cells, and macrophages themselves. Among these products are prostaglandins, leukotriene B4 (a potent chemoattractant), monocyte chemoattractant protein,[151] PDGF,[152] transforming growth factor,[153-155] fibroblast growth factor (a mitogen for endothelial cells, as well as fibroblasts),[156] tumor necrosis factor ("cachectin"),[157,158] and interleukin 1. Tumor necrosis factor and interleukin 1 have multiple effects, including induction of tissue factor and adhesive molecule expression in endothelial cells; interleukin 1 also induces PDGF secretion in smooth muscle cells. Macrophages do not express the full gamut of potential activities at any one time; rather, complex feedback mechanisms involving multiple stimulatory and inhibitory molecules released by other cells ensure expression of specific sets of activities suited to the particular purpose at hand. T lymphocytes play a particularly important role in this process, modulating macrophage differentiation by release

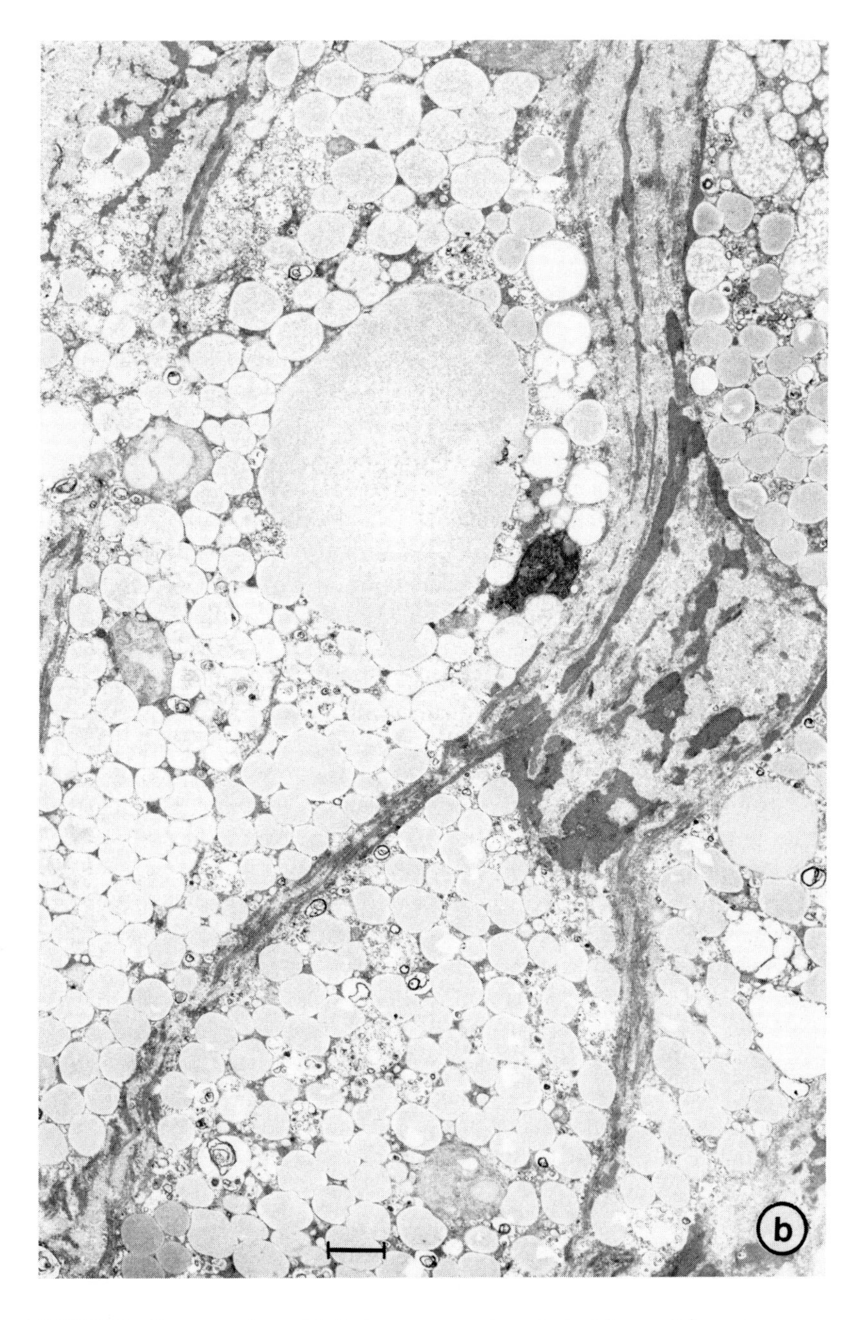

FIGURE 9b. The macrophage. (b) illustrates a macrophage-derived foam cell from an atherosclerotic lesion from human femoral artery. (Bar: 2 μm.)

of various lymphokines. The sensitivity of the macrophage's response to all these signals is made possible by an unusually diverse collection of plasma membrane receptors.[159]

B. Macrophage Receptors — The Key to Macrophage Activity and Foam Cell Transformation

All activities of the macrophage, including its versatile differentiation and its migratory, secretory, and endocytotic activities, are regulated by its receptors.[160] Macrophage receptors include three main superfamilies of structurally related proteins: the integrins, the immunoglobulin-interacting receptors, and a group of complement component (C3b/C4b)-interacting receptors. In addition, there are various lectin-like receptors. More than 60 ligands have been reported to bind to macrophage receptors; some of these ligands bind to more than one receptor, and some of the receptors recognize more than one ligand. This makes it difficult to be precise about how many receptors the macrophage has, but there can be no doubt that the number is very large.

The receptors that mediate phagocytosis recognize the F_c portion of immunoglobulins (F_c receptor), and complement components (CR3 and CR1).[161-163] Particulate debris and foreign matter destined for internalization are coated with immunoglobulin and complement molecules, and upon receptor-ligand binding, a spreading of the macrophage's plasma membrane over the object is initiated. Spreading is then perpetuated by further sequential receptor-ligand interactions, allowing the cell progressively to engulf the object. After the object has been pinched off into the cytoplasm as a phagosome, fusion of hydrolase-containing primary lysosomes results in digestion. Phagocytotic and secretory activities are influenced by many other receptors, e.g., integrins, which also influence migratory behavior.

Cell spreading, movement, and shape changes induced by ligand receptor interactions are effected by the cytoskeleton, which comprises a three-dimensional network of actin filaments cross-linked by filamin.[164] The cytoskeleton is a dynamic structure subject to rapid disassembly, reassembly, and redistribution within the cell.[165] Upon stimulation of macrophage receptors, assembly of filamentous actin from actin monomers occurs at the cell cortex, enabling initiation of cell movement. Restructuring of the cytoskeleton involves the activities of two actin modulating proteins, gelsolin and profilin.[164]

Macrophage receptors of particular importance in the context of atherosclerosis are those concerned with lipoprotein binding and uptake.[166] Although macrophages are reported to have LDL receptors, which facilitate the uptake of LDL by the usual tightly regulated process of receptor-mediated endocytosis, these receptors are not abundant. However, activated macrophages express two further types of lipoprotein receptor: one that recognizes another lipoprotein, β-migrating very-low-density lipoprotein (β-VLDL, see Chapter 10), and one referred to as the scavenger receptor.[166-168] These receptors profoundly influence the fate of lipoproteins in the artery wall.

The β-VLDL receptor is immunologically related to the LDL receptor,[166] but it has only a low affinity for LDL, binding avidly to β-VLDL and chylomicron remnants, lipoproteins that contain apolipoprotein E (see section VII B). The scavenger receptor mediates the uptake of LDL that has been experimentally acetylated or that has undergone oxidative modification during or after its passage through the endothelial cell[169-171] — changes that render the LDL unrecognizable to the LDL receptor. Two types of macrophage scavenger receptor have been identified by complementary DNA cloning; both have an unusual α-helical coiled coil domain joined to a collagen-like triple helix exposed on the external surface of the membrane, and one or both of these regions appear to be responsible for ligand binding.[172,173] Unlike the LDL receptor, both the scavenger receptor and the β-VLDL receptor are not subject to downregulation, and as a result, the macrophage will take up as much oxidatively modified LDL (oxLDL) and β-VLDL as are available to it. In addition, LDL that has formed complexes with proteoglycans is internalized by yet other poorly regulated receptor pathways.[139] In these ways, macrophages can become filled with cholesteryl esters, forming the "foam cells"[169,174] that are such a conspicuous feature of fatty streaks and established atherosclerotic lesions (Figure 9b). That these processes actually occur in cells isolated from atherosclerotic-type lesions, and not only in model systems, is demonstrated by the work of Olli Jaakkola (Chapter 9).

In vivo some lipid-laden foam cells appear to be able to migrate back to the blood stream, thereby removing lipid from the lesion.[149] Others, however, remain trapped in the arterial wall, releasing chemoattractants and mitogens. Macrophages are thus a major source of the stimuli that entice smooth muscle cells to migrate to the intima, proliferate, and engage in intense synthetic activity. Immobilized with their massive overload of lipid, many eventually die, contributing to the accumulation of necrotic material and lipid in the core of the plaque.

VII. CELL INTERACTIONS AND HYPOTHESES OF ATHEROGENESIS

Basic descriptions of the structure and function of vascular cells, as we have seen so far, are impossible without repeated reference to interactions between different cell types. This is just as true for the maintenance of normal vessel function as it is for the events that lead to vascular disease. Not surprisingly, therefore, the theme of cell interactions has been central in the development of hypotheses of atherosclerosis. Although a number of hypotheses of atherogenesis have been advanced, these are not, for the most part, mutually exclusive, tending to differ more in the emphasis given to particular events or aspects than to fundamentally opposite points of view. By examining these hypotheses briefly, we shall take a closer look at how various sets of interactions may operate in the pathogenesis of atherosclerosis. In doing so, we shall examine not only the interactions between endothelial cells, smooth muscle cells, and macrophages, but also consider those involving blood-borne lipoprotein particles, platelets, coagulation factors, and lymphocytes, with these cells.

A. The Response to Injury Hypothesis — Endothelium at Center Stage

A widely accepted explanation of atherosclerosis is based on the "response to injury hypothesis" attributed originally to Virchow[175] and developed and elaborated over the last two decades by Ross and colleagues.[5,55,176,177] This hypothesis, in its modern form, proposes that the initiating event in atherogenesis is injury to the endothelium.[5] Endothelial injury may take a variety of forms, but however mediated, the resultant effect is release of growth factors into the intima and subintimal zone. The growth factors act as chemoattractants and mitogens on smooth muscle cells, stimulating them to migrate from the media into the intima, proliferate, and adopt synthetic properties. The source of the growth factors may be platelets, macrophages, endothelial cells, smooth muscle cells, or any combination of these cell types.

The concept of endothelial injury is deliberately broad to encompass a spectrum of changes ranging from the stripping of the entire lining of a vessel, at one extreme, to retraction of still-adherent adjacent cells, through to more subtle, morphologically imperceptible, functional alterations, at the other extreme. The widely established risk factors — high blood pressure, cigarette smoking, and cholesterol — are all plausible causes of such damage via mechanical stress, toxic effects (e.g., due to carbon monoxide[178]), and altered membrane properties, respectively. Endotoxins, homocystinemia,[179] immune complexes,[180] viruses,[181] and other factors reported to promote atherogenesis are all envisaged, according to the response to injury hypothesis, as primary mediators of endothelial dysfunction.

Experimental studies on animals, using transluminal balloon and other nylon catheters, have shown that very extensive erosion of the endothelium, or repetitive desquamation in the same area, will induce smooth muscle cell proliferation, but a single episode of limited endothelial cell loss without deeper damage does not, in itself, do so.[5,145,182] As we have previously seen (section IV. B), removal of endothelial cells leads to regeneration from the border of the injured zone. If accomplished rapidly enough, the "normal" endothelial control over smooth muscle function is maintained. However, if endothelial cells in a given area are repeatedly stimulated to divide, they may eventually lose their ability to replicate, resulting in ultimate failure of the repair mechanism, or other functional alterations. Whatever the precipitating factors, the net outcome of such endothelial dysfunction is hypothesized to include the secretion of growth factors by endothelial cells, and increased expression of adhesive molecules on their luminal surfaces.

Where injury does involve removal or retraction of endothelium, platelets will immediately adhere to, and spread across, the exposed connective tissue surface, forming the primary source of growth factors (PDGF and transforming growth factor beta) to which smooth muscle cells respond. Where the endothelium is "injured", but remains morphologically intact, as happens, for example, in experimental animals fed a cholesterol-rich diet, the expression of adhesive properties (section IV. B) causes circulating monocytes to stick to the luminal surface. Subsequent migration between the endothelial cells into the intimal connective tissue ensues, with transformation to macrophages that then become the major source of growth factors and cytokines. Lipoproteins accumulate in the subintimal area, either by endothelial-mediated transport or, in the absence of endothelium, by directly binding to exposed connective tissue components and subsequently becoming trapped after endothelial regeneration. Uptake of lipoproteins, especially oxidized LDL by macrophages via the scavenger receptor system, then leads to the formation of foam cells and the appearance of the fatty streak.

Intimal smooth muscle cells, whether newly migrated from the media or resident in the normal thickened intima of human coronary arteries, respond to the growth factors released by endothelial cells, platelets, and/or macrophages by modulating to the synthetic phenotype. As a result, large quantities of extracellular matrix components are manufactured, resulting in accumulation of fibrous material and bulk growth of the plaque. Further growth of the established plaque may occur by the same paracrine processes or may, in part, become a self-perpetuating process enhanced by autocrine mechanisms. Although smooth muscle cell proliferation contributes to these processes, the rate of proliferation observed is relatively low, in line with the slow progression typical of most atherosclerotic lesions.[183] Growth of the plaque, or macrophage activity in earlier lesions, may be associated with loss of the overlying endothelium. Thus, even if repaired to restore an intact covering following earlier insults, additional episodes of platelet adherence and release of PDGF may occur, further stimulating smooth muscle cell proliferation and synthetic activity. As the plaque grows still larger, this process may be augmented yet further by the development of cracks in weakened regions of the plaque, causing repeated episodes of intraplaque hemorrhage, thrombosis, and release of PDGF from platelets.

According to the response to injury hypothesis, then, endothelial injury can initiate a host of different, but interrelated, events, the precise sequence and relative importance of which is left open, allowing for a certain plasticity in the genesis of different atherosclerotic plaques. Apart from initiation, endothelial damage may also play a role in progression of the plaque. One pathway of pathogenesis leading from endothelial injury involves macrophages and the fatty streak, but because of the ability of endothelial cells to secrete growth factors directly, this pathway, according to the refined version of the endothelial injury hypothesis, need not necessarily be obligatory for all plaques.

With the focus on endothelial injury in the sense of dysfunction, it is important, however, not to overlook the importance of effects arising from frank mechanical damage, especially deep injury to the vessel wall, as occurs in invasive intervention procedures. In animal models, under conditions in which endothelial denudation alone does not provoke a proliferative response, deeper injury to the subintima and media may do so. Deeper injury of this nature, associated with massive platelet adherence to exposed connective tissue with consequent release of PDGF, is suggested to be a major factor underlying restenosis following angioplasty and atherectomy, and accelerated atherosclerosis in bypass grafts and transplanted hearts.[184,185] The effects of direct mechanical injury to the vessel wall remain a major problem limiting the success of invasive procedures, and this forms a major stimulus for current work on the effects of growth factors and extracellular matrix components on the responses of vascular cells to injury (Chapter 6).

B. Cholesterol and the Lipid Infiltration Hypothesis

The lipid infiltration hypothesis shares a long history with the response to injury hypothesis, dating back to discussion of the role of insudation of plasma constituents into the intima.[186] In its modern form, this hypothesis envisages the key event in atherosclerosis as infiltration of lipids (lipoproteins) from the serum through the endothelial barrier, with deposition into the subendothelial connective tissue. As emphasized by Steinberg,[48] the lipid infiltration and endothelial injury hypotheses cannot be entirely separated from one another; there are many features of overlap and reinforcement between the two. There is, however, a distinction between the two hypotheses with respect to the relative importance of the roles ascribed to lipids and endothelial injury, and their timing. Implicit in the lipid hypothesis is the assumption that advanced lesions evolve only from fatty streaks; the response to injury hypothesis, by contrast, does not preclude other origins. Furthermore, in the lipid hypothesis, endothelial injury is envisaged as important in the progression of the fatty streak to the fibrous plaque, rather than in initiation of the early lesion.

Of the major risk factors for atherosclerosis, elevated serum cholesterol and other lipid abnormalities are the most widely discussed and thoroughly researched.[5,10,31,34,187-189] The transport of cholesterol to and from cells is mediated by various species of lipoprotein particles, cholesterol levels in the blood being determined principally by the rates of uptake and release of these lipoproteins by cells. Lipoprotein binding and uptake is mediated by a number of different lipoprotein receptors, which recognize specific types of protein, called apolipoproteins (apos), that stabilize the surface of the lipoprotein particles.[190] Two of these receptors, the LDL receptor and the macrophage scavenger receptor, have already been referred to in the earlier descriptions of the cell types of the vessel wall. LDL is a cholesterol-rich lipoprotein derived by the action of endothelial lipoprotein lipase on circulating VLDL particles released by the liver. The LDL receptor, extensively characterized through the highly acclaimed studies of Brown and Goldstein,[34,67,147,191,192] recognizes two apolipoproteins: apo B-100 (the sole lipoprotein of LDL) and apo E. Individuals who have inherited defects in the LDL receptor gene or in the apo B-100 molecule (through which the receptor recognizes LDL) have raised serum cholesterol levels owing to their inability to clear adequate quantities of LDL from the blood, and this is associated with a high incidence of early atheroma.[34,193] Two further types of lipoprotein receptor for which candidate proteins have been identified are the high-density lipoprotein receptor[194] and the apo E receptor.[167,195] The former mediates the uptake of high-density lipoproteins (HDL) — lipoproteins containing apo A, which appear to be involved in removal of cholesterol from cells in peripheral tissues, facilitating its transport back to the liver. The latter, identified as LRP (LDL receptor-related protein, equivalent to the α_2 macroglobulin receptor) mediates the clearance by the liver of lipoproteins that contain apo E.[196-200] Principal among the apo-E-containing lipoproteins are chylomicrons, the lipoproteins produced upon absorption of dietary lipids in the intestine, and chylomicron remnants, the residual lipoproteins derived by the action of endothelial lipoprotein lipase on chylomicrons. A more detailed discussion of these lipoproteins, their receptors, and cholesterol homeostasis is presented in Chapter 10.

The discoveries that LDL can undergo oxidative modification in the artery wall and that macrophages possess a scavenger receptor through which they avidly take up oxidatively modified LDL (oxLDL) have been pivotal to the development of the cholesterol hypothesis in its current form.[48] Oxidative modification of LDL, which may take place via a number of chemical pathways, results in dramatic alterations to its properties; oxLDL is no longer recognized by the LDL receptor. It acts as a chemoattractant for monocytes, inhibits macrophage migration, and is cytotoxic.[48] From this standpoint atherogenesis can be envisaged as starting with elevated levels of LDL in the serum, which through the nonsaturable endothelial transport process leads to delivery of large amounts of LDL to the intima. Here the LDL interacts with, and forms complexes with, proteoglycans and other matrix components, resulting in retention and accumulation of lipid in the intima (for a spectacular visual demonstration of this process, see Chapter 2 by Frank and Nievelstein-Post). In the microenvironment of the vessel wall, protected from natural antioxidants such as occur in serum, oxLDL is formed. This process may involve the action of cellular oxygenase or superoxide anion generated by cells, formation of fatty acid lipid peroxides, followed by covalent attachment of oxidized lipids to apo B so that new epitopes are created. The entry of monocytes into the intima can then be attributed to the chemoattractant property of oxLDL; once in the intima, the monocyte/macrophages rapidly take up oxLDL and LDL-proteoglycan complexes via scavenger receptors. Instead of exiting from the vessel wall with their lipid load, the macrophages are immobilized by the inhibitory effect of oxLDL and thus become "foam cells". HDL can stimulate the efflux of cholesterol from foam cells; however, its capacity to do so is diminished by similar oxidative changes to that befalling LDL.[201] The concentration of oxLDL may increase still further by the ability of the accumulated macrophages to convert LDL to oxLDL. With the consequent buildup of foam cells, a fatty streak is formed. The endothelium, intact through the inital stages of this process, may become damaged owing to the cytotoxic effect of oxLDL, and focal desquamation may thus result. The cycles of growth factor production, described under the response to injury hypothesis, then swing into full flow, resulting in conversion of the fatty streak to the fibrous plaque.

A detailed update of the role of oxLDL in atherosclerosis is given by Palinski in Chapter 8.

C. The Monoclonal Hypothesis and Smooth Muscle Cell Neoplasia

This hypothesis centers on the smooth muscle cell, rather than the endothelial cell or the macrophage, and originates from evidence interpreted to suggest that the smooth muscle cells within atherosclerotic lesions are often monoclonal in origin. Smooth muscle cell proliferation in atherosclerosis, according to this view, would represent a neoplastic process, possibly involving a mutational event, rather than a response to arterial injury. The etiological factors responsible could include viruses[181,202] or carcinogens.[203]

The suggestion that populations of smooth muscle cells within a plaque might derive from a single transformed cell originally came from the study of glucose-6-phosphate dehydrogenase (G-6-PD) isozyme types.[204] Samples taken from atherosclerotic plaques from a group of Afro-American females heterozygous for G-6-PD were found frequently to be monotypic, i.e., they displayed just one of the two G-6-PD types. There is, however, a distinction to be drawn between monotypism and monoclonality. Whereas a monoclonal origin is one possible explanation for monotypism, it is not the only one; G-6-PD type could, for example, be an indirect marker for other characteristics of selective survival value in the proliferative response. Moreover, interpretation is complicated by the presence of ditypism in some samples taken elsewhere from the same plaques as those in which monotypic samples were found.

Evidence for the monoclonal hypothesis has, however, been strengthened by the detection of a transforming gene (oncogene) in human atherosclerotic plaques. Penn and colleagues[205] reported that transfection of a cultured mouse fibroblast line (NIH 3T3) with DNA from plaques results in transformation, the transformed cells being capable of producing a low incidence of slow-growing tumors in athymic "nude" mice (which have an impaired cell-mediated immune response). DNA isolated from cultured smooth muscle cells derived from plaques was subsequently shown to be positive in the same transfection-nude mouse assay.[206] The cultured plaque smooth muscle cells show enhanced expression of the protooncogene *myc*, but this does not itself appear to be the transforming gene. Many oncogenes are thought to encode proteins that interfere with the normal growth factor-mediated mechanisms that control cell proliferation.[207] Among the mechanisms identified are enhanced synthesis of a growth factor already being produced, synthesis of growth factor(s) not normally produced, and alterations in the properties of growth factor receptors or their associated signal transduction systems.[207] Expression of oncogenes in smooth muscle cells of the atherosclerotic plaque is thus fully consistent with the pattern of changes identified with the synthetic phenotype.

D. "Encrustation" and the Thrombogenic Hypothesis

The original "thrombus encrustation" hypothesis[208] predates the early version of the injury/insudation hypothesis with which, historically, it has been portrayed as opposing. Further elaborated in the post-war years,[209] the essence of this hypothesis was that atherosclerotic lesions are derived from small mural thrombi that collect (e.g., on areas of injury) and, through subsequent organization by smooth muscle cells, eventually become transformed into fibrotic lesions. As we have seen, it is now generally recognized that plaques consist of far more than transformed thrombi, and there is little evidence to suggest that thrombi act as nucleation sites for the initial development of lesions. Thrombus formation does, however, make an important contribution to the growth and progression of atherosclerotic plaques, and is the major factor in acute clinical sequelae.[7-9,210-212] Once atherosclerosis has become established, repeated episodes of mural thrombosis, associated with endothelial loss or plaque fissure, may occur during the evolution of advanced lesions. On each occasion, thrombotic material may become incorporated into the mass of the plaque as further cycles of smooth muscle cell proliferation and connective tissue production occur in response to the consequent PDGF release. Moreover, as noted at the outset, plaque rupture resulting in major intraluminal thrombosis is the critical event precipitating myocardial infarction and unstable angina. Thus, thrombosis has an important, though variable, part to play in atherosclerosis, which can be integrated into the framework of events proposed by the other hypotheses.

Just as recent work in the lipid field links in with the network of interactions involved in vascular injury and smooth muscle proliferation, so too can it be linked to thrombogenesis. Lipoprotein (a),[213] a highly atherogenic variant of LDL, binds to the plasminogen binding sites on endothelial cells, thereby potentially inhibiting plasminogen activation of the fibrinolytic process and promoting coagulation.[214-216] This lipoprotein could therefore conceivably promote atherogenesis by a dual "lipid accumulation" and "thrombotic" effect. In similar vein, recent evidence suggests that oxLDL has procoagulant effects as an inducer of endothelial tissue factor activity and as a suppressor of protein C activation.[217]

Notwithstanding the general consensus on the role of thrombosis, a minority view effectively turns the tables, suggesting that hypoxia, possibly linked to thrombosis, may be a critical early event and that the problem may not start at the luminal surface at all.[218] This provocative proposal has been developed from observations on an animal model in which intimal proliferation is induced by placing a nonconstricting mechanical collar around the outer surface of the carotid. The essence of the suggestion is that hypoxia resulting from restriction of blood flow in the vasa vasorum (which could similarly arise from thrombosis in the vasa vasorum) leads to PDGF release in the media and, thus, smooth muscle cell proliferation. The parallels of this system to the atherosclerotic disease process in humans remains to be more fully assessed.

E. The Lymphocyte and Immune Mechanisms

Apart from the monocyte/macrophage, which as we have seen, plays multiple complex roles in atherosclerosis, lymphocytes are found in atherosclerotic lesions, and these cells potentially provide another major source of cytokines and mediators for events important in lesion development. Macrophages and lymphocytes are closely interdependent in their functions, and from their presence, atherosclerosis shows all the hallmarks of a chronic inflammatory response to irritation or injury.[47,219-221] Both "suppressor/cytolytic" (CD8-type) and "helper/inducer" (CD4-type) lymphocytes have been identified immunohistochemically in atherosclerotic lesions, and up to one third of these cells may be in an immunologically activated state.[222-225] Some reports suggest that CD8-type cells predominate in early lesions, and CD4-type cells become more abundant in advanced plaques. Recent evidence indicates that the plaque T-cell population is polyclonal, suggesting activation of the cells prior to entry, or activation within the plaque without clonal proliferation.[226]

Reports that chronic serum sickness[227,228] and elevated levels of circulating immune complexes[180,229] may act as promoting factors in atherogenesis have added weight to the idea that immune mechanisms can form an important component of the disease. LDL-immune complexes have been detected in some groups of coronary disease patients, and in experimental studies, these complexes profoundly alter cholesterol metabolism. Evidence is emerging that high levels of circulating immune complexes may, in some instances, be associated with a particular form of lesion featuring a concentric, rather than eccentric, growth pattern, involvement of the media, and particularly pronounced inflammatory cell invasion.[49] These features are suggested to be sufficiently distinct from the usual form of lesion to warrant a separate term, "atheroarteritis", and lesions of this nature have been proposed to account for a proportion of the cases of severe disease in individuals who do not have the usual risk factors.

How do lymphocytes arrive in the lesion and how do they participate in, and augment, the atherosclerotic process? One hypothesis[230] suggests monocytes could be activated by circulating immune complexes; the resulting release of monokines would increase adhesivity of the endothelial plasma membrane and set in train the now familiar events of monocyte migration, macrophage transformation, foam cell formation, and endothelial damage. Once macrophages are established within a lesion, whether by this or other mechanisms, release of interleukin-1 from these cells could lead to adherence and activation of lymphocytes. A key product of the activated lymphocyte is interferon gamma, the presence of which has been demonstrated in lesions.[225] Interferon gamma, among multiple effects, induces expression of major histocompatibility class II antigens, in particular, HLA-DR, in endothelial cells,[231] increasing their adhesivity to lymphocytes. Another important effect of interferon gamma is to stimulate release of interleukin-2 from lymphocytes. This lymphokine has the capacity to augment still further adhesion to the endothelium and ultimately cause its lysis, as well as triggering release of PDGF, interleukin-1, and other growth factors, from macrophages. These growth factors and mediators, in turn, are potent inducers of smooth muscle cell migration and proliferation. Although interferon-gamma itself has an inhibitory effect on smooth muscle cell proliferation, this effect would, in all likelihood, be overwhelmed by the magnitude of the effects of the other lymphokines and growth factors released.

Studies using the immunosuppressive agent cyclosporine A have been undertaken to examine the role of T lymphocytes in more detail. Cyclosporine A, which specifically inhibits the activation of T lymphocytes, has variously been reported to reduce, or have no effect upon, smooth muscle cell proliferation following experimentally induced endothelial injury.[232,233] In a recent investigation of the proliferative response to experimental (balloon-catheter) injury in congenitally athymic nude rats (which lack T lymphocytes), no difference in extent of intimal thickening or of leukocyte population was observed between homozygous and heterozygous animals.[234] Thus, T lymphocytes do not appear to have a significant role in the acute smooth muscle cell proliferative response in this system, though this does not preclude their participation in later stages of lesion development or in human atherosclerosis. Recent evidence does, however, suggest that immune mechanisms are a major factor in the accelerated atherosclerosis ("arteriopathy") associated with transplantation. Here, coronary endothelium expresses high levels of the class II major histocompatibility complex HLA-DR, associated with extensive T-cell (and macrophage) infiltration and proliferation immediately below the endothelium.[235] By contrast, atherosclerotic plaques from nontransplant patients were reported to show HLA-DR expression infrequently, and the distribution pattern of lymphocytes in these lesions appeared different.[235]

Immune mechanisms thus seem to play a variable role in atherosclerosis, which again can be integrated with the framework provided by the preceding hypotheses. In "atheroarteritis" and "transplant arteriopathy" smooth muscle cell proliferation, matrix production and the other events of lesion growth may well be the outcome of a primary severe immune reaction. In the more common forms of atherosclerosis, immune mechanisms may become involved as lesions develop, but, overall, probably play a lesser role. The key importance of the macrophage is well established, and lymphocytes that penetrate the lesion will have local modulating effects on macrophage function. The article by Vollmer and Roessner in Chapter 3 discusses in more detail the parallels between transplant arteriopathy and atherosclerosis from the histopathologist's viewpoint.

VIII. CONCLUDING COMMENTS

Our excursion through selected topical aspects of atherosclerosis research has highlighted the fact that there is no single straightforward answer to the question "what causes atherosclerosis?" Current hypotheses allow for the interplay of multiple causes and effects. That a substantial body of epidemiological observation now appears explicable at the cellular and molecular level increases confidence that current thinking is on the right track. The links established with the lipid/lipoprotein/oxidative modification/macrophage receptor story stand out as a particularly persuasive example in this respect. However, in the process of fashioning unifying theories, evidence that does not fit the argument perfectly is often disregarded, only later assuming importance as new findings come to light. Some of the confident opinions on the nature of atherosclerosis expressed in past decades[236] remind us that the evolution of understanding is a continuing process and that current interpretation represents no more than a transitory "best fit" of the evidence available to date.

From the point of view of those populations afflicted with coronary artery disease, the major outstanding questions center on prevention and treatment. Is lesion regression feasible and can disease progression be halted or slowed? And once invasive intervention has become necessary, can restenosis and accelerated atherosclerosis be prevented or delayed? The answers and practical solutions to these problems represent formidable challenges for the combined interdisciplinary resources of basic and clinical scientists for the 1990s.

ACKNOWLEDGMENTS

The authors are indebted to the British Heart Foundation (N.J. Severs) and the Deutsche Forschungsgemeinschaft (SFB 223, SFB 310; H. Robenek) for continuing support of their work. Our recent joint studies are supported by a NATO Collaborative Research Grant (CRG.910122). We thank all our colleagues for their various contributions. Particular thanks are due to Stephen Rothery for help with photography, artwork, microscopy, and reference organization. The manusucript has been substantially improved by constructive critical comment, and access to human pathological material for microscopy, and for these we are indebted to Dr. Colin R. Green, Dr. Nicholas S. Peters, Jill Reckless, Dr. Kevin Pritchard, and Dr. Jessica Mann.

REFERENCES

1. **Working Group on Arteriosclerosis of the NHLBI,** *Report of the Working Group on Arteriosclerosis of the National Heart, Lung and Blood Institute, Vol.* 2, U.S. Government Printing Office, Washington, D.C., 1981.
2. **The Principal Investigators of the MONICA Project,** WHO MONICA Project, Geographic variation in mortality from cardiovascular diseases, *World Health Stat. Q.*, 40, 171, 1987.
3. **Uemura, K. and Pisa, Z.,** Trends in cardiovascular mortality in industrialized countries since 1950, *World Health Stat. Q.*, 41, 155, 1988.
4. **Woolf, N.,** *Pathology of Atherosclerosis*, Butterworths, London, 1982.
5. **Ross, R.,** The pathogenesis of atherosclerosis — an update, *N. Engl. J. Med.*, 314, 488, 1986.
6. **Woolf, N.,** Pathology of atherosclerosis, *Br. Med. Bull.*, 46, 960, 1990.
7. **Davies, M.J. and Thomas, A.C.,** Plaque fissuring — the cause of acute myocardial infarction, sudden ischaemic death, and crescendo angina, *Br. Heart. J.*, 53, 363, 1985.
8. **Davies, M.J.,** A macro and micro view of coronary vascular insult in ischemic heart disease, *Circulation*, 82 (Suppl. II), 38, 1990.
9. **Fuster, V., Stein, B., Ambrose, J.A., Badimon, L., Badimon, J.J., and Chesebro, J.H.,** Atherosclerotic plaque rupture and thrombosis. Evolving concepts, *Circulation*, 82 (Suppl. II), 47, 1990.
10. **Dzau, V. and Braunwald, E.,** Resolved and unresolved issues in the prevention and treatment of coronary artery disease: a workshop consensus statement, *Am. Heart J.*, 121, 1244, 1991.
11. **Feinleib, M., Thom, T., and Havlik, R.J.,** Decline in coronary heart disease mortality in the United States, *Atheroscler. Rev.*, 9, 29, 1982.
12. **Stallones, R.A.,** Mortality due to ischaemic heart disease: observations and explanations, *Atheroscler. Rev.*, 9, 43, 1982.
13. **Marmot, M.G., Booth, M., and Beral, V.,** International trends in heart disease mortality, *Atheroscler. Rev.*, 9, 19, 1982.
14. **Shaper, A.G.,** Ischaemic heart disease: risk factors and prevention, in *Diseases of the Heart*, Julian, D.G., Camm, A.J., Fox, K.M., Hall, R.J.C., and Poole-Wilson, P.A., Eds., Bailliere Tindall, London, 1989, 1001.
15. **Waller, B.F.,** "Crackers, breakers, stretchers, drillers, scrapers, shavers, burners, welders and melters": the future treatment of atherosclerotic coronary artery disease? A clinico-morphological assessment, *J. Am. Coll. Cardiol.*, 13, 969, 1989.
16. **Serruys, P.W., Strauss, B.H., Beatt, K.J., Bertrand, M.E., Puel, J., Rickards, A.F., Meier, B., Goy, J.J., Vogt, P., Kappenberger, L., and Sigwart, U.,** Angiographic follow-up after placement of a self expanding coronary artery stent, *N. Engl. J. Med.*, 324, 13, 1991.
17. **Dorros, G., Cowley, M.J., Simpson, J., Bentivvoglio, L.G., Block, P.C., Bourassa, M., Detre, K., Grsselin, A.J., Grüntzig, A.R., Kelsey, S.F., Kent, K.M., Mock, M.B., Mullin, S.M., Myler, R.K., Passamani, E.R., Stertzer, S.H., and Williams, D.O.,** Percutaneous transluminal coronary angioplasty: report of complications from the National Heart, Lung and Blood Institute PTCA Registry, *Circulation*, 67, 723, 1983.
18. **Bredlau, C.E., Roubin, G.S., Leimgruber, P.P., Douglass, J.S., King, S.B., III, and Gruntzig, A.R.,** In-hospital morbidity and mortality in patients undergoing elective coronary angioplasty, *Circulation*, 72, 1044, 1985.
19. **Nobuyoshi, M., Kimura, T., Nosaka, H., Mioka, S., Ueno, K., Yokoi, H., Hamasaki, N., Horiuchi, H., and Ohishi, H.,** Restenosis after successful percutaneous transluminal coronary angioplasty: serial angiographic follow-up of 229 patients, *J. Am. Coll. Cardiol.*, 12, 616, 1988.
20. **Lange, R.A., Flores, E.D., and Hillis, L.D.,** Restenosis after coronary balloon angioplasty, *Ann. Rev. Med.*, 42, 127, 1991.

21. **Karas, S.P., Santoian, E.C., and Gravanis, M.B.,** Restenosis following coronary angioplasty, *Clin. Cardiol.*, 14, 791, 1991.
22. **Stewart, R.W., Cosgrove, D.M., Loop, F.D., and Lytle, B.W.,** Current status of coronary artery surgery, *Heart Transplant*, 3, 210, 1984.
23. **Atkinson, J.B., Forman, M.B., Vaughn, W.F., Robinowitz, M., McAlister, H.A., and Virani, R.,** Morphologic changes in long-term saphenous vein grafts, *Chest*, 88, 341, 1985.
24. **Billingham, M.E.,** Cardiac transplant atherosclerosis, *Transplant Proc.*, 19, 19, 1987.
25. **Barnhart, G.R., Pascoe, E.A., Mills, S.A., Szentpetery, S., Eich, D.M., Mohanakumar, T., Hastillo, A., Thompson, J.A., Hess, M.L., and Lower, R.R.,** Accelerated coronary arteriosclerosis in cardiac transplant recipients, *Transplant Rev.*, 1, 31, 1987.
26. **Braunlin, E.A., Hunter, D.W., Canter, C.E., Gutierrez, F.R., Ring, W.S., Olivari, M.T., Titus, J.L., Spray, T.L., and Bolman, R.M., III,** Coronary artery disease in pediatric cardiac transplant recipients receiving triple-drug immunosuppression, *Circulation*, 84 (Suppl. 3), 303, 1991.
27. **Hunt, S. and Billingham, M.,** Long-term results of cardiac transplantation, *Ann. Rev. Med.*, 42, 437, 1991.
28. **Kuwahara, M., Jacobsson, J., Kagan, E., Ramwell, P.W., and Foegh, M.L.,** Coronary artery ultrastructural changes in cardiac transplant atherosclerosis in the rabbit, *Transplantation*, 52, 759, 1991.
29. **Davies, M.J., Krikler, D.M., and Katz, D.,** Atherosclerosis: inhibition or regression as therapeutic possibilities, *Br. Heart J.*, 65, 302, 1991.
30. **De Feyter, P.J., Serruys, P.W., Davies, M.J., Richardson, P., Lubsen, J., and Oliver, M.F.,** Quantitative coronary angiography to measure progression and regression of coronary atherosclerosis: value, limitations and implications for clinical trials, *Circulation*, 84, 412, 1991.
31. **Tunstall-Pedoe, H.,** Epidemiology and prevention of ischaemic heart disease, *Curr. Opinion Cardiol.*, 5, 379, 1990.
32. **Gorlin, R.,** Hypertension and ischaemic heart disease: the challenge of the 1990s, *Am. Heart J.*, 121 (Suppl. 2, part 2), 658, 1991.
33. **Chamberlain, J.C. and Galton, D.J.,** Genetic susceptibility to atherosclerosis, *Br. Med. Bull.*, 46, 917, 1990.
34. **Goldstein, J.L. and Brown, M.S.,** Familial hypercholesterolemia, in *The Metabolic Basis of Inherited Disease*, Stanbury, J.B., Wyngaarden, J.B., Fredrickson, D.S., Goldstein, J.L., and Brown, M.S., Eds., McGraw-Hill, New York, 1983, 672.
35. **Thompson, G.R.,** Primary hyperlipidaemia, *Br. Med. Bull.*, 46, 986, 1990.
36. **White, R.A.,** *Atherosclerosis & Arteriosclerosis: Human Pathology and Experimental Animal Methods and Models*, CRC Press, Boca Raton, FL, 1989.
37. **Davies, M.J. and Woolf, N.,** *Atheroma. Atherosclerosis in Ischaemic Heart Disease, Vol. 1, The Mechanisms*, Science Press, London, 1990.
38. **Poole-Wilson, P.A. and Sheridan, D.J.,** *Atheroma. Atherosclerosis in Ischaemic Heart Disease, Vol. 2, Myocardial Consequences*, Science Press, London, 1990.
39. **Rhodin, J.A.G.,** Architecture of the vessel wall, in *Handbook of Physiology, The Cardiovascular System*, Bohr, D.F., Somlyo, A.P., and Sparks, H.V., Eds., American Physiological Society, Bethesda, MD, 1980, 1.
40. **Maseri, A., Crea, F., Kaski, J.C., and Crake, T.,** Mechanisms of angina pectoris in syndrome X, *J. Am. Coll. Cardiol.*, 17, 499, 1991.
41. **Hamby, R.I.,** Cardiac anatomy in coronary atherosclerosis, in *Clinical-Anatomical Correlates in Coronary Artery Disease*, Futura Publications, Mount Kisco, New York, 1979, 80.
42. **Stary, H.,** Evolutional progression of atherosclerotic lesions in coronary arteries of children and young adults, *Arteriosclerosis*, 9 (Suppl.1), 19, 1989.
43. **McConnell, J., Benson, G.S., and Schmidt, W.A.,** The vasculature of the human penis: a reexamination of the morphological basis for the bolster theory of erection, *Anat. Rec.*, 203, 475, 1982.

44. **Banya, Y., Ushiki, T., Ishida, K., Tohyama, K., Kurosawa, T., Goto, Y., Aoki, H., Ide, C., and Kubo, T.,** Wall structure of arteriovenous anastomoses in the human corpus cavernosum penis as revealed by light, scanning and transmission microscopy, *Int. J. Impotence Res.*, 2, 21, 1990.
45. **Ross, R., Wight, T.N., Strandness, E., and Thiele, B.,** Human atherosclerosis. I. Cell constitution and characteristics of advanced lesions of the superficial femoral artery, *Am. J. Pathol.*, 114, 79, 1984.
46. **Woolf, N.,** The morphology of atherosclerotic lesions, in *Pathology of Atherosclerosis*, Butterworths, London, 1982, 47.
47. **Munro, J.M. and Cotran, R.S.,** Biology of disease — the pathogenesis of atherosclerosis: atherogenesis and inflammation, *Lab. Invest.*, 58, 249, 1988.
48. **Steinberg, D., Parthasarathy, S., Carew, T.E., Khoo, J.C., and Witztum, J.L.,** Beyond cholesterol. Modifications of low-density lipoprotein that increase its atherogenicity, *N. Engl. J. Med.*, 320, 915, 1989.
49. **Wissler, R.W.,** Update on the pathogenesis of atherosclerosis, *Am. J. Med.*, 91 (Suppl. 1B), 3S, 1991.
50. **Watanabe, T., Tokunaga, O., Fan, J., and Shimokama, T.,** Atherosclerosis and macrophages, *Acta Pathol. Jpn.*, 39, 473, 1989.
51. **Wolfbauer, G., Glick, J.M., Minor, L.K., and Rothblat, G.H.,** Development of the smooth muscle foam cell: uptake of macrophage lipid inclusions, *Proc. Natl. Acad. Sci. U.S.A.*, 83, 7760, 1986.
52. **Davies, M.J., Woolf, N., Rowles, P.M., and Pepper, J.,** Morphology of the endothelium over atherosclerotic plaques in human coronary arteries, *Br. Heart J.*, 60, 459, 1988.
53. **Faggiotto, A. and Ross, R.,** Studies of hypercholesterolemia in the non-human primate. II. Fatty streak conversion into fibrous plaque, *Arteriosclerosis*, 4, 341, 1984.
54. **Rosenfeld, M.E., Tsukada, T., Chait, A., Bierman, E.L., Gown, A.M., and Ross, R.,** Fatty streak expansion and maturation in Watanabe heritable hyperlipidemic and comparably hypercholesterolemic fat-fed rabbits, *Arteriosclerosis*, 7, 24, 1987.
55. **Ross, R.,** Endothelial injury and atherosclerosis, in *Endothelial Cell Biology in Health and Disease*, Simionescu, N. and Simionescu, M., Eds., Plenum Press, New York, 1988, 371.
56. **Simionescu, M. and Simionescu, N.,** Functions of the endothelial cell surface, *Ann. Rev. Physiol.*, 48, 279, 1986.
57. **Simionescu, M.,** Receptor-mediated transcytosis of plasma molecules by vascular endothelium, in *Endothelial Cell Biology in Health and Disease*, Simionescu, N. and Simionescu, M., Eds., Plenum Press, New York, 1988, 69.
58. **Wagner, D.D., Olmsted, J.B., and Marder, V.J.,** Immunolocalization of von Willebrand protein in Weibel-Palade bodies of human endothelial cells, *J. Cell Biol.*, 95, 355, 1982.
59. **Warhol, M.D. and Sweet, J.M.,** The ultrastructural localization of von Willebrand factor in endothelial cells, *Am. J. Pathol.*, 117, 310, 1984.
60. **Severs, N.J.,** Caveolae: static inpocketings of the plasma membrane, dynamic vesicles or plain artifact? *J.Cell Sci.*, 90, 341, 1988.
61. **Wagner, R.C. and Chen, S.-C.,** Transcapillary transport of solute by the endothelial vesicular system: evidence from thin serial section analysis, *Microvasc. Res.*, 42, 139, 1991.
62. **Clough, G.,** Relationship between microvascular permeability and ultrastructure, *Prog. Biophys. Molec. Biol.*, 55, 47, 1991.
63. **Bearer, E.L. and Orci, L.,** Endothelial fenestral diaphragms: a quick-freeze deep-etch study, *J. Cell Biol.*, 100, 418, 1985.
64. **Wagner, R.C. and Casley-Smith, J.R.,** Endothelial vesicles. Review, *Microvasc. Res.*, 21, 267, 1981.
65. **Simionescu, N.,** Transcytosis and traffic of membranes in the endothelial cell, in *International Cell Biology*, Schweiger, H.G., Ed., Springer-Verlag, Berlin, 1981, 657.
66. **Vasile, E., Simionescu, M., and Simionescu, N.,** Visualization of the binding, exocytosis and transcytosis of low density lipoprotein in the arterial endothelium in situ, *J. Cell Biol.*, 96, 1677, 1983.

67. **Goldstein, J.L., Brown, M.S., Anderson, R.B., Russel, D.W., and Scheider, W.J.,** Receptor-mediated endocytosis: concepts emerging from the LDL receptor system, *Ann. Rev. Cell. Biol.*, 1, 1, 1985.
68. **Lin, H.C., Moore, M.S., Sanan, D.A., and Anderson, R.G.W.,** Reconstitution of clathrin-coated pit budding from plasma membranes, *J. Cell Biol.*, 114, 881, 1991.
69. **Pearse, B.M.F. and Robinson, M.S.,** Clathrin, adaptors, and sorting, *Ann. Rev. Cell. Biol.*, 6, 151, 1990.
70. **Hajjar, D.P., Falcone, D.J., Fowler, S., and Minick, C.R.,** Endothelium modifies the altered metabolism of the injured aortic wall, *Am. J. Pathol.*, 102, 28, 1981.
71. **Davies, P.F., Truskey, G.A., Warren, H.B., O'Connor, S.E., and Eisenhaure, B.H.,** Metabolic cooperation between vascular endothelial cells and smooth muscle cells in co-culture: changes in low density lipoprotein metabolism, *J. Cell Biol.*, 101, 871, 1985.
72. **Simionescu, M., Simionescu, N., and Palade, G.E.,** Segmental differentiations of cell junctions in the vascular endothelium. The microvasculature, *J. Cell Biol.*, 67, 863, 1975.
73. **Simionescu, M., Simionescu, N., and Palade, G.E.,** Segmental differentiations of cell junctions in the vascular endothelium. Arteries and veins, *J. Cell Biol.*, 68, 705, 1976.
74. **Claude, P.,** Morphological factors influencing transepithelial permeability: a model for the resistance of the zonula occludens, *J. Membr. Biol.*, 39, 219, 1978.
75. **Kachar, B. and Reese, T.S.,** Evidence for the lipidic nature of tight junction strands, *Nature*, 296, 464, 1982.
76. **Stevenson, B.R., Siliciano, J.D., Mooseker, M.S., and Goodenough, D.A.,** Identification of ZO-1: a high molecular weight polypeptide associated with the tight junction (zonula occludens) in a variety of epithelia, *J. Cell Biol.*, 103, 755, 1986.
77. **Citi, S., Sabanay, H., Jakes, R., Geiger, B., and Kendrick-Jones, J.,** Cingulin, a new peripheral component of tight junctions, *Nature*, 333, 272, 1988.
78. **Bennett, M.V.L., Barrio, L.C., Bargiello, T.A., Spray, D.C., Hertzberg, E., and Sáez, J.C.,** Gap junctions: new tools, new answers, new questions, *Neuron*, 6, 305, 1991.
79. **Willecke, K., Hennemann, H., Dahl, E., Jungbluth, S., and Heynkes, R.,** The diversity of connexin genes encoding gap-junctional proteins, *Eur. J. Cell Biol.*, 56, 1, 1991.
80. **Larson, D.M. and Sheridan, J.D.,** Intercellular junctions and transfer of small molecules in primary vascular endothelial cultures, *J. Cell Biol.*, 92, 183, 1982.
81. **Larson, D.M. and Sheridan, J.D.,** Junctional transfer in cultured vascular endothelium. II. Dye nucleotide transfer, *J. Membr. Biol.*, 83, 157, 1985.
82. **Pepper, M.S., Spray, D.C., Chanson, M.C., Montesano, R., Orci, L., and Meda, P.,** Junctional communication is induced in migrating capillary endothelial cells, *J. Cell Biol.*, 109, 3027, 1989.
83. **Franke, W.W., Cowen, P., Grund, C., Kuhn, C., and Kapprell, H.-P.,** The endothelial junction: The plaque and its components, in *Endothelial Cell Biology in Health and Disease*, Simionescu, N. and Simionescu, M., Eds., Plenum Press, New York, 1988, 147.
84. **Takeichi, M.,** Cadherin cell adhesion receptors as a morphogenetic regulator, *Science*, 251, 1451, 1991.
85. **Magee, A.I. and Buxton, R.S.,** Transmembrane molecular assemblies regulated by the greater cadherin family, *Curr. Opinion Cell Biol.*, 3, 854, 1991.
86. **Heimark, R.L., Degner, M., and Schwartz, S.M.,** Identification of Ca^{2+}-dependent cell-cell adhesion molecule in endothelial cells, *J. Cell Biol.*, 110, 1745, 1990.
87. **Hynes, R.O.,** Integrins: a family of cell surface receptors, *Cell*, 48, 549, 1987.
88. **Defilippi, P., van Hinsbergh, V., Bertolotto, A., Rossino, P., Silengo, L., and Tarone, G.,** Differential distribution and modulation of expression of alpha1/beta1 integrin on human endothelial cells, *J. Cell Biol.*, 114, 855, 1991.
89. **Lampugnani, M.G., Resnati, M., Dejana, E., and Marchisio, P.C.,** The role of integrins in the maintenance of endothelial monolayer integrity, *J. Cell Biol.*, 112, 479, 1991.
90. **Hemler, M.E.,** VLA proteins in the integrin family: structures, functions, and their role on leukocytes, *Ann. Rev. Immunol.*, 8, 365, 1990.

91. **Albelda, S.M., Oliver, P.D., Romer, L.H., and Buck, C.A.,** EndoCAM: a novel endothelial cell-cell adhesion molecule, *J. Cell Biol.*, 110, 1227, 1990.
92. **Newman, P.J., Berndt, M.C., Gorsky, J., White, G.C., II, Lyman, S., Paddock, C., and Muller, W.A.,** PECAM-1 (CD 31) cloning and relation to adhesion molecules of the immunoglobulin gene superfamily, *Science*, 247, 1219, 1990.
93. **Albelda, S.M., Muller, W.A., Buck, C.A., and Newman, P.J.,** Molecular and cellular properties of PECAM-1 (endoCAM/CD31): a novel vascular cell-cell adhesion molecule, *J. Cell Biol.*, 114, 1058, 1991.
94. **Gottlieb, A.I., Langille, B.L., Wong, M.K.K., and Kim, D.W.,** Biology of disease — structure and function of the endothelial cytoskeleton, *Lab. Invest.*, 65, 123, 1991.
95. **Kachar, B. and Pinto da Silva, P.,** Rapid massive assembly of tight junctional strands, *Science*, 213, 541, 1981.
96. **Moncada, S., Gryglewski, R., Bunting, S., and Vane, J.R.,** An enzyme isolated from arteries transforms prostaglandin endoperoxides to an unstable substance that inhibits platelet aggregation, *Nature*, 263, 663, 1976.
97. **Marcum, J.A. and Rosenberg, R.D.,** The biochemistry and physiology of anticoagulantly active heparin-like molecules, in *Endothelial Cell Biology in Health and Disease*, Simionescu, N. and Simionescu, M., Eds., Plenum Press, New York, 1988, 207.
98. **Esmon, C.T.,** Assembly and function of the protein C anticoagulant pathway on endothelium, in *Endothelial Cell Biology in Health and Disease*, Simionescu, N. and Simionescu, M., Eds., Plenum Press, New York, 1988, 191.
99. **Stern, D.M., Handley, D.A., and Nawroth, P.P.,** Endothelium and the regulation of coagulation, in *Endothelial Cell Biology in Health and Disease*, Simionescu, N. and Simionescu, M., Eds., Plenum Press, New York, 1988, 275.
100. **Gimbrone, M.A., Jr. and Bevilacqua, M.P.,** Vascular endothelium: functional modulation at the blood interface, in *Endothelial Cell Biology in Health and Disease*, Simionescu, N. and Simionescu, M., Eds., Plenum Press, New York, 1988, 255.
101. **Schleef, R.R., Loskutoff, D.J., and Podor, T.J.,** Immunoelectron microscopic localization of type 1 plasminogen activator inhibitor on the surface of activated endothelial cells, *J. Cell Biol.*, 113, 1413, 1991.
102. **Furchgott, R.F.,** Role of endothelium in responses of vascular smooth muscle, *Circ. Res.*, 53, 537, 1983.
103. **Yanagisawa, M., Kurihara, H., Kimura, S., Tomobe, M., Kobayashi, M., Mitsui, Y., Yazaki, Y., Goto, K., and Masaki, T.,** A novel potent vasoconstrictor peptide produced by vascular endothelial cells, *Nature*, 332, 411, 1988.
104. **Palmer, R.M.J., Ferridge, A.G., and Moncada, S.,** Nitric oxide release accounts for the biological activity of endothelium-derived relaxing factor, *Nature*, 327, 524, 1987.
105. **Palmer, R.M.J., Ashton, D.S., and Moncada, S.,** Vascular endothelial cells synthesize nitric oxide from L-arginine, *Nature*, 333, 664, 1988.
106. **Furchgott, R.F., Khan, M.T., and Jothianandan, D.,** I. Endothelium-derived relaxing factors and nitric oxide. Comparison of properties of nitric oxide and endothelium-derived relaxing factor: some cautionary findings, in *Endothelium-Derived Relaxing Factors*, Rubanyi, G.M. and Vanhoutte, P.M., Eds., S. Karger, Basel, 1990, 8.
107. **Garg, U.C. and Hassid, A.,** Nitric oxide-generating vasodilators and 8-bromo-cyclic guanosine monophosphate inhibit mitogenesis and proliferation of cultured rat vascular smooth muscle cells, *J. Clin. Invest.*, 83, 1774, 1989.
108. **Whittle, B.J.R. and Moncada, S.,** The endothelin explosion. a pathophysiological reality or a biological curiosity? *Circulation*, 81, 2022, 1990.
109. **Ross, R., Raines, E.W., and Bowen-Pope, D.F.,** The biology of platelet-derived growth factor, *Cell*, 46, 155, 1986.
110. **Ross, R.,** Peptide regulatory factors: platelet-derived growth factor, *Lancet*, 1, 1179, 1989.
111. **Heldin, C.-H. and Westermark, B.,** Platelet-derived growth factors: a family of isoforms that bind to two distinct receptors, *Br. Med. Bull.*, 45, 453, 1989.

112. **Larochelle, W.J., Giese, N., May-Siroff, M., Robbins, K.C., and Aaronson, S.A.,** Chimeric molecules map and dissociate the potent transforming and secretory properties of PDGF A and PDGF B, *J. Cell Sci. Suppl.*, 13, 31, 1990.
113. **Dicorleto, P.E. and Bowen-Pope, D.F.,** Cultured endothelial cells produce a platelet-derived growth factor-like protein, *Proc. Natl. Acad. Sci. U.S.A.*, 80, 1919, 1983.
114. **Castellot, J.J., Jr., Addonizio, M.L., Rosenberg, R., and Karnovsky, M.J.,** Cultured endothelial cells produce a heparin-like inhibitor of smooth muscle cell growth, *J. Cell Biol.*, 90, 372, 1981.
115. **Gabella, G.,** Structural apparatus for force transmission in smooth muscles, *Physiol. Rev.*, 64, 455, 1984.
116. **Bagby, R.M.,** Ultrastructure, cytochemistry and organization of myofilaments in vertebrate smooth muscle cells, in *Ultrastructure of Smooth Muscle*, Motta, P.M., Ed., Kluwer, Boston, 1990, 23.
117. **Travo, P., Weber, K., and Osborn, M.,** Co-existence of vimentin and desmin type intermediate filaments in a subpopulation of adult rat vascular smooth muscle cells growing in primary culture, *Exp. Cell Res.*, 139, 87, 1982.
118. **Kargacin, G.J., Cooke, P.H., Abramson, S.B., and Fay, F.S.,** Periodic organization of the contractile apparatus in smooth muscle revealed by the motion of dense bodies in single cells, *J. Cell Biol.*, 108, 1465, 1989.
119. **Draeger, A., Amos, W.B., Ikebe, M., and Small, J.V.,** The cytoskeletal and contractile apparatus of smooth muscle: contraction bands and segmentation of the contractile elements, *J. Cell Biol.*, 111, 2463, 1990.
120. **Small, J.V., Fürst, D.O., and De Mey, J.,** Localization of filamin in smooth muscle, *J. Cell Biol.*, 102, 210, 1986.
121. **Geiger, B., Dutton, A.H., Tokuyasu, K.T., and Singer, S.J.,** Immunoelectron microscope studies of membrane-microfilament interactions: distributions of α-actin, tropomyosin and vinculin in intestinal brush border and chicken gizzard smooth muscle cells, *J. Cell Biol.*, 91, 614, 1981.
122. **Forbes, M.S., Rennels, M.L., and Nelson, E.,** Caveolar systems and sarcoplasmic reticulum in coronary smooth muscle cells of the mouse, *J. Ultrastruct. Res.*, 67, 325, 1979.
123. **Burridge, K., Faith, K., Kelly, T., Nuckolls, G., and Turner, C.,** Focal adhesions: transmembrane junctions between the extracellular matrix and cytoskeleton, *Ann. Rev. Cell. Biol.*, 4, 487, 1988.
124. **Moore, L.K., Beyer, E.C., and Burt, J.M.,** Characterization of gap junction channels in A7r5 vascular smooth muscle cells, *Am. J. Physiol. Cell Physiol.*, 260, C975, 1991.
125. **Forbes, M.S.,** Ultrastructure of vascular smooth muscle cells in mammalian heart, in *The Coronary Artery*, Kalsner, S., Ed., Croom Helm, London, 1982, 3.
126. **Walker, J.W., Somlyo, A.V., Goldman, Y.E., Somlyo, A.P., and Trentham, D.R.,** Kinetics of smooth and skeletal muscle activation by laser pulse photolysis of caged inositol 1,4,5-triphosphate, *Nature*, 327, 249, 1987.
127. **Marston, S.B.,** What is latch? New ideas about tonic contraction in smooth muscle, *J. Musc. Res. Cell Motil.*, 10, 97, 1989.
128. **Marston, S.B. and Redwood, C.S.,** The molecular anatomy of caldesmon. Review article, *Biochem. J.*, 279, 1, 1991.
129. **Small, J.V., Herzog, M., Barth, M., and Draeger, A.,** Supercontracted state of vertebrate smooth muscle cell fragments reveals myofilament lengths, *J. Cell Biol.*, 111, 2451, 1990.
130. **Burgeson, R.E.,** New collagens, new concepts, *Ann. Rev. Cell. Biol.*, 4, 551, 1988.
131. **Kjllén, L. and Lindahl, U.,** Proteoglycans: structure and interactions, *Ann. Rev. Biochem.*, 60, 443, 1991.
132. **Campbell, J.H., Black, M.J., and Campbell, G.R.,** Replication of smooth muscle cells in atherosclerosis and hypertension, in *Blood Cells and Arteries in Hypertension and Atherosclerosis*, Meyer, P. and Marche, P., Eds., Raven Press, New York, 1989, 15.

133. **Campbell, G.R. and Campbell, J.H.,** The phenotypes of smooth muscle expressed in human atheroma, *Ann. N.Y. Acad. Sci.*, 598, 143, 1990.
134. **Gabbiani, G., Kocher, O., Bloom, W.S., Vandekerckhove, J., and Weber, K.,** Actin expression in smooth muscle cells of rat aortic intimal thickening, human atheromatous plaque, and cultured rat aortic media, *J. Clin. Invest.*, 73, 148, 1984.
135. **Glukhova, M.A., Kabakov, A.E., Frid, M.G., Ornatsky, O.I., Belkin, A.M., Mukhin, D.N., Orekhov, A.N., Koteliansky, V.E., and Smirnov, V.N.,** Modulation of human aorta smooth muscle cell phenotype: a study of muscle-specific variants of vinculin, caldesmon, and actin expression, *Proc. Natl. Acad. Sci. U.S.A.*, 85, 9542, 1988.
136. **Kocher, O. and Gabbiani, G.,** Cytoskeletal features of normal and atheromatous human arterial smooth muscle cells, *Hum. Pathol.*, 17, 875, 1986.
137. **Kocher, O., Gabbiani, F., Gabbiani, G., Reidy, M.A., Cokay, M.S., Peters, H., and Hüttner, I.,** Phenotypic features of smooth muscle cells during the evolution of experimental carotid artery intimal thickening. Biochemical and morphologic studies, *Lab. Invest.*, 65, 459, 1991.
138. **Wagner, W.D.,** Modification of collagen and elastin in the human atherosclerotic plaque, in *Pathobiology of the Human Atherosclerotic Plaque*, Glasgov, S., Neumann, W.P., and Schaffer, S.A., Eds., Springer-Verlag, New York, 1990, 167.
139. **Srinivasan, S.R., Radhakrishnamurthy, B., Vijayagopal, P., and Berenson, G.S.,** Proteoglycans, lipoproteins, and atherosclerosis, *Adv. Exp. Med. Biol.*, 285, 373, 1991.
140. **Amento, E.P., Ehsani, N., Palmer, H., and Libby, P.,** Cytokines and growth factors positively and negatively regulate interstitial collagen gene expression in human vascular smooth muscle cells, *Arterioscler. Thromb.*, 11, 1223, 1991.
141. **Walker, L.N., Bowen-Pope, D.F., Ross, R., and Reidy, M.A.,** Production of platelet-derived growth factor-like molecules by cultured arterial smooth muscle cells accompanies proliferation after arterial injury, *Proc. Natl. Acad. Sci. U.S.A.*, 83, 7311, 1986.
142. **Libby, P., Warner, S.J.C., Salomon, R.N., and Birinyi, L.K.,** Production of platelet-derived growth factor-like mitogen by smooth-muscle cells from human atheroma, *N. Engl. J. Med.*, 318, 1493, 1988.
143. **Sejersen, T., Betshultz, C., Sjölund, M., Heldin, C.-H., Westermark, B., and Thyberg, J.,** Rat skeletal myoblasts and arterial smooth muscle cells express the gene for the B chain (c-sis) of platelet-derived growth factor (PDGF) and produce a PDGF-like protein, *Proc. Natl. Acad. Sci. U.S.A.*, 83, 6844, 1986.
144. **Clowes, A.W. and Schwartz, S.M.,** Significance of quiescent smooth muscle migration in the injured rat carotid artery, *Circ. Res.*, 56, 139, 1985.
145. **Clowes, A.W., Clowes, M.M., Fingerle, J., and Reidy, M.A.,** Kinetics of cellular proliferation after arterial injury. V. Role of acute distension in the induction of smooth muscle proliferation, *Lab. Invest.*, 60, 360, 1989.
146. **Massmann, J.,** Mononuclear cell infiltration of the aortic intima in domestic swine, *Exp. Pathol.*, 17, 110, 1979.
147. **Brown, M.S. and Goldstein, J.L.,** Lipoprotein metabolism in the macrophage: implications for cholesterol deposition in atherosclerosis, *Ann. Rev. Biochem.*, 52, 223, 1983.
148. **Gerrity, R.G.,** The role of the monocyte in atherogenesis. I. Transition of blood-borne monocytes into foam cells in fatty lesions, *Am. J. Pathol.*, 103, 181, 1981.
149. **Gerrity, R.G.,** The role of the monocyte in atherosclerosis. II. Migration of foam cells from atherosclerotic lesions, *Am. J. Pathol.*, 103, 191, 1981.
150. **Fan, J., Yamada, T., Tokunaga, O., and Watanabe, T.,** Alterations in the functional characteristics of macrophages induced by hypercholesterolemia, *Virchows Arch. [B]*, 61, 19, 1991.
151. **Wolpe, S.D. and Cerami, A.,** Macrophage inflammatory proteins 1 and 2: members of a novel superfamily of cytokines, *FASEB J.*, 3, 2565, 1989.

152. **Shimokado, K., Raines, E.W., Madtes, D.K., Barrett, T.B., Benditt, E.P., and Ross, R.,** A significant part of macrophage-derived growth factor consists of at least two forms of PDGF, *Cell*, 43, 277, 1985.
153. **Massagué, J.,** The transforming growth factor family, *Ann. Rev. Cell. Biol.*, 6, 597, 1990.
154. **Hsuan, J.J.,** Transforming growth factors β, *Br. Med. Bull.*, 45, 425, 1989.
155. **Boyd, F.T., Cheifetz, S., Andres, J., Laiho, M., and Massagué, J.,** Transforming growth factor-β receptors and binding proteoglycans, *J. Cell Sci. Suppl.*, 13, 131, 1990.
156. **Baird, A. and Walicke, P.A.,** Fibroblast growth factors, *Br. Med. Bull.*, 45, 438, 1989.
157. **Balkwill, F.R.,** Tumour necrosis factor, *Br. Med. Bull.*, 45, 389, 1989.
158. **Jones, E.Y. and Stuart, D.I.,** The structure of tumour necrosis factor — implications for biological function, *J. Cell Sci. Suppl.*, 13, 11, 1990.
159. **Gordon, S.,** *Macrophage Plasma Membrane Receptors: Structure and Function, J. Cell Sci. Suppl.* 9, The Company of Biologists, Cambridge, 1988.
160. **Gordon, S., Perry, V.H., Rabinowitz, S., Chumg, L.-P., and Rosen, H.,** Plasma membrane receptors of the mononuclear phagocyte system, *J. Cell Sci. Suppl.*, 9, 1, 1988.
161. **Mellman, I., Koch, T., Healey, G., Hunziker, W., Lewis, V., Plutner, H., Miettinen, H., Vaux, D., Moore, K., and Stuart, S.,** Structure and function of Fc receptors on macrophages and lymphocytes, *J. Cell Sci. Suppl.*, 9, 45, 1988.
162. **Law, S.K.A.,** C3 receptors on macrophages, *J. Cell Sci. Suppl.*, 9, 67, 1988.
163. **Ravetch, J.V. and Kinet, J.-P.,** Fc receptors, *Ann. Rev. Immunol.*, 9, 457, 1991.
164. **Yin, H.L. and Hartwig, J.H.,** The structure of the macrophage actin skeleton, *J. Cell Sci. Suppl.*, 9, 169, 1988.
165. **Greenberg, S., El Khoury, J., Di Virgilio, F., Kaplan, E.M., and Silverstein, S.C.,** Ca^{2+} - independent F-actin assembly and disassembly during Fc receptor-mediated phagocytosis in mouse macrophages, *J. Cell Biol.*, 113, 757, 1991.
166. **Fogelman, A.M., van Lenten, B.J., Warden, C., Haberland, M.E., and Edwards, P.A.,** Macrophage lipoprotein receptors, *J. Cell Sci. Suppl.*, 9, 135, 1988.
167. **Herz, J., Hamann, U., Rogne, S., Myklebost, O., Gausepohl, H., and Stanley, K.K.,** Surface location and high affinity for calcium of a 500kD liver membrane protein closely related to the LDL-receptor suggest a physiological role as lipoprotein receptor, *EMBO J.*, 7, 4119, 1988.
168. **Naito, M., Kodama, T., Matsumoto, A., Doi, T., and Takahashi, K.,** Tissue distribution, intracellular localization, and *in vitro* expression of bovine macrophage scavenger receptors, *Am. J. Pathol.*, 139, 1411, 1991.
169. **Kita, T., Nagano, Y., Yokode, M., Ishii, K., Kume, N., Ooshima, A., Yoshida, H., and Kawai, C.,** Probucol prevents the progression of atherosclerosis in Watanabe heritable hyperlipidemic rabbit, an animal model for familial hypercholesterolemia, *Proc. Natl. Acad. Sci. U.S.A.*, 84, 5928–5931, 1987.
170. **Carew, T.E., Schwenke, D.C., and Steinberg, D.,** Antiatherogenic effect of probucol unrelated to its hypocholesterolemic effect: evidence that antioxidants in vivo can selectively inhibit low density lipoprotein degradation in macrophage-rich fatty streaks and slow the progression of atherosclerosis in Watanabe heritable hyperlipidemic rabbit, *Proc. Natl. Acad. Sci. U.S.A.*, 84, 7725, 1987.
171. **Palinski, W., Rosenfeld, M.E., Ylä-Herttuala, S., Gurtner, G.C., Scoher, S.S., Butler, S.W., Parthasarathy, S., Carew, T.E., Steinberg, D., and Witztum, J.L.,** Low density lipoprotein undergoes oxidative modification in vivo, *Proc. Natl. Acad. Sci. U.S.A.*, 86, 1372, 1989.
172. **Kodama, T., Freeman, M., Rohrer, L., Zabrecky, J., Matsudaira, P., and Krieger, M.,** Type I macrophage scavenger receptor contains α-helical and collagen-like coiled coils, *Nature*, 343, 531, 1990.
173. **Rohrer, L., Freeman, M., Kodama, T., Penman, M., and Krieger, M.,** Coiled-coil fibrous domains mediate ligand binding by macrophage scavenger receptor type II, *Nature*, 343, 570, 1990.

174. **Henricksen, T., Mahoney, E.M., and Steinberg, D.,** Enhanced macrophage degradation of biologically modified low density lipoprotein, *Arteriosclerosis*, 3, 149, 1983.
175. **Virchow, R.,** Der atheromatöse Prozess der Arterien, *Wien. Med. Wochenschr.*, 6, 825, 1856.
176. **Ross, R. and Glomset, J.A.,** The pathogenesis of atherosclerosis, *N. Engl. J. Med.*, 295, 369, 1976.
177. **Ross, R.,** Cellular interactions in atherosclerosis — the role of growth factors, in *Atherosclerosis VIII*, Crepaldi, G., Gotto, A.M., Manzato, E., and Baggio, G., Eds., Excerpta Medica, Amsterdam, 1989, 13.
178. **Mustard, J.F. and Murphy, E.N.,** Effect of smoking on blood coagulation and platelet survival in man, *Br. Med. J.*, 1, 846, 1963.
179. **Harker, L.A., Ross, R., Slichter, S.J., and Scott, C.R.,** Homocystine-induced arteriosclerosis: the role of endothelial cell injury and platelet response in its genesis, *J. Clin. Invest.*, 58, 731, 1976.
180. **Füst, G., Szondy, E., Szekely, J., Nanai, I., and Gerö, S.,** Studies on the occurrence of circulating immune complexes in vascular disease, *Atherosclerosis*, 29, 181, 1978.
181. **Hajjar, D.P.,** Viral pathogenesis of atherosclerosis: impact of molecular mimicry and viral genes, *Am. J. Pathol.*, 139, 1195, 1991.
182. **Walker, L.N., Ramsey, M.M., and Bowyer, D.E.,** Endothelial healing following defined injury to rabbit aorta: depth of injury and mode of repair, *Atherosclerosis*, 47, 123, 1983.
183. **Gordon, D. and Schwartz, S.M.,** Cell proliferation in human atherosclerosis, *Trends Cardiovasc. Med.*, 1, 24, 1991.
184. **Ip, J.H., Fuster, V., Badimon, L., Badimon, J.J., Taubman, M.B., and Chesebro, J.H.,** Syndromes of accelerated atherosclerosis: role of vascular injury and smooth muscle cell proliferation, *J. Am. Coll. Cardiol.*, 15, 1667, 1990.
185. **Miller, W.P. and Liedtke, A.J.,** Atheromatous coronary artery disease, *Curr. Opinion Cardiol.*, 6, 489, 1991.
186. **Virchow, R.,** Phlagose und Thrombose im Gefäßsystem, in *Gesamelte Abhandlungen zur Wissenschaftlichten Medizin*, Meidinger Sohn., Frankfurt am Main, 458, 1856.
187. **Pekkanen, J., Linn, S., Heiss, G., Suchindran, C.M., Leon, A., Rifkind, B.M., and Tyroler, H.A.,** Ten-year mortality from cardiovascular disease in relation to cholesterol level among men with and without preexisting cardiovascular disease, *N. Engl. J. Med.*, 322, 1700, 1990.
188. **Galton, D.J. and Thompson, G.R.,** *Lipids and Cardiovascular Disease, Vol.46*, Churchill Livingstone, London, 1990.
189. **McMurry, M.P., Cerqueira, M.T., Connor, S.L., and Connor, W.E.,** Changes in lipid and lipoprotein levels and body weight in tarahumara Indians after consumption of an affluent diet, *N. Engl. J. Med.*, 325, 1704, 1991.
190. **Gotto, A.M., Jr.,** Structural and functional aspects of HDL and LDL: influences of apolipoproteins, *Atheroscler. Rev.*, 17, 39, 1988.
191. **Brown, M.S. and Goldstein, J.L.,** Receptor-mediated control of cholesterol metabolism, *Science*, 191, 150, 1976.
192. **Brown, M.S., Kovanen, P.T., and Goldstein, J.L.,** Regulation of plasma cholesterol by lipoprotein receptors, *Science*, 212, 628, 1981.
193. **Breslow, J.L.,** Lipoprotein transport gene abnormalities underlying coronary heart disease susceptibility, *Ann. Rev. Med.*, 42, 357, 1991.
194. **Oram, J.F., McKnight, G.L., and Hart, C.E.,** The high density lipoprotein receptor, *Atheroscler. Rev.*, 20, 103, 1990.
195. **Beisiegel, U., Weber, W., Ihrke, G., Herz, J., and Stanley, K.K.,** The LDL-receptor-related protein, LRP, is an apolopoprotein E-binding protein, *Nature*, 341, 162, 1989.
196. **Strickland, D.K., Ashcom, J.D., Williams, S., Burgess, W.H., Migliorini, M., and Argraves, W.S.,** Sequence identity between the α_2-macroglobulin receptor and low density lipoprotein receptor-related protein suggests that this molecule is a multifunctional receptor, *J. Biol. Chem.*, 265, 17,401, 1990.

197. **Kristensen, T., Moestrup, S.K., Gliemann, J., Bendsten, L., Sand, O., and Sottrup-Jensen, L.,** Evidence that the newly cloned low-density-lipoprotein receptor related protein (LRP) is the α_2-macroglobulin receptor, *FEBS Lett.*, 276, 151, 1990.
198. **Herz, J., Goldstein, J.L., Strickland, D.K., Ho, Y.K., and Brown, M.S.,** 39-kDa protein modulates binding of ligands to low density lipoprotein receptor-related protein/α_2-macroglobulin receptor, *J. Biol. Chem.*, 266, 21,232, 1991.
199. **Beisiegel, U., Weber, W., and Bengtsson-Olivecrona, G.,** Lipoprotein lipase enhances the binding of chylomicrons to low density lipoprotein receptor-related protein, *Proc. Natl. Acad. Sci. U.S.A.*, 88, 8342, 1991.
200. **Kowal, R.C., Herz, J., Goldstein, J.L., Esser, V., and Brown, M.S.,** Low density lipoprotein receptor-related protein mediates uptake of cholesteryl esters derived from apoprotein E-enriched lipoproteins, *Proc. Natl. Acad. Sci. U.S.A.*, 86, 5810, 1989.
201. **Nagano, Y., Arai, H., and Kita, T.,** High density lipoprotein loses its effect to stimulate efflux of cholesterol from foam cells after oxidative modification, *Proc. Natl. Acad. Sci. U.S.A.*, 88, 6457, 1991.
202. **Benditt, E.P., Barrett, T., and McDougall, J.K.,** Viruses in the etiology of atherosclerosis, *Proc. Natl. Acad. Sci. U.S.A.*, 80, 6386, 1983.
203. **Majesky, M.W., Reidy, M.A., Benditt, E.P., and Jachau, M.R.,** Focal smooth muscle proliferation in the aortic intima produced by an initiation-promotion sequence, *Proc. Natl. Acad. Sci. U.S.A.*, 82, 3450, 1985.
204. **Benditt, E.P. and Benditt, J.M.,** Evidence for a monoclonal origin of human atherosclerotic plaques, *Proc. Natl. Acad. Sci. U.S.A.*, 70, 1753, 1973.
205. **Penn, A., Garte, S.J., Warren, L., Nesta, D., and Mindich, B.,** Transforming gene in human atherosclerotic plaque DNA, *Proc. Natl. Acad. Sci. U.S.A.*, 83, 7951, 1986.
206. **Parkes, J.L., Cardell, R.R., Hubbard, F.C., Hubbard, D., Meltzer, A., and Penn, A.,** Cultured human atherosclerotic plaque smooth muscle cells retain transforming potential and display enhanced expression of the *myc* protooncogene, *Am. J. Pathol.*, 138, 765, 1991.
207. **Marshall, C.J.,** Oncogenes, *J. Cell Sci. Suppl.*, 4, 417, 1986.
208. **Rokitansky, C.,** *Manual of Pathological Anatomy, Vol. 4,* Sydenham Society, London, 1852.
209. **Duguid, J.B.,** Thrombosis as a factor in the pathogenesis of aortic atherosclerosis, *J. Pathol. Bacteriol.*, 60, 57, 1948.
210. **Davies, M.J. and Thomas, A.,** Thrombosis and acute coronary artery lesions in sudden cardiac ischaemic death, *N. Engl. J. Med.*, 310, 1137, 1984.
211. **Richardson, P.D., Davies, M.J., and Born, G.V.R.,** Influences of plaque configuration and stress distribution on fissuring of coronary atherosclerotic plaques, *Lancet*, ii, 941, 1989.
212. **Forrester, J.S., Litvack, F., and Grundfest, W.,** Initiating events of acute coronary arterial occlusion, *Ann. Rev. Med.*, 42, 35, 1991.
213. **Utermann, G.,** The mysteries of lipoprotein (a), *Science*, 246, 904, 1989.
214. **Scott, J.,** Thrombogenesis linked to atherogenesis at last? *Nature*, 341, 22, 1989.
215. **Gonzalez-Gronow, M., Edelberg, J.M., and Pizzo, S.M.,** Further characterization of the cellular plasminogen binding sites: evidence that plasminogen 2 and lipoprotein (a) compete for the same site, *Biochemistry*, 28, 2374, 1989.
216. **Rees, A., Bishop, A., and Morgan, R.,** The Apo(a) gene: structure/function relationships and the possible link with thrombotic atheromatous disease, *Br. Med. Bull.*, 46, 873, 1990.
217. **Weis, J.R., Pitas, R.E., Wilson, B.D., and Rodgers, G.M.,** Oxidized low-density lipoprotein increases cultured human endothelial cell tissue factor activity and reduces protein C activation, *FASEB J.*, 5, 2459, 1991.
218. **Martin, J.F., Booth, R.F.G., and Moncada, S.,** Arterial wall hypoxia following thrombosis of the vasa vasorum is an initial lesion in atherosclerosis, *Eur. J. Clin. Invest.*, 21, 355, 1991.
219. **Hansson, G.K., Jonasson, L., Seifert, P.S., and Stemm, E.S.,** Immune mechanisms in atherosclerosis, *Arteriosclerosis*, 9, 567, 1989.

220. **Clerc, G.,** Atherosclerosis as an immune disease, *Med. Hypotheses*, 36, 24, 1991.
221. **Libby, P. and Hansson, G.K.,** Involvement of the immune system in human atherogenesis: current knowledge and unanswered questions, *Lab. Invest.*, 64, 5, 1991.
222. **Jonasson, L., Holm, J., Skalli, O., Bondjers, G., and Hansson, G.K.,** Regional accumulation of T cells, macrophages, and smooth muscle cells in the human atherosclerotic plaque, *Arteriosclerosis*, 6, 131, 1986.
223. **Emeson, E.E. and Robertson, A.L.,** T lymphocytes in aortic and coronary intimas: their potential role in atherogenesis, *Am. J. Pathol.*, 130, 369, 1988.
224. **Munro, J.M., van der Walt, J.D., Munro, C., Chalmers, J.A.C., and Cox, E.L.,** An immunocytochemical analysis of human aortic fatty streaks, *Hum. Pathol.*, 18, 375, 1987.
225. **Hansson, G.K., Holm, J., and Jonasson, L.,** Detection of activated T lymphocytes in the human atherosclerotic plaque, *Am. J. Pathol.*, 135, 169, 1989.
226. **Stemme, S., Rymo, L., and Hansson, G.K.,** Polyclonal origin of T lymphocytes in human atherosclerotic plaques, *Lab. Invest.*, 65, 654, 1991.
227. **Roberts, J.R.,** The effects of serum sickness on cholesterol atherosclerosis in the rabbit, *Circulation*, 22, 657, 1960.
228. **Minick, C.R. and Murphy, G.E.,** Experimental induction of atherosclerosis by synergy of allergic injury to arteries and lipid rich diet. II. Effect of repeatedly injected foreign protein in rabbits fed a lipid rich, cholesterol poor diet, *Am. J. Pathol.*, 73, 265, 1973.
229. **Szondy, E., Horvath, M., Mezey, Z., Szekely, J., Legyel, E., and Gerö, S.,** Free and complexed anti-lipoprotein antibodies in vascular disease, *Atherosclerosis*, 49, 69, 1983.
230. **Lopes-Virella, M.F. and Virella, G.,** Immune mechanisms in the pathogenesis of atherosclerosis, in *Hypercholesterolemia, Hypocholesterolemia, Hypertriglyceridemia*, Malmendier, C.L., Ed., Plenum Press, New York, 1990, 383.
231. **Pober, J.S. and Gimbrone, M.A., Jr.,** Expression of la-like antigens by human vascular endothelial cells is inducible *in vitro*: demonstration by monoclonal antibody binding and immunoprecipitation, *Proc. Natl. Acad. Sci. U.S.A.*, 79, 6641, 1982.
232. **Jonasson, L., Holm, J., and Hansson, G.K.,** Cyclosporin A inhibits smooth muscle cell proliferation in the vascular response to injury, *Proc. Natl. Acad. Sci. U.S.A.*, 136, 967, 1990.
233. **Ferns, G.A.A., Reidy, M.A., and Ross, R.,** Vascular effects of cyclosporine A in vivo and in vitro, *Am. J. Pathol.*, 137, 403, 1990.
234. **Ferns, G.A.A., Reidy, M.A., and Ross, R.,** Balloon catheter de-endothelialization of the nude rat carotid, *Am. J. Pathol.*, 138, 1045, 1991.
235. **Salomon, R.N., Hughes, C.C.W., Schoen, F.J., Payne, D.D., Pober, J.S., and Libby, P.,** Human coronary transplantation-associated atherosclerosis. Evidence for a chronic immune reaction to activated graft endothelial cells, *Am. J. Pathol.*, 138, 791, 1991.
236. **Montague Murray, H.,** *Quain's Dictionary of Medicine*, Longmans, Green and Co., London, Bombay and Calcutta, 1910.
237. **Branton, D., Bullivant, S., Gilula, N.B., Karnovsky, M.J., Moor, H., Mühlethaler, K., Northcote, D.H., Packer, L., Satir, B., Satir, P., Speth, V., Staehelin, L.A., Steere, R.L., and Weinstein, R.S.,** Freeze-etching nomenclature, *Science*, 190, 54, 1975.

Chapter 2

THREE-DIMENSIONAL ORGANIZATION OF THE INTIMA OF RABBIT AORTA IN EARLY ATHEROSCLEROTIC LESION DEVELOPMENT AS VIEWED BY QUICK-FREEZE, DEEP-ETCH ELECTRON MICROSCOPY

Joy S. Frank and Patricia F.E.M. Nievelstein-Post

TABLE OF CONTENTS

ISBN 0-8493-5505-2

I. INTRODUCTION

Extracellular lipid deposition in the intima of the artery wall is an important step in the development of atherosclerotic lesions. The stages from the increased uptake of low-density lipoprotein (LDL) across the endothelial cell, to the retention and possible alteration of LDL in the intima of the artery wall, to the formation of a fatty streak lesion, are still incompletely understood.[1] Schwenke and Carew[2] demonstrated a marked and focal increase in the concentration of intact ^{131}I-labeled LDL in lesion-prone sites of the rabbit aorta after cholesterol feeding. Recent ultrastructural studies demonstrated the presence of unesterified cholesterol-rich extracellular liposomes in the intima of rabbits on a high-cholesterol diet before the onset of monocyte infiltration in the intima.[3,4] Mora et al.[5] used histochemical techniques to colocalize apolipoprotein B, unesterified cholesterol, and extracellular liposomes, in the subendothelial space of hyperlipidemic rabbits.

The subendothelial-intimal matrix is a complex meshwork of various proteoglycans, various collagen types, and other proteins, like elastin and fibronectin.[6,7] It is well known from several *in vitro* studies that LDL has a high affinity for proteoglycans.[8-11] LDL also binds to fibronectin and, to some degree, to collagen[12,13] and elastin.[14] Thus, the extracellular matrix may be very important for retaining LDL in the intima and actively involved in its alteration into a form where it is more avidly taken up by macrophages to form foam cells. Ultrastructural studies that have attempted to visualize early lipid deposition in the intimal meshwork have been hampered by the fact that extracellular lipid and proteoglycans are poorly preserved by routine electron microscopy. No single conventional electron microscopic technique allows the simultaneous visualization of all matrix components at once and retains lipid deposits. Most techniques involve fixation and chemical staining procedures, which can induce alterations in molecular and cellular structure. Even when additives to preserve lipids are utilized, it has been difficult to obtain a three-dimensional image.

In this chapter we discuss two recent studies[15,16] where we have applied quick-freeze, deep-etch, rotary replication techniques to visualize lipid deposition in the rabbit intimal extracellular matrix.[15,16] This technique provides a three-dimensional view of the macromolecular structure of the intima. As a result, we are provided with a more realistic representation of the physical relationships between the complex matrix components and the lipid deposits in the intima. The ultrastructure of the aortic intima of hypercholesterolemic rabbits will be compared with control rabbits. We concentrate on prelesional stages of atherosclerosis by selecting sites of the aorta that are free of fatty streaks or lesions. The results will be discussed in the context of the current concepts in atherogenesis.

II. METHODS

A. Ultrarapid Freezing

All micrographs in this chapter were prepared by ultrarapid freezing of rabbit aorta followed by freeze-etch replication.

Samples of aortic tissue were taken from control rabbits and rabbits with a high level of plasma cholesterol. Hypercholesterolemia was obtained in three different ways: (1) endogenously by using Watanabe Heritable Hyperlipidemic (WHHL) rabbits;[15] (2) exogenously by using rabbits on a high cholesterol diet for 10 or 21 d;[15] and (3) by infusion of a high concentration (320 mg) purified human LDL in a normal rabbit.[16] The animals were sacrificed by an overdose of pentobarbital. A needle was inserted in the left ventricle, and blood was flushed from the aorta by perfusion with oxygenated physiological buffer (133 m*M* NaCl, 3.6 m*M* KCl, 1.0 m*M* $CaCl_2$, 5.0 m*M* Trizma maleate, 16 m*M* dextrose, pH 7.4, 37°C) and allowed to drain into the abdominal cavity. The heart and attached aorta, while being perfused with oxygenated buffer, was placed in a dissection dish. Rings of aortic arch were cut into 3-mm^2 pieces that were immediately placed on moist filter paper glued to aluminum support discs, for ultrarapid freezing.[16]

The initial freezing is the key step. It has to be done fast enough to prevent the formation of ice crystals in and around the intima. The discs were plunged onto an ultrapure copper block that had been precooled by liquid helium (–269°C, 4°K). Because the tissue sample has a high water content and water has a low thermal conductivity, the zone of good freezing is limited to a thickness of approximately 10 μm. While this is definitely a limitation of ultrarapid freezing technology, in our case it proved to be quite helpful. Since we were particularly interested in the superficial intima and subendothelial region of the aorta, it was relatively easy to ensure that we had generated such a superficial fracture by the quality of freezing.

B. Freeze-Fracture and Replication

Fracturing was performed at –150°C and at a vacuum of 1×10^{-7} Torr. For deep etching, the stage of the Balzers 301 freeze-fracture apparatus was warmed gradually over a period of 8 min to –95°C. For 3 min the stage was maintained at –110°C, followed by 3 min at –100°C and 2 min at –95°C. The fractured and etched surface of the aorta was replicated by depositing 2 nm of platinum/carbon from the electron-beam gun mounted at a 25°angle while the tissue was rotated through 360°. This was followed by two 4-sec bursts of carbon. Tissue was digested from the replica with household bleach, rinsed in distilled water, and picked up on formvar-coated grids.

The replicas were examined in a JEOL 100 CX electron microscope. The electron micrographs are printed as negative images. This makes all structures coated with metal look white, so they appear to be highlighted against a background that looks dark and shadowy. Images have a three-dimensional appearance, like low-resolution scanning electron micrographs of cells, but at the resolution of the transmission microscope, in which individual macromolecules can be resolved.

C. Schematic Presentation of Freeze-Fracture Plane

Figure 1 shows a schematic cross section through the superficial layers of a rabbit aorta. The area where most fracture planes are derived is indicated by the arrow and is situated in the superficial intima (also referred to as the subendothelial-intimal space). The relative thickness of the intimal layer in control rabbits (including the LDL-injected rabbits) is shown on the left (under A) and is limited to a few micrometers. In Watanabe Heritable Hyperlipidemic (WHHL) rabbits, as shown on the right (under B), the intima is thickened. The amount of collagen is increased, and there are smooth muscle cells present that have migrated from the media.

III. RESULTS AND DISCUSSION

A. Interpretation of Freeze-Etched Rabbit Intima: General Structure

The type of preservation of cell architecture that we can obtain with quick freezing and deep etching is illustrated in Figure 2. The fracture plane has passed through the luminal surface of an endothelial cell membrane, revealing a densely packed cytoplasm with numerous microfilaments and vesicles of varying shapes and sizes. The fractured plasma membrane has small openings of the endothelial vesicles. Because of the deep etching, the true surface of the endothelial cell can be seen. It consists of the glycocalyx with some of its extensions inserting into the outer surface of the plasmalemma. When the fracture plane moves to the abluminal side of the endothelial cell and into the intima, deep etching provides us with a unique view of this complex matrix network (Figure 3). It consists of filaments with extensive lateral associations that form a honeycomb-like lattice. The lattice is seen to consist of thick filaments with a metal-coated diameter of 6.7 to 7.9 nm (2 nm deposited platinum) and thinner filaments of 2.2 to 2.3 nm in diameter, which branch at irregular intervals. At the branch points variously sized granules of 5 to 10 nm are evident. The uniform distribution of this matrix with its polymeric lattice structure is consistent with the idea that it is composed of proteoglycans and glycosaminoglycans. Embedded within the matrix are abundant collagen and elastin fibers (Figures 4 and 5). The 30 nm-wide collagen fibrils form bundles that extend over large distances. In the lipid-rich intima of WHHL and cholesterol-fed rabbits, the collagen courses up to the basement membrane of the endothelial cells (Figures 6 and 7). Deep etching reveals the presence of periodically spaced (67.4 nm) transverse cuffs of material on the surface of the collagen fibrils, which results in cross banding that is characteristic of type I collagen (Figure 8). Within the bundles of collagen, short filaments

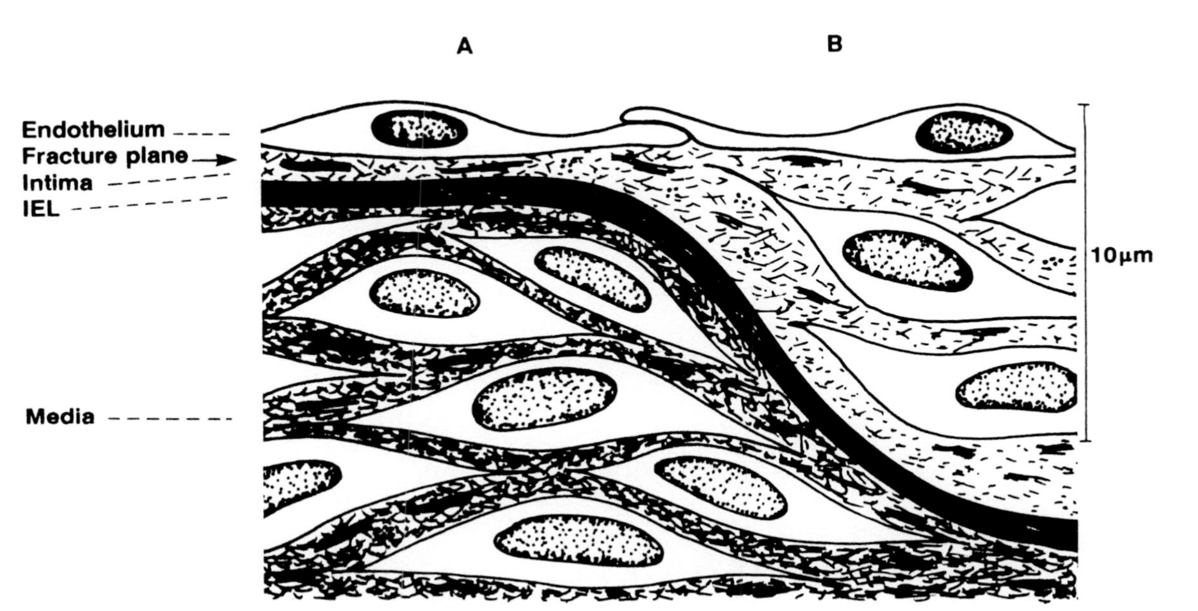

FIGURE 1. Level of fracture plane. The arrow indicates the superficial fracture plane generated in the intima just underneath, and sometimes through, an endothelial cell. The area of good freezing (10 μm) is indicated. On the left (under A) is a cross section through control rabbit aorta. A thickened intima, such as found in WHHL rabbits, is shown on the right (under B). Smooth muscle cells that have migrated from the media into the intima are present in the area of the fracture plane.

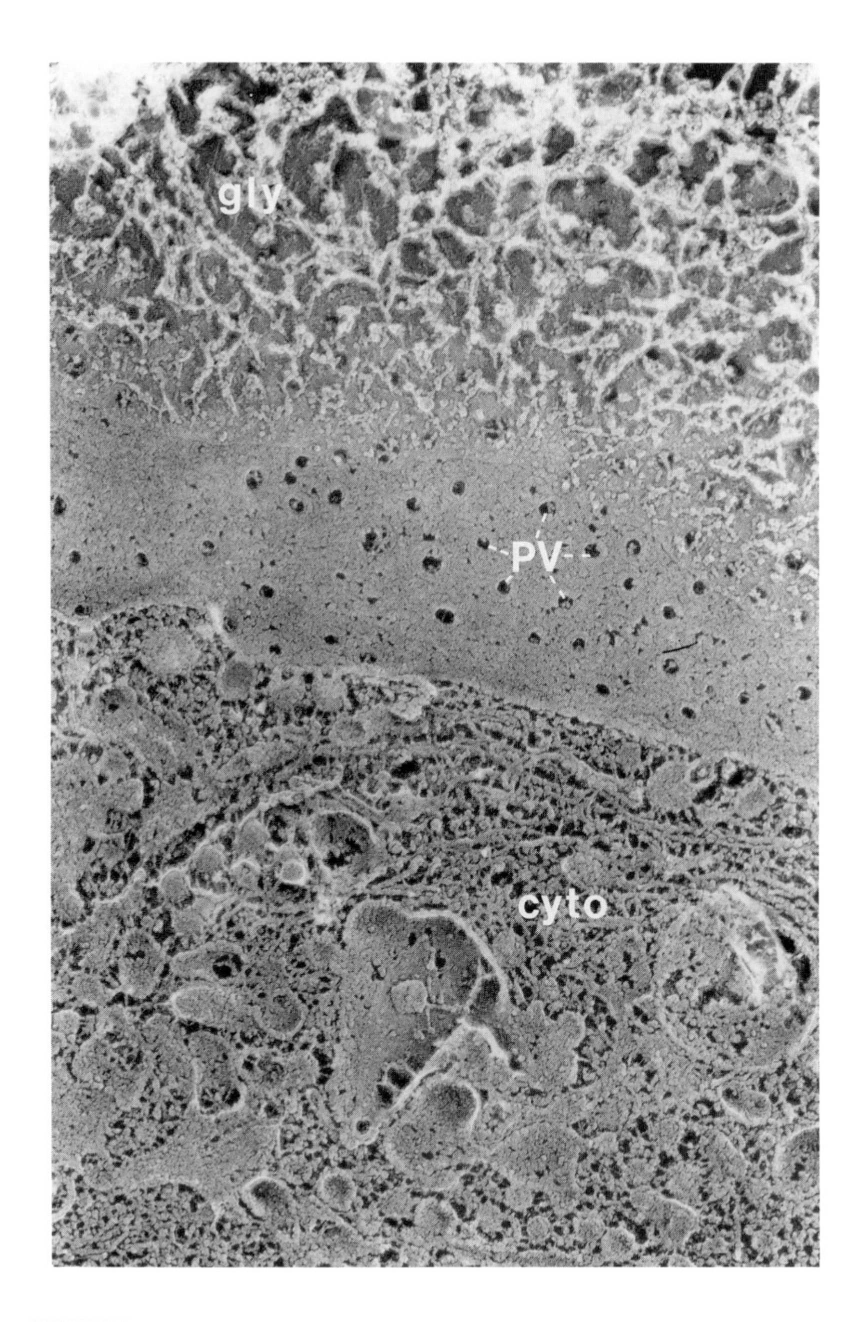

FIGURE 2. Fracture plane through an endothelial cell from control rabbit aorta. Deep etching reveals the glycocalyx (gly) on the outer surface of the cell, the P face of the plasmalemma with numerous pinocytotic (endothelial) vesicles (PV), and the cytoplasm (cyto) with intracellular matrix and organelles. (Magnification × 58,500.)

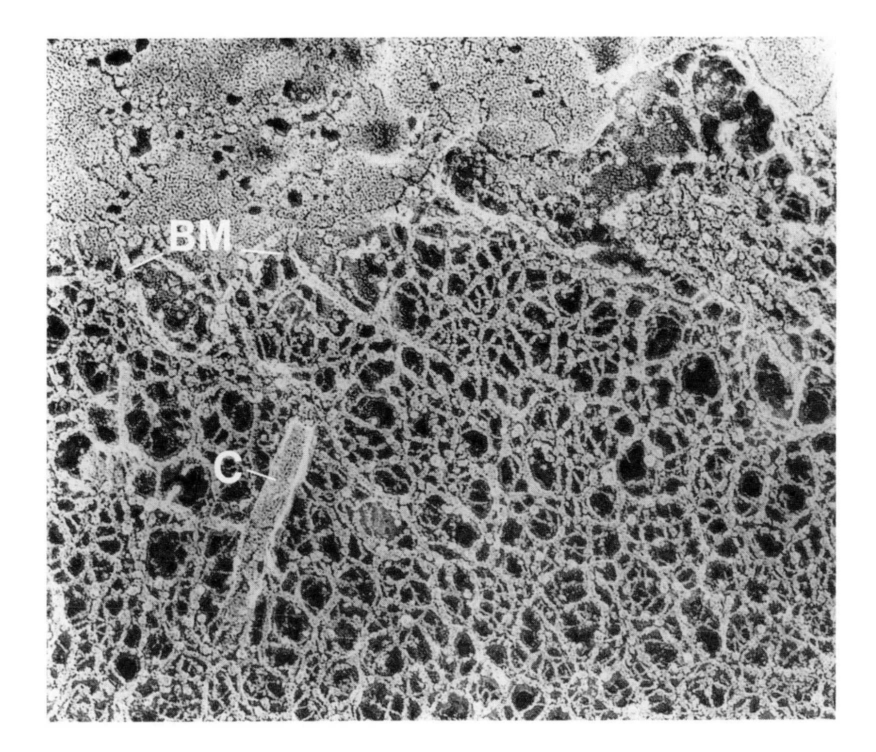

FIGURE 3. Electron micrograph of a freeze-etch replica from a control rabbit aorta just underneath the endothelium. The basement membrane (BM) of an endothelial cell is shown. The extracellular matrix consists of fine filaments arranged in a honeycomb-like organization. A collagen fibril is linked to the fine fibrillar meshwork. (Magnification × 101,700.)

connect the fibrils to each other and to the matrix network. Granules are present along the length of the collagen, especially at junctions between the fibril and the matrix filaments.

B. Lipid Deposition in the Intima

The fatty streak is widely accepted to be the precursor of the fibrous plaque. Foam cells that make up the fatty streak are mostly formed as monocyte/ macrophages take up lipid. One of the earliest events in experimental animals on a high-cholesterol diet is an increase in the adherence of circulating monocytes.[17] Since circulating monocytes contain almost no lipid, they must accumulate lipid when they are in the subendothelial-intimal space. However, Goldstein et al.[18] showed that LDL, the major atherogenic lipoprotein generating foam cell lesions, when incubated with monocytes or macrophages, is not taken up in sufficient rates to generate foam cells. It is now known that LDL must undergo some form of modification in its structure and biological properties (i.e., aggregation and/or oxidation).[19-22] The importance of being able to visualize the earliest lipid deposition in the intima, while maintaining its structural integrity and that of the surrounding matrix, is obvious.

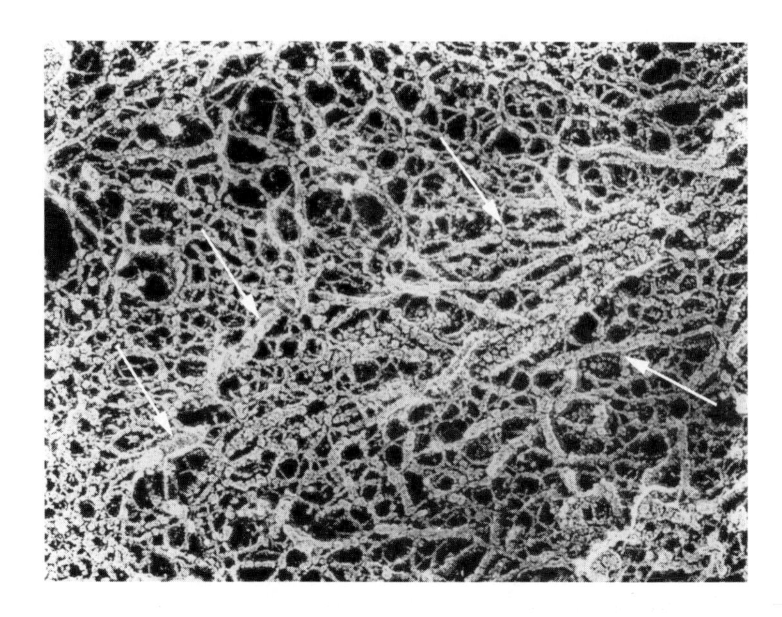

FIGURE 4. Electron micrograph of a freeze-etch replica from a control rabbit aortic intima. Demonstration of elastin fibrils (arrows) in the extracellular matrix. (Magnification × 101,700.)

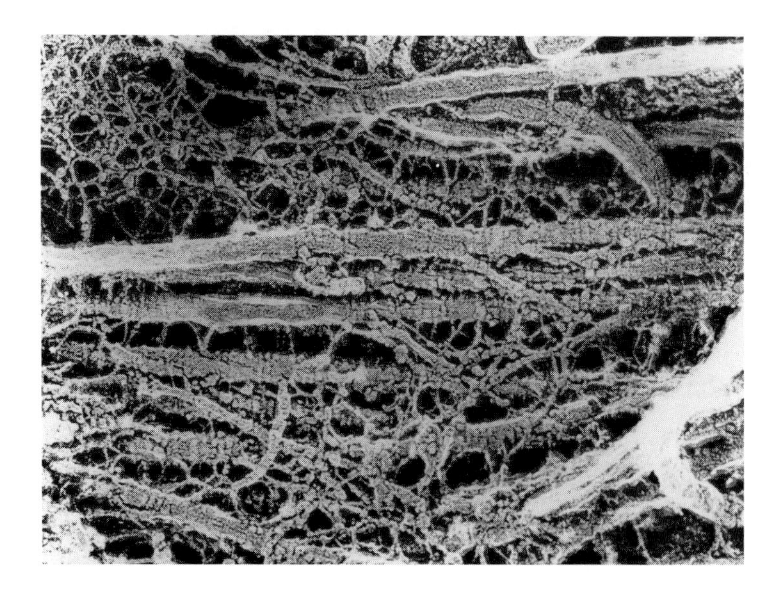

FIGURE 5. Electron micrograph of a freeze-etch replica from a control rabbit aortic intima. Demonstration of collagen fibrils that are connected to finer extracellular matrix fibrils. The typical banding pattern of collagen is visible. (Magnification × 101,700.)

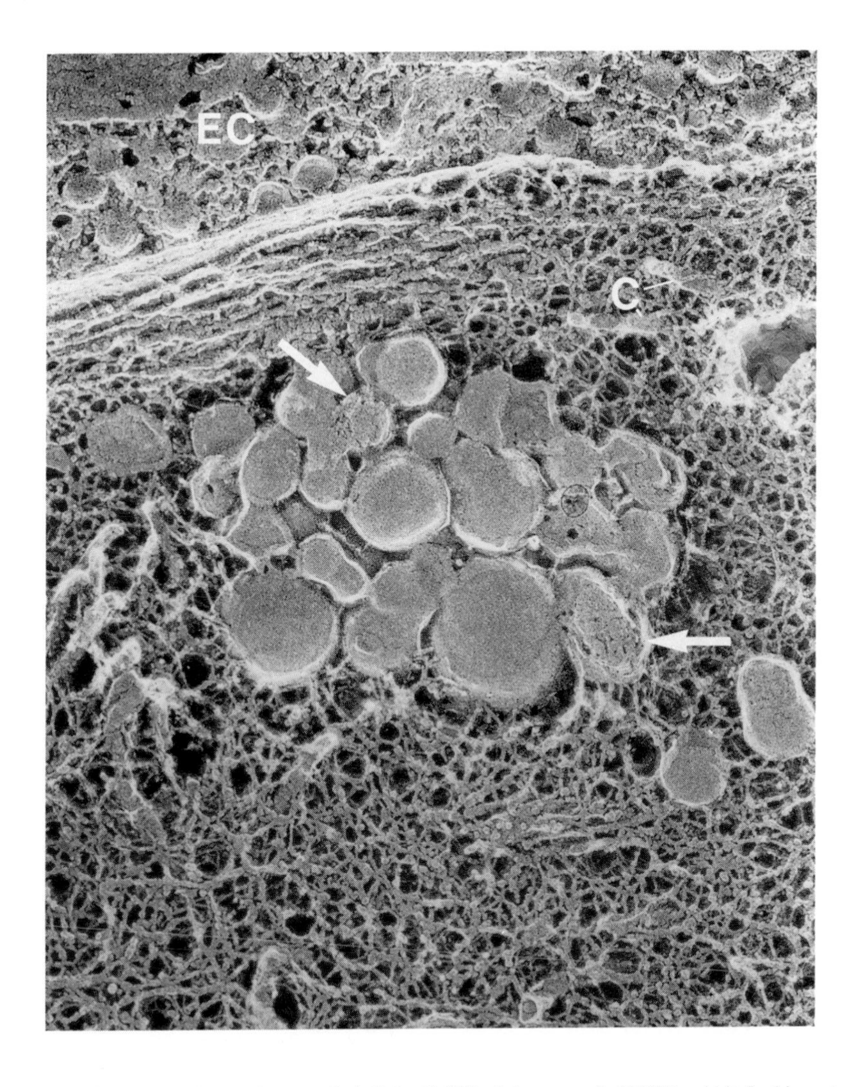

FIGURE 6. Area just under an endothelial cell (EC) of the aorta of a WHHL rabbit. In this part of the subendothelial-intimal space is a tightly clustered group of lipid particles. This lipid appears to be aggregated. Note that the matrix filaments appear to attach to the lipid holding it in the lattice. The surface of most particles is smooth; however, some have a knobby surface (arrows). (Magnification × 75,600).

1. Watanabe Heritable Hyperlipidemic Rabbits

As would be expected, these rabbits had varying amounts of lipid deposition in the intima, from small clusters of liposomes to fully developed plaques. For ultrarapid freezing we selected areas of the WHHL rabbits that were free of fatty streaks or lesions. Figure 9 shows an area typical of the superficial intima. It is characterized by extensive accumulation of lipid particles and collagen, which

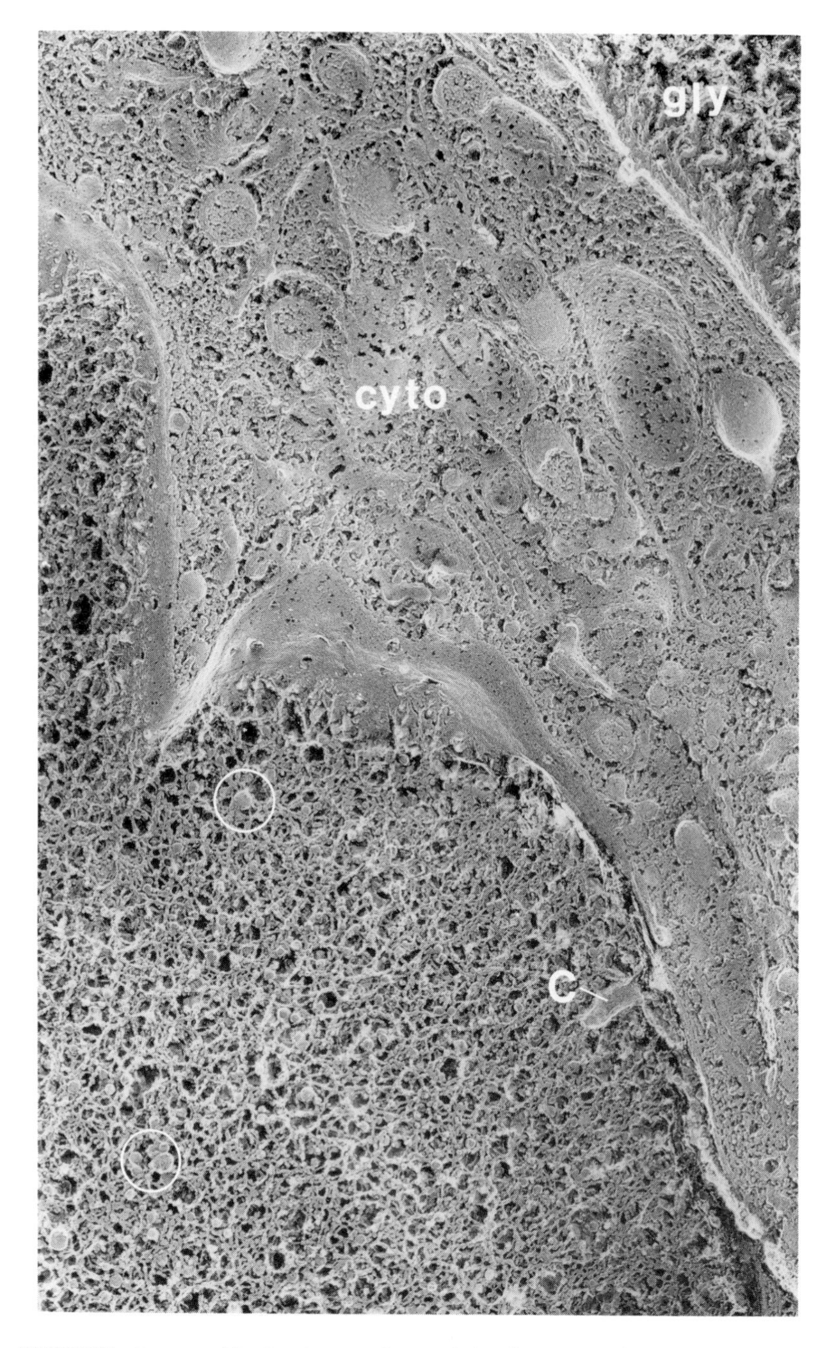

FIGURE 7. Low magnification electron micrograph that gives an overview of a fracture running from the luminal surface of the endothelial cell with its glycocalyx (gly) on the outside, through the cytoplasm (cyto) to the abluminal side of the cell and into its matrix. Enclosed in the circles are individual lipid particles of 40 to 60 nm diameter enmeshed in the filaments of the matrix. This tissue was from the aortic arch of a rabbit after 10 d on a high-cholesterol diet. (Magnification × 35,750.)

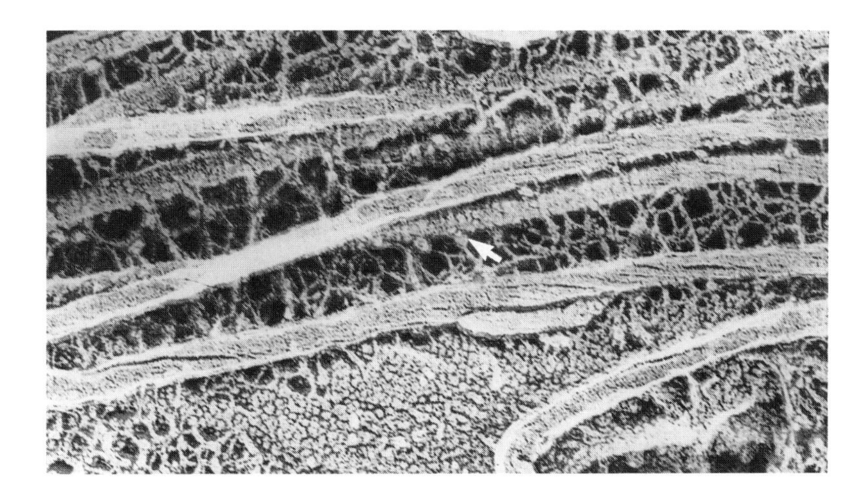

FIGURE 8. Collagen type I fibrils in control rabbit aortic intima. The cross striation typical for collagen fibrils is seen. Part of the fibrils show the internal helical structure. The collagen fibrils are interconnected by fine short filaments. Granules are present at junctions between the fibril and matrix filaments (arrow). (Magnification × 106,300.)

are enmeshed within the matrix filaments. While lipid particles as small as 23 nm could be resolved, the majority of the particles (80%) were between 70 and 169 nm. Many of the lipid particles were in clusters and appeared to be fusing into larger-sized vesicles (Figure 9).

2. Cholesterol-Fed and LDL-Injected Rabbits

Lipid deposition in the intima of cholesterol-fed rabbits occurs as early as 10 d after starting the cholesterol-rich diet.[5,15] These deposits (termed "extracellular liposomes") were found to contain unesterified cholesterol.[3] We were able to see such lipid deposition and its relationship to the matrix in great detail, after 10 d of cholesterol feeding (Figures 7 and 10). The diameter of most of the lipid was between 23 and 68 nm (75%), which is within the range of LDL and β-VLDL. Another feature was that the most frequent configuration of the lipid deposits was in clusters linked directly to the matrix filaments (Figure 10). Collagen and elastin fibers were usually present as well. Within the clusters the smaller particles appeared to be fusing. It is striking that in the 10-d and 21-d cholesterol-fed animals, individual lipid particles between 30 and 40 nm could be seen just below the endothelial cells (Figures 7 and 11). These small lipid particles were enmeshed in the basement membrane matrix complex. Mora et al.[5] demonstrated (in this same area of rabbit aorta) the presence of apolipoprotein B, after 1 week of cholesterol feeding.

In a recent study we injected intravenously 320 mg of human LDL into normal rabbits. We were able to demonstrate with immunofluorescence and immunogold labeling the presence of apolipoprotein B, 2 hours after the bolus injection.[16]

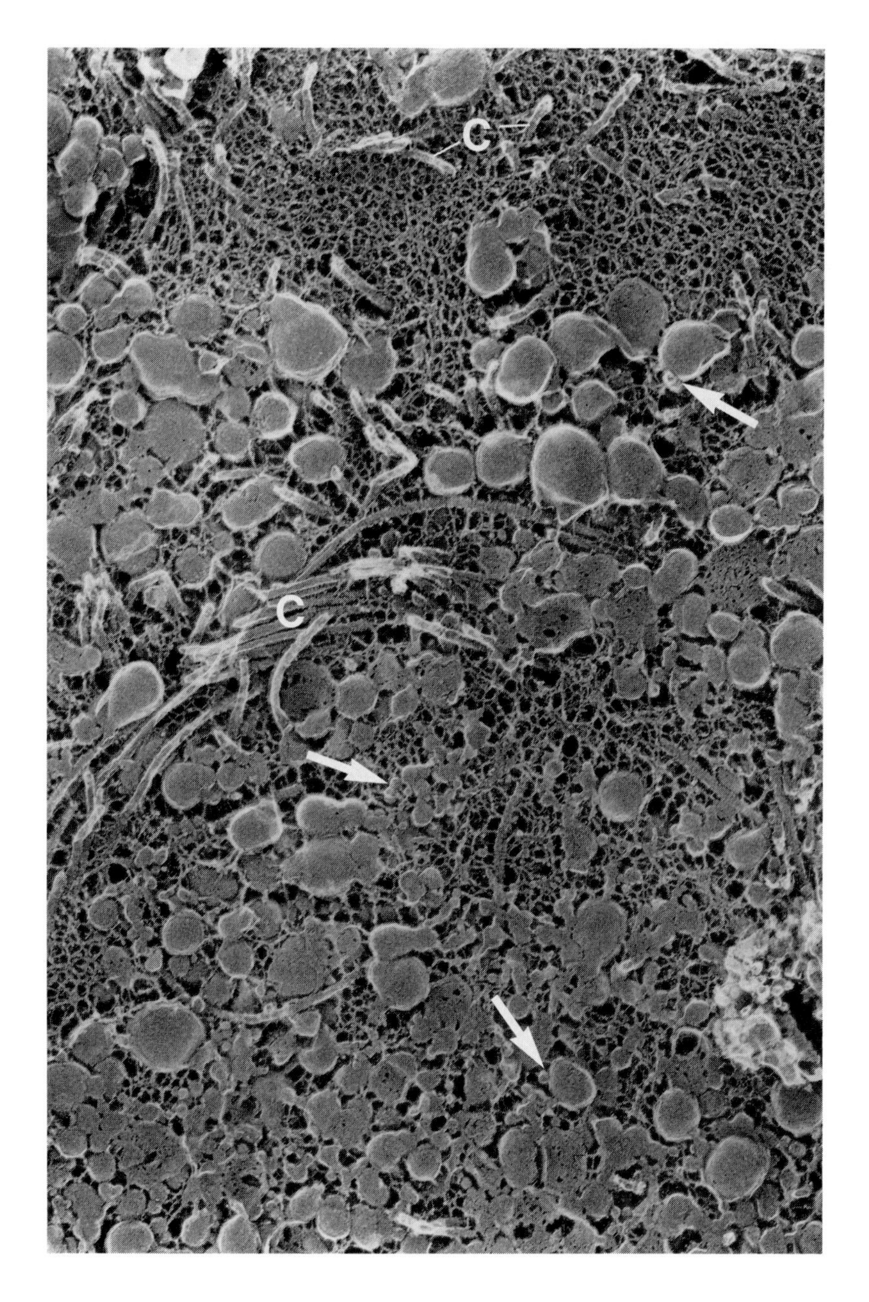

FIGURE 9. Area of intima from aorta of WHHL rabbit. Note large amounts of lipid surrounded by collagen fibrils (C) and enmeshed in the matrix filaments. The lipid particles are in grape-like clusters. The smallest particles appear to be fusing to form the larger particles (examples indicated by arrows). (Magnification × 43,600.)

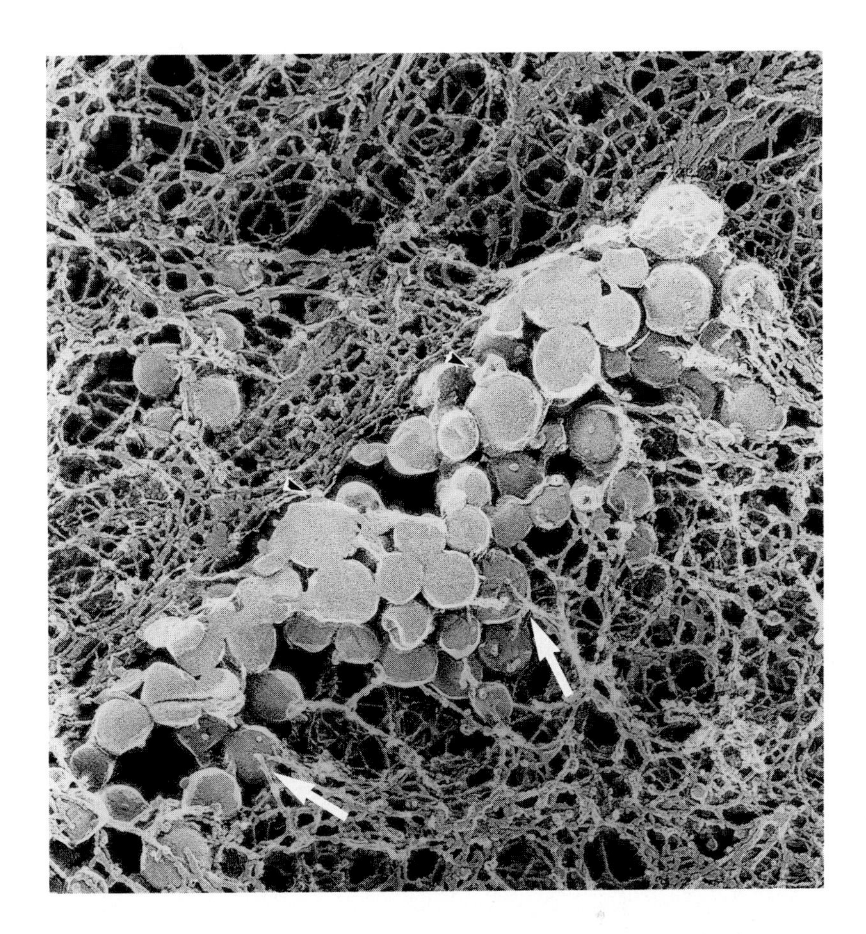

FIGURE 10. Area of the intima from a rabbit aortic arch taken 10 d after a high-cholesterol diet. Clusters of lipid particles are enmeshed in the matrix filaments. Attachments of the matrix to the lipid forming the cluster are indicated by the arrows. Examples of close association of small lipid particles with larger lipid particles, which suggests fusion, is indicated by arrowheads. (Magnification × 72,600.)

With ultrarapid freezing and etching, these LDL-injected rabbits showed the deposition of individual 23 nm-sized lipid particles enmeshed in the matrix, which corresponded in size to the injected LDL. In a few discrete areas there was deposition of lipid aggregates (Figure 12), and sometimes there were clusters of 23 nm-sized lipid particles alongside larger lipid particles (Figure 13). These larger particles were similar in size to the lipid particles described in the WHHL and fat-fed rabbits.[15] The close association of the clustered LDL-sized particles to the larger lipid structures, as seen in Figure 13, suggests that fusion of the 23 nm-sized lipid particles may have occurred to form the larger lipid structures. On higher magnification, 23 nm-sized lipid particles showed a knobby surface

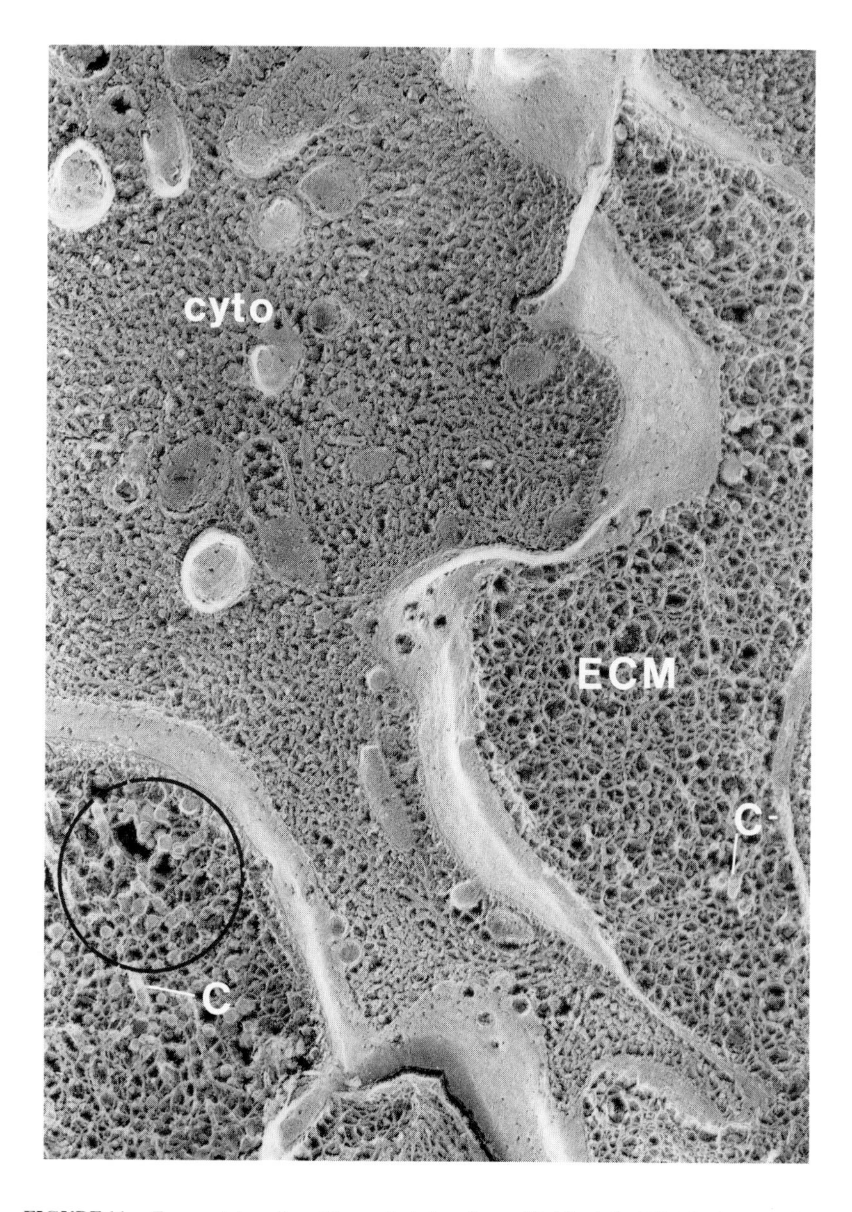

FIGURE 11. Freeze-etch replica of the aortic intima from a 21-d fat-fed rabbit. The fracture plane runs through the cytoplasm (cyto) of the endothelial cell and into the extracellular matrix (ECM). Small lipid particles which are within the size of β-VLDL (30 to 60 nm) are dispersed throughout the extracellular matrix. A group of such particles just underneath the endothelium is indicated within the black circle. (C=Collagen). (Magnification × 42,500.)

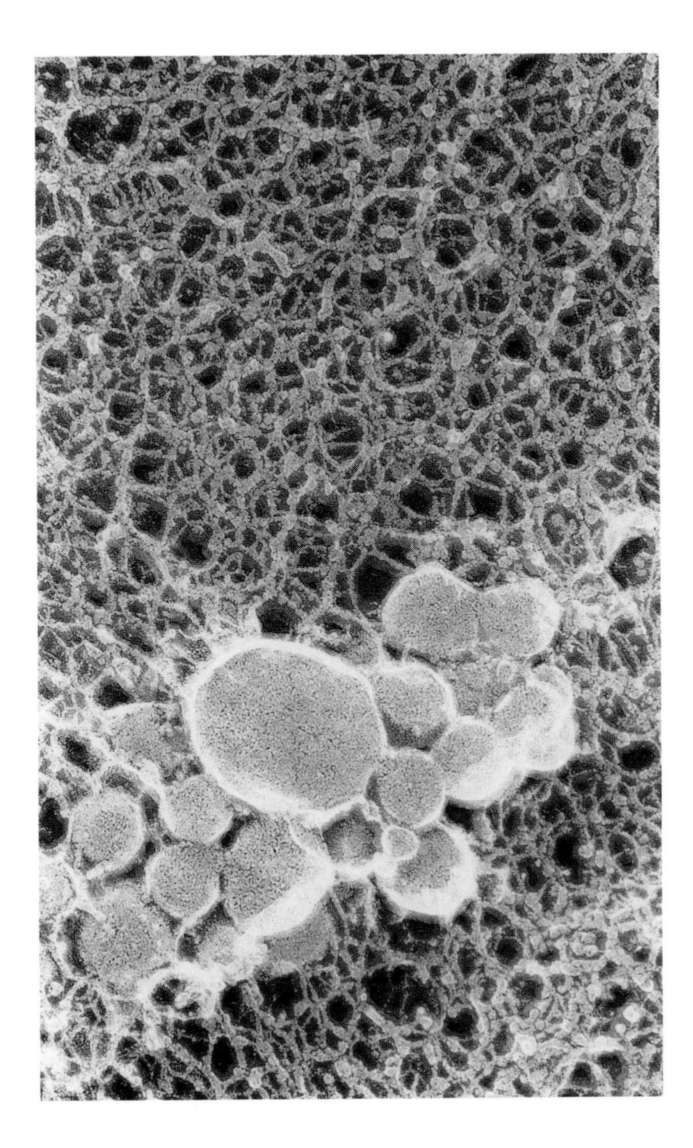

FIGURE 12. Electron micrograph of a freeze-etch replica from the aortic intima of a rabbit 2 h after bolus injection of human LDL. It shows the deposition of a lipid aggregate embedded in the extracellular matrix meshwork. (Magnification × 136,900.)

topography, typical for freeze-etched LDL,[23,24] whereas the larger lipid structures had a smooth surface (Figure 14). All the lipid particles, 23 nm-sized particles, and larger particles were enmeshed in the matrix filaments that were generally associated with collagen fibrils. There was no evidence of aggregates in the injected pool of human LDL, suggesting that the clustering of the LDL-

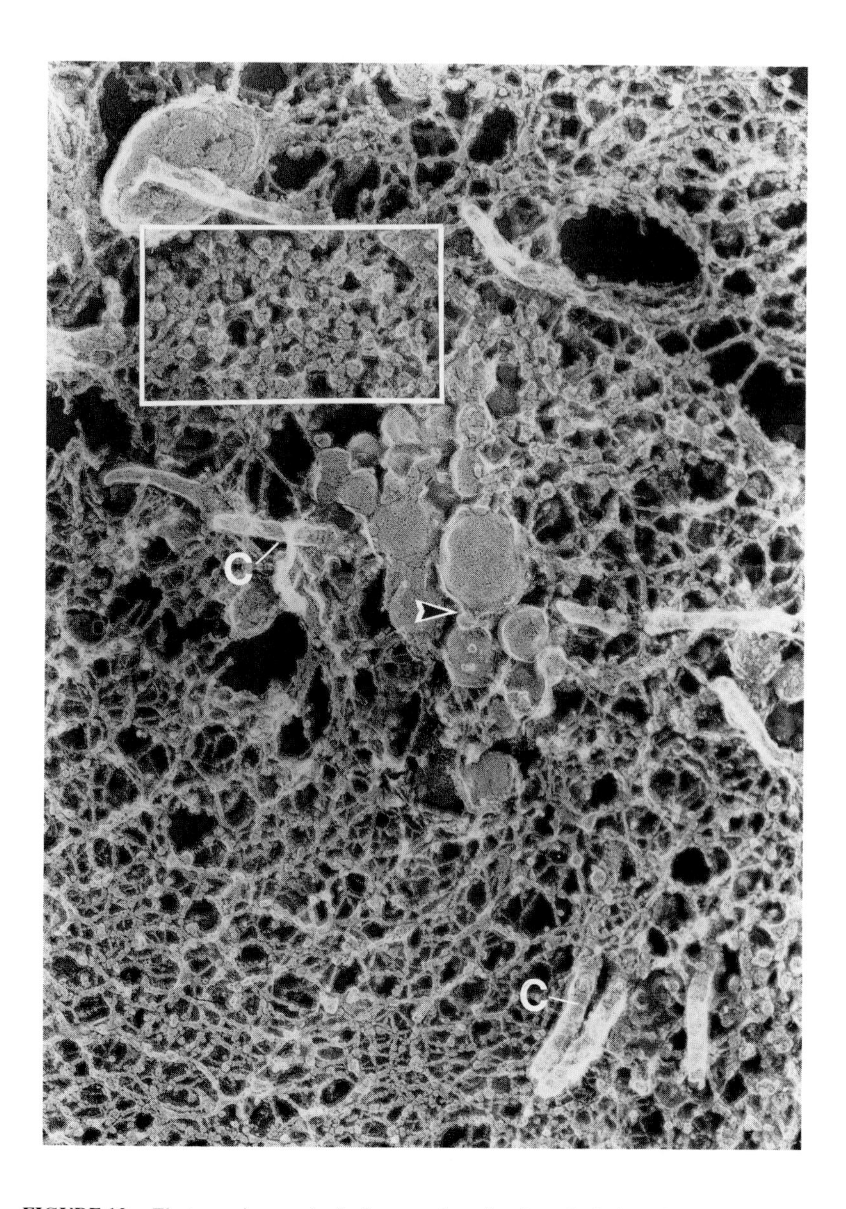

FIGURE 13. Electron micrograph of a freeze-etch replica from the intima of the aortic arch of a rabbit 2 h after it had been injected with a high concentration of human LDL. In this part of the intima, clusters of 23 nm-sized lipid-like particles are seen (within white rectangle) alongside larger lipid structures. These larger lipid particles appear to have formed from fusion of 23 nm-sized lipid particles (arrowhead). Typical collagen fibrils are present (C). (Magnification × 76,500.). (From Nievelstein et al., *Atherosclerosis Thromb.*, 11, 1795–1805, 1991.)

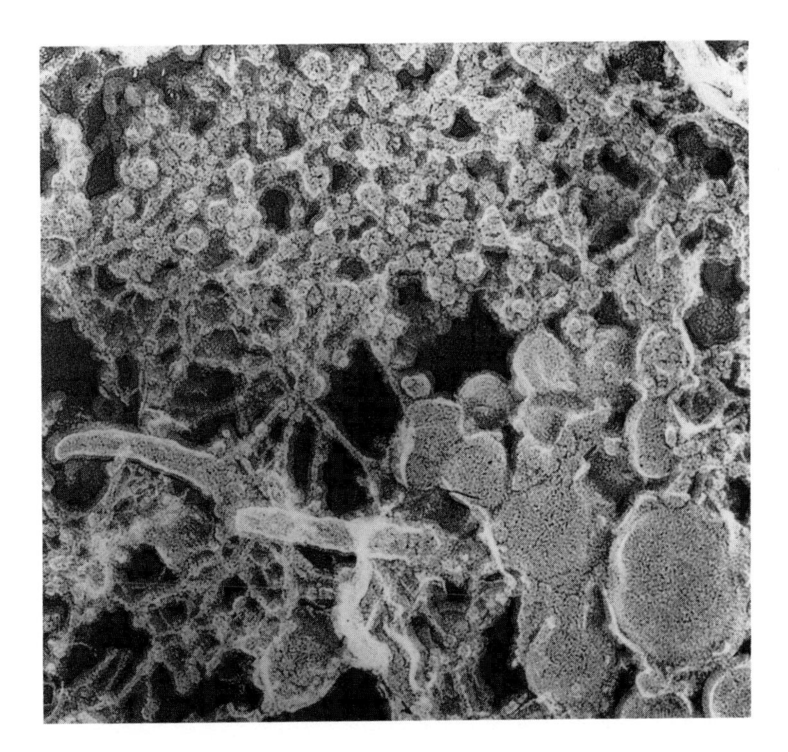

FIGURE 14. Freeze-etch electron micrograph that shows an enlarged area of Figure 13. The 23 nm-sized particles have a knobby surface. This is in contrast to the larger lipid structures (up to 161 nm) nearby that have a smooth topography. (Magnification × 153,000.). (From Nievelstein et al., *Atherosclerosis Thromb.*, 11, 1795–1805, 1991.)

sized particles occurred *in vivo*. Neither the clusters of 23 nm-sized lipid particles nor the clusters of larger lipid-like structures were present in any of the replicas from control rabbits injected only with buffer.

C. Implications of Freeze-Etch Studies

The unique feature of the freeze-etch technique is that it can retain the organization of lipids and extracellular matrix components in a state close to that *in vivo*. The replicas show the specific linkage of the lipid particles to fine fibrils in the extracellular matrix surrounded by other structures, such as collagen. The demonstration of lipid aggregates in the bolus-injected rabbits is the first structural observation to suggest that self-aggregation of lipid occurs *in vivo*. This is important because self-aggregation of LDL may be a necessary alteration of LDL for its uptake by monocyte/macrophages. The ultrastructural observation that individual LDL-sized particles are closely associated with the matrix filaments supports the importance of the interaction of LDL with proteoglycans and other extracellular components. The presence of collagen surrounding lipid clusters suggests that this matrix protein might play a role in aggregate formation. Lipid particles immobilized in the matrix may function as a nucleation site for other lipid particles to form aggregates. Once entrapped in the matrix, changes in the surface charge and/or chemisty of the lipid might occur.

In summary, we are able, with the freeze-etch technique, to retain features of the native organization of the intima in a detailed three-dimensional form. However, the specific identification of the lipid particles awaits further study using combinations of freeze-etch technology with immunogold labeling.

ACKNOWLEDGMENT

We thank Giuliano Mottino for his enthusiastic and excellent technical assistance.

REFERENCES

1. **Steinberg, D. and Witztum, J.L.,** Lipoproteins and atherogenesis, *JAMA*, 264, 3047–3052, 1990.
2. **Schwenke, D.C. and Carew, T.E.,** Initiation of atherosclerotic lesions in cholesterol fed rabbits. I. Focal increases in arterial LDL concentrations precede development of fatty streak lesions, *Arteriosclerosis* 9, 895–907, 1991.
3. **Simionescu, N., Vasile, E., Lupu, F., Popescu, G., and Simionescu, M.,** Prelesional events in atherogenesis. Accumulation of extracellular cholesterol-rich liposomes in the arterial intima and cardiac valves of the hyperlipidemic rabbit, *Am. J. Pathol.*, 123, 1109–1125, 1986.
4. **Kruth, H.S.,** Subendothelial accumulation of unesterified cholesterol. An early event in the atherosclerotic lesion development. *Atherosclerosis*, 57, 337–341, 1985.
5. **Mora, R., Lupu, F., and Simionescu, N.,** Prelesional events in atherogenesis. Colocalization of apolipoprotein B, unesterified cholesterol and extracellular phospholipid liposomes in the aorta of hyperlipidemic rabbit, *Atherosclerosis*, 67, 143–154, 1987.
6. **Wight, T.N.,** Cell biology of arterial proteoglycans, *Arteriosclerosis*, 9, 1–20, 1989.
7. **Mayne, R.,** Collagenous components of blood vessels, *Arteriosclerosis*, 6, 585–593, 1986.
8. **Iverious, P.H.,** The interaction between human plasma lipoproteins and connective tissue glycosaminoglycans, *J. Biol. Chem.*, 247, 2607–2613, 1972.
9. **Srinevasan, S.R., Dolan, P., Radhakrischnamurthy, B., and Berenson, G.S.,** Isolation of lipoprotein-acid mucopolysaccharide complexes from fatty streaks of human aorta, *Atherosclerosis*, 16, 95–104, 1972.
10. **Camejo, G.,** The interaction of lipids and lipoproteins with the extracellular matrix of arterial tissue: its possible role in atherogenesis, *Adv. Lipid Res.*, 19, 1–55, 1982.
11. **Camejo, G., Linden, T., Olsson, U., Wiklund, O., Lopez, F., and Bondjers, G.,** Binding parameters and concentration modulate formation of complexes between LDL and arterial proteoglycans in serum, *Atherosclerosis*, 79, 121–128, 1989.
12. **Le Loux, M., Boudin, D., Salmon, S., and Polonovski, J.,** The affinity of type I collagen for lipid in vitro, *Biochim. Biophys. Acta*, 708, 26–32, 1982.
13. **Cseh, K., Karadi, I., Rischak, K., Szollar, L., Janaki, G., Jakab, L., and Romics, L.,** Binding of fibronectin to human lipoproteins, *Clin. Chim. Acta*, 182, 75–86, 1989.

14. **Podet, E.J., Shaffer, D.R., Gianturco, S.H., Bradley, W.A., Yang, C.Y., and Guyton, J.R.,** Interaction of low density lipoproteins with human aortic elastin, *Arterioscler. Thromb.*, 11, 116–122, 1991.
15. **Frank, J.S. and Fogelman, A.M.,** Ultrastructure of the intima in WHHL and cholesterol-fed rabbit aortas prepared by ultrarapid freezing and freeze-etching, *J. Lipid Res.*, 30, 967–978, 1989.
16. **Nievelstein, P.F.E.M., Fogelman, A.M., Mottino, G., and Frank, J.S.,** Lipid accumulation in rabbit aortic intima two hours after bolus infusion of low density lipoproteins: a deep-etch and immunolocalization study of ultrarapidly frozen tissue, *Arterioscler. Thromb.*, 11, 1795–1805, 1991.
17. **Gerrity, R.G.,** The role of the myocyte in atherogenesis. I. Transition of blood-borne monocytes into foam cells and fatty streak lesions, *Am. J. Pathol.*, 103, 181–190, 1981.
18. **Goldstein, J.L., Ho, Y.K., Basu, S.K., and Brown, M.S.,** Binding site on marcophages that mediates uptake and degradation of acetylated low-density lipoproteins producing massive cholesterol deposition, *Proc. Natl. Acad. Sci. U.S.A.*, 76, 333–337, 1979.
19. **Steinberg, D., Parthasarathy, S., Carew, T.E., Khoo, J.C., and Witztum, J.L.,** Beyond cholesterol: modifications of low-density lipoproteins that increases its atherogenicity, *N. Engl. J. Med.*, 320, 915–924, 1989.
20. **Fogelman, A.M., Schechter, J.S., Hokam, M., Child, J.S., and Edwards, P.A.,** Malondialdehyde alteration of low-density lipoproteins leads to cholesterol accumulation in human monocyte macrophages, *Proc. Natl. Acad. Sci. USA*, 77, 2214–2218, 1980.
21. **Khoo, J.C., Miller, E., McLaughlin, P., and Steinberg, D.,** Enhanced macrophage uptake of low density lipoprotein after self aggregation, *Arteriosclerosis*, 8, 348–358, 1988.
22. **Guyton, J.R., Klemp K.F., and Mims, M.P.,** Altered ultrastructural morphology of self-aggregated low density lipoproteins: coalescence of lipid domains forming droplets and vesicles, *J. Lipid Res.*, 32, 953–962, 1991.
23. **Gulik-Krzywicki, T., Yates, M., and Aggerbeck, L.P.,** Structure of serum low-density lipoprotein, *J. Mol. Biol.*, 131, 475–484, 1979.
24. **Aggerbeck, L.P., Yates, M., and Gulik-Krzywicki, T.,** Freeze-etching electron microscopy of serum lipoproteins, *Electronmicroscopy*, 348, 352–362, 1980.

Chapter 3

RENAL TRANSPLANT ARTERIOPATHY: SIMILARITIES TO ATHEROSCLEROSIS

Ekkehard Vollmer and Albert Roessner

TABLE OF CONTENTS

ISBN 0-8493-5505-2

I. INTRODUCTION

Progress in transplant surgery and the increasing number of allogenic transplantations have led to a parallel increase in obliterative arteriopathy. Although this is a general problem, renal transplantation is, numerically, particularly affected. Since Hamburger et al.,[1] in 1953, made the first attempt to prolong the survival of a girl with renal insufficiency, by transplanting a kidney from the girl's mother, renal transplantation has become accepted clinical practice. Successful surgery and adequate acceptance by the recipient will ensure satisfactory rehabilitation. Transplants taken from an HLA (human leukocyte antigen)-identical relative have an especially favorable prognosis, but even the success of transplanting autopsy kidneys has been considerably improved in recent years.

The classification of transplant rejections, i.e., the different reactions of the recipient that lead to rejection, may be based on the time period between transplantation and rejection, on the causative immunological pathomechanism, or on the resulting histologic changes. Interstitial and glomerular reactions observed in the transplanted kidney, as well as vascular and, in particular, arterial lesions, present specific problems that affect the patient and intrigue the pathologist. Acute or chronic obliterative transplant arteriopathy after allogenic organ transplantation are grave complications whose pathogenesis still, in many ways, remains unexplained. It is often suggested that the morphology of transplanted vessels, both macroscopically and in the histological analysis of its cellular makeup, is morphogenetically related to atherosclerosis.[2-6] The morphological pattern observed comprises fibrosis with an increase in the number of spindle-shaped cells, as well as intimal aggregates of so-called foam cells (Figures 1a, b). Other authors have discussed disorders of lipid metabolism in transplant patients,[7] and also the influence of serum factors as direct correlates of the prevalence and progression of atherosclerotic vessel changes in patients with renal insufficiency.[8] Clearly, possible iatrogenic side effects (e.g., the impact of immunosuppressive therapy on cholesterol) play their role in the progression of such lesions.[9-11] Besides these, the established risk factors of atherosclerosis bear an additional and crucial risk for renal transplant rejection.[12]

The principal cause of transplant arteriopathy is seen today as an initial dysfunction of vascular endothelia leading to accelerated vascular changes.[13] These, in turn, will trigger the transformation and activation of immunocompetent cells, with drastic changes of morphology, function, and metabolism of the cells involved in the sclerotic process, in a manner very similar to that observed in atherosclerosis.

In the following sections the different cell types of the vascular wall of renal transplants are analyzed with respect to pathomorphological, functional, and cytogenetic aspects. In addition, the results of immunohistochemical and cytochemical studies on the topographical distribution of apolipoproteins (Apo), and on the characterization of the cells involved in Apo metabolism, are evaluated at the light and electron microscopical level. These investigations of rejected

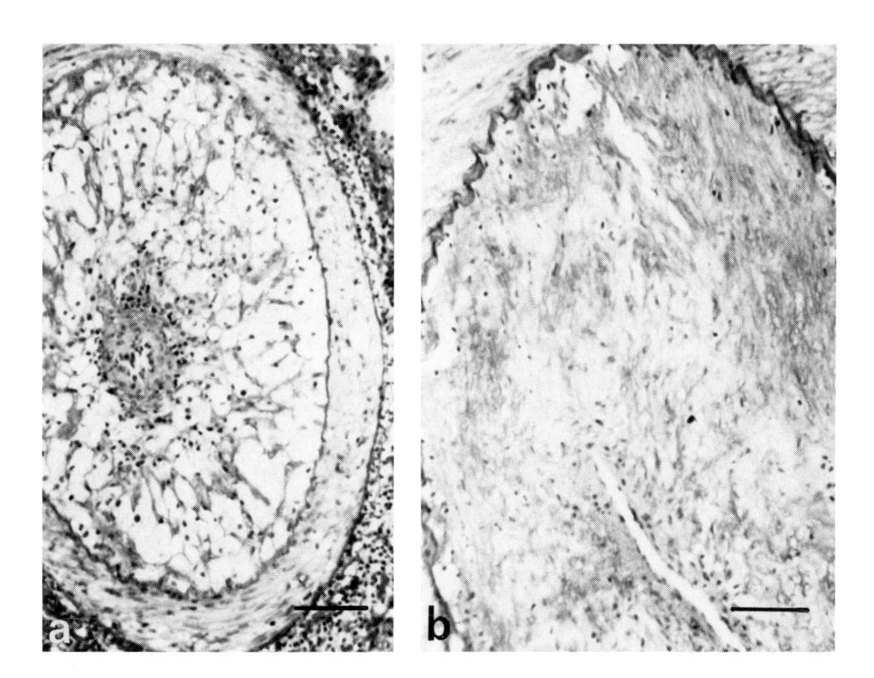

FIGURE 1. Renal transplant arteriopathy in arteriae interlobulares with high-grade luminal obliteration. (a) Numerous foam cell sheets and (b) fibrosis and proliferation of spindle cells in the intima; (Paraffin). (Bars: 100 μm.)

renal transplants in chronic transplant arteriopathy were intended to establish the possible parallels with atherosclerosis, with a view to providing a better understanding of the pathogenesis of arteriosclerotic vessel changes in general.

II. MATERIALS AND METHODS

Histologic investigations were performed on nephrectomy specimens after chronic rejection of 30 kidney transplants; the intervening phase in the recipients ranged from 3 to 96 months. The age of recipients was between 12 and 64, and that of donors between 5 and 50 years. No positive correlation was found between the duration of transplant acceptance and the degree of eventual obliteration of the arterial vessels. Furthermore, the pathoanatomical vascular findings failed to correlate with available clinical parameters, i.e., sex, blood pressure, smoking history, original renal disease, laboratory data for diabetes, fat metabolism, etc., or data on immunosuppressive therapy.

For conventional light microscopy, the surgical specimens, after appropriate cutting, were fixed for 24 h in buffered 3% formol and routinely embedded in paraffin. Three-μm sections were stained with hematoxylin eosin and elastica van Gieson. Immunolabeling was performed by a modified alkaline-phosphatase/antialkaline-phosphatase (APAAP) technique.[14] Alkaline-phosphatase/

antialkaline-phosphatase complexes were purchased from Dianova (Hamburg), and linking antibodies were purchased from Dako (Hamburg). Stains were developed in 30 min by incubation with naphthol-AS-biphosphate/ dimethylformamide together with a neofuchsin solution. Simultaneously, endogenous alkaline phosphatase was suppressed by levamisole.

Ultrastructural examination was performed either on chemically fixed material (glutaraldehyde or periodate-lysine-paraformaldehyde) or after cryofixation.[15] Physical fixation by freezing provides a method for combining adequate ultrastructural information with preservation of hypersensitive antigenic linkage sites that may otherwise be destroyed, even with weak chemical fixation. In order to ensure optimal preparation of our samples, freeze substitution was performed after cryofixation.[16,17] Our substitution medium was methanol containing 0.5% uranyl acetate; embedding was carried out in apolar, hydrophilic resin (Lowicryl) at temperatures between –50 and –35°C. We also used the modified ultracryosection method according to Tokuyasu,[18] with 2.3-M sucrose as cryoprotectant. Immunolabeling was performed with secondary antibodies coupled to 10-nm gold. Material embedded conventionally in Epon was also included in our investigations.

Primary antibodies against alpha and gamma actin were obtained from Ortho (Heidelberg). Primary antibodies against the intermediate filament proteins desmin and vimentin, those distinguishing the different lymphocyte subtypes (CD45RO, UCHLI; CD22, L26; T helper/inducer, OPD4), and antibodies for the detection of proliferating cell nuclear antigen (PCNA) were obtained from Dako (Hamburg). Markers for endothelial cells, antibodies against factor-VIII-related antigen (F VIII), and Ulex europaeus lectin 1 (UE1) were purchased from the same source. Specific antibodies to discriminate between monocytes (27-E-10) and mature macrophages (25-F-9) were kindly provided by Prof. Dr. C. Sorg and his team from the Institute of Experimental Dermatology, University of Münster (FRG). Antibodies against Apo A_1, A_2, and B were purchased from Boehringer (Mannheim).

III. ENDOTHELIUM

A. General Remarks

The endothelium, an integrative element between blood and tissue, plays a key role in the maintenance of structural and functional integrity of the circulation. With respect to atherogenesis, the well-known risk factors (chronic hypercholesterolemia, hypertension, and cigarette smoking) are presumed to provoke endothelial damage by functional alteration of endothelial cells, mostly resulting in the clinical manifestation of inflammation.[19-21] The role of the endothelium in lipid metabolism, e.g., in the modification of LDL, is an additional important aspect of its function.[22]

The central regulatory part played by endothelia in acute or chronic processes of inflammation has been defined and verified in recent years. The endothelium is activated under inflammatory conditions and will react by expressing a number of new functions, including specific structures for the recognition of leukocyte adherence, an increased rate of mitosis, and the secretion of growth factors and cytokines. The mechanisms steering the rise of inflammatory infiltration and vascularization also provoke the onset of various disease patterns. Just as in atherosclerosis, the endothelial cell has been identified as the primary target for the initiation of renal obliterative transplant arteriopathy.[4,13,23,24]

B. Results

In the arteries of minor and major caliber, our study of chronically rejected renal transplants revealed endothelial cells of spindle shape and of the cuboid phenotype (Figures 2a, b), as described in atherosclerosis.[25] The cuboid type, comparatively the more frequent form in renal transplant arteriopathy, is associated mainly with microthrombi, but also with mononucleate cells. The latter are identified either as monocyte/macrophages or belong to the T-lymphocyte populations (see below).

As in atherosclerosis, endothelial cells in renal transplant arteriopathy are often observed detached from the basal lamina. There are also some areas of complete endothelial denudation.

Histochemistry reveals that arterial endothelial cells in rejected renal transplants are poorly labeled with UE1 and intensely labeled by the antibody against F VIII (Figures 2d, e). Erythrocytes, however, are also labeled by UE1. Both these endothelial markers are well suited for outlining an increased subendothelial accumulation of partly foamy leukocytes (like those seen in the fatty streaks of atherosclerosis) and reveal a multiple loosening and lifting of the endothelial cells. The mononuclear leukocytes often appear integrated into the texture of the endothelium. The markers often allow distinct endothelial layers to be discriminated in which the individual endothelial cells show a variable intensity of labeling (Plate 2b, following page 118). Proliferating endothelial cells expressing PCNA are encountered (Figure 2c).

Immunoelectron microscopy permits ultrastructural localization of the binding sites for the endothelial cell markers. F VIII occurs in an intracytoplasmic location mainly in storage granules (Weibel-Palade bodies) (Figure 3a), and UE1 is localized in electron-dense areas of the plasma membrane and within the cytoplasm (Figure 3b). These findings accord with previous reports on cell cultures.[26]

IV. SMOOTH MUSCLE CELLS

A. General Remarks

Smooth muscle cells play a particularly important part in the development of the atherosclerotic plaque.[27] Being the predominant cell type of the fibrous plaque, their proliferation is the key to further expansion of the lesion. It is generally supposed that plaque smooth muscle cells originate from the tunica

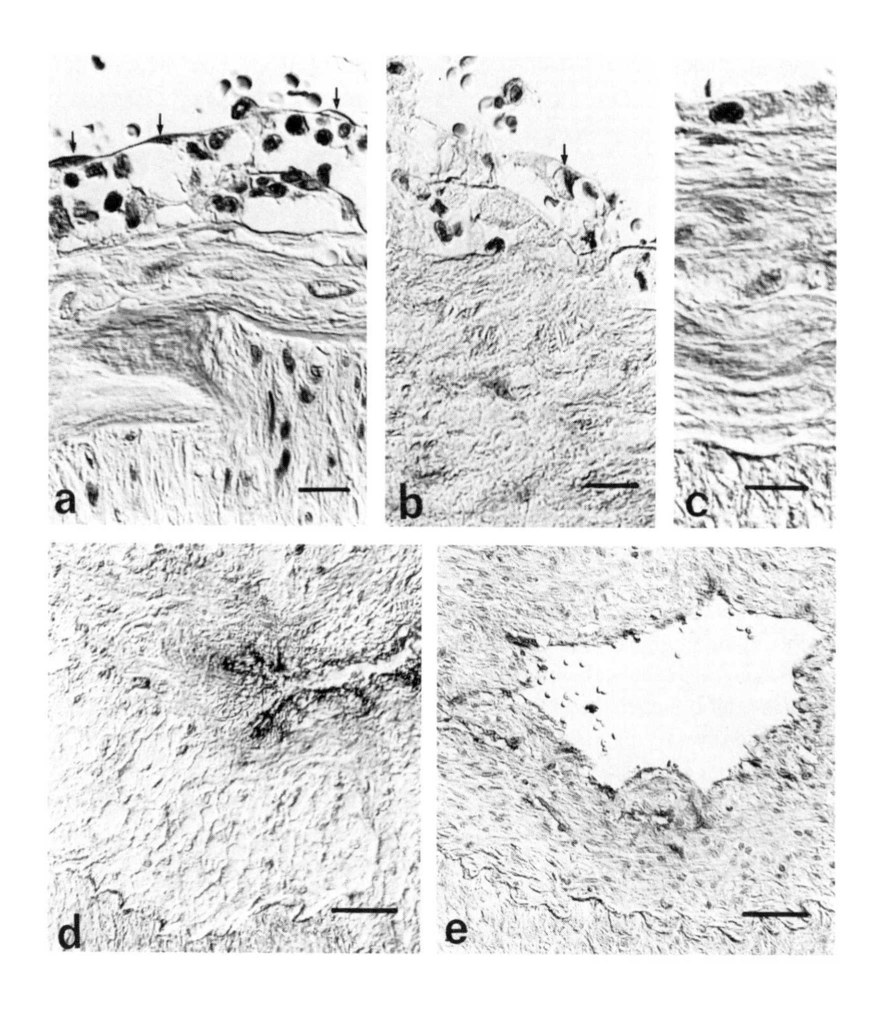

FIGURE 2. Renal transplant arteriopathy with microthrombi and mononuclear cells adhering to flat (a) as well as cuboid (b) endothelium (arrows) and mononuclear cells migrating into the subendothelial space. Endothelial cell (c) expressing the proliferating cell nuclear antigen. Endothelial cells express Factor-VIII-related antigen more strongly (d) than the binding site for Ulex europaeus 1 (e); (Paraffin). (Bars: (a–c) 20 μm; (d, e) 50 μm.)

media,[28] but their actual migration from media into intima has not been conclusively proven in humans. In fact, some smooth muscle cells are known to occur physiologically, now and then, in the normal human vascular intima. Two principal (sub)types of the smooth muscle cell are present in human vessels: (1) the synthesizing, metabolically active form and (2) the contractile phenotype. Synthesizing cells contain only a few bundles of myofilaments, but abundant organelles such as free ribosomes, granular endoplasmic reticulum, and mitochondria. Synthesizing smooth muscle cells react strongly to growth factors and other matrix stimuli to which contractile cells do not respond.[29] Much remains

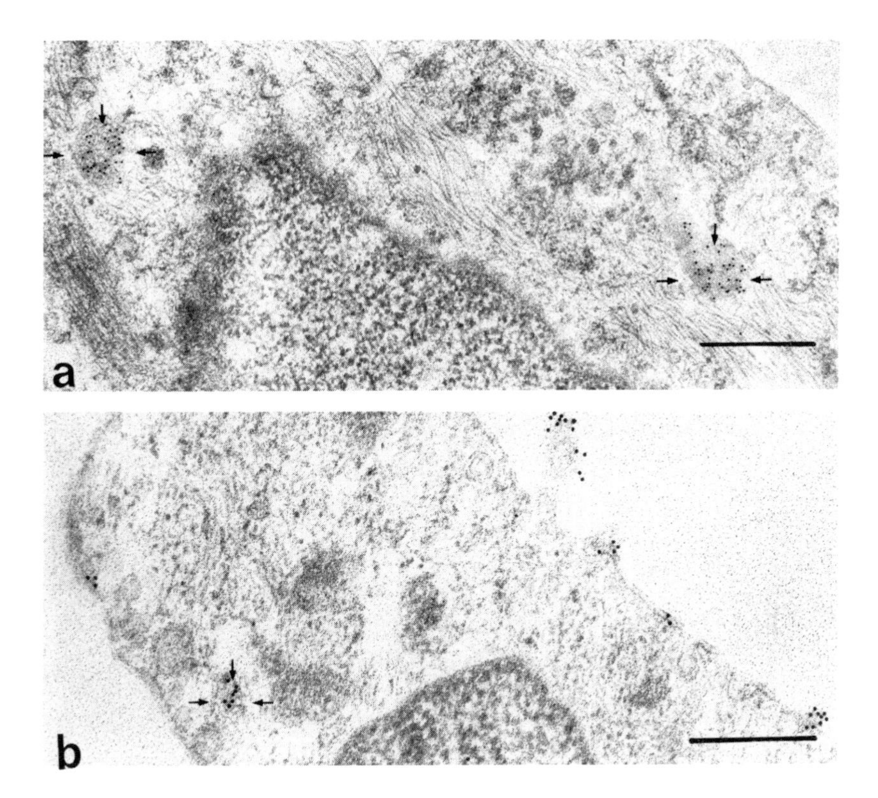

FIGURE 3. Endothelial cells with intracytoplasmic expression of Factor-VIII-related antigen (a), and labeling with Ulex europaeus 1 (b). The latter occurs in intracytoplasmic locations as well as at dense areas of the membrane surface; (Lowicryl). (Bars: 0.5 μm.)

to be done to define the morphological and biochemical differences in the smooth muscle cells of media and intima. The heterogeneity of these cells has become a topic of current interest, especially with regard to atherosclerosis.[30,31] The spindle-shaped cells observed in transplant arteriopathy, and their cytogenetic origin, have also attracted recent attention,[4-6,24] since these cells may play a similar key role in the progression of these lesions.

B. Results

Our histological studies of chronically rejected renal transplants revealed the presence of some smooth muscle cells in the arterial intima of minor sclerotic lesions, similar to those found in normal vessels. Also observed were areas of distinct intimal fibrosis in which central parts were depleted of smooth muscle cells and contained abundant extracellular matrix, and other areas where numerous smooth muscle cells occurred in the shoulder part of fibrous caps. In these areas the internal elastic lamina is less often spliced and destroyed than in the picture of atherosclerosis.

Immunohistochemical tests in the renal artery showed muscle-specific actin in both intima and media, manifest as widely uniform and intense staining of smooth muscle cells (Figure 4a). Desmin is expressed in medial smooth muscle cells to a much lower extent than actin and is hardly ever detected in the intimal layer (Figure 4b). Vimentin (Figure 4c) is expressed in spindle-shaped cells of the intima and in the endothelium, but as with desmin, the expression of this intermediate filament protein in medial smooth muscle cells is less strong than is the expression of actin.

In medium-sized renal arteries with particularly distinct vascular lesions, a strong actin expression is found in spindle-shaped cells of the intima as well as in smooth muscle cells of the media, even where intimal fibrosis is marked. Desmin expression, by contrast, is mostly negative, especially in the intima and, except for some single smooth muscle cells, in the media. Expression of vimentin is less strong than that of actin, but still somewhat stronger than that of desmin. The actual size of the vessels made no difference in these patterns.

At the ultrastructural level, the progress of vascular lesions was paralleled with an increase of synthesizing-type smooth muscle cells. Immunoelectron microscopy demonstrated muscle-specific actin rather evenly on almost all myocytes, whereas desmin expression on single smooth muscle cells varied in intensity (Figure 4d). Stronger and more frequent staining of smooth muscle cells was observed with the antibody against vimentin, especially in a perinuclear location and in a net-like distribution around lipid droplets — a pattern also found in atherosclerosis.

In some cases immunohistochemical evidence of PCNA was found in nuclei of the spindle-shaped cells of medium-caliber renal arteries that showed distinct signs of intimal fibrosis (Figure 5).

V. MONOCYTE/MACROPHAGES AND LYMPHOCYTES

A. General Remarks

The so-called foam cells characterizing atherosclerotic plaques had for many years been assumed to originate from smooth muscle cells; it is now generally believed, however, that they derive predominantly from cells of the monocyte/macrophage system. This concept is emphasized in the latest morphologic studies on the pathogenesis of atherosclerosis, especially in early stages of the

---

FIGURE 4. Immunohistochemical demonstration of the cytoskeleton in consecutive sections of an arteria renalis (lumen = upper margin) with minor transplant arteriopathy (a–c). (a) shows marked actin expression in myocytes of intimal (I) and medial (M) layers. Desmin expression (b) is by comparison lower, visible only in the media. Vimentin expression (c) is weaker than that of actin, but it occurs in medial smooth muscle cells in the intima, and, additionally, in endothelial cells (arrows). The example in (d) is an electron micrograph of smooth muscle cells, showing individual differences of desmin expression; the cell at the left margin shows dense gold labeling, in contrast to that at the right margin where a portion of the nucleus is visible; (a–c, Paraffin; d, cryosubstitution with Lowicryl). (Bars: (a–c) 50 μm; (d) 0.5 μm.)

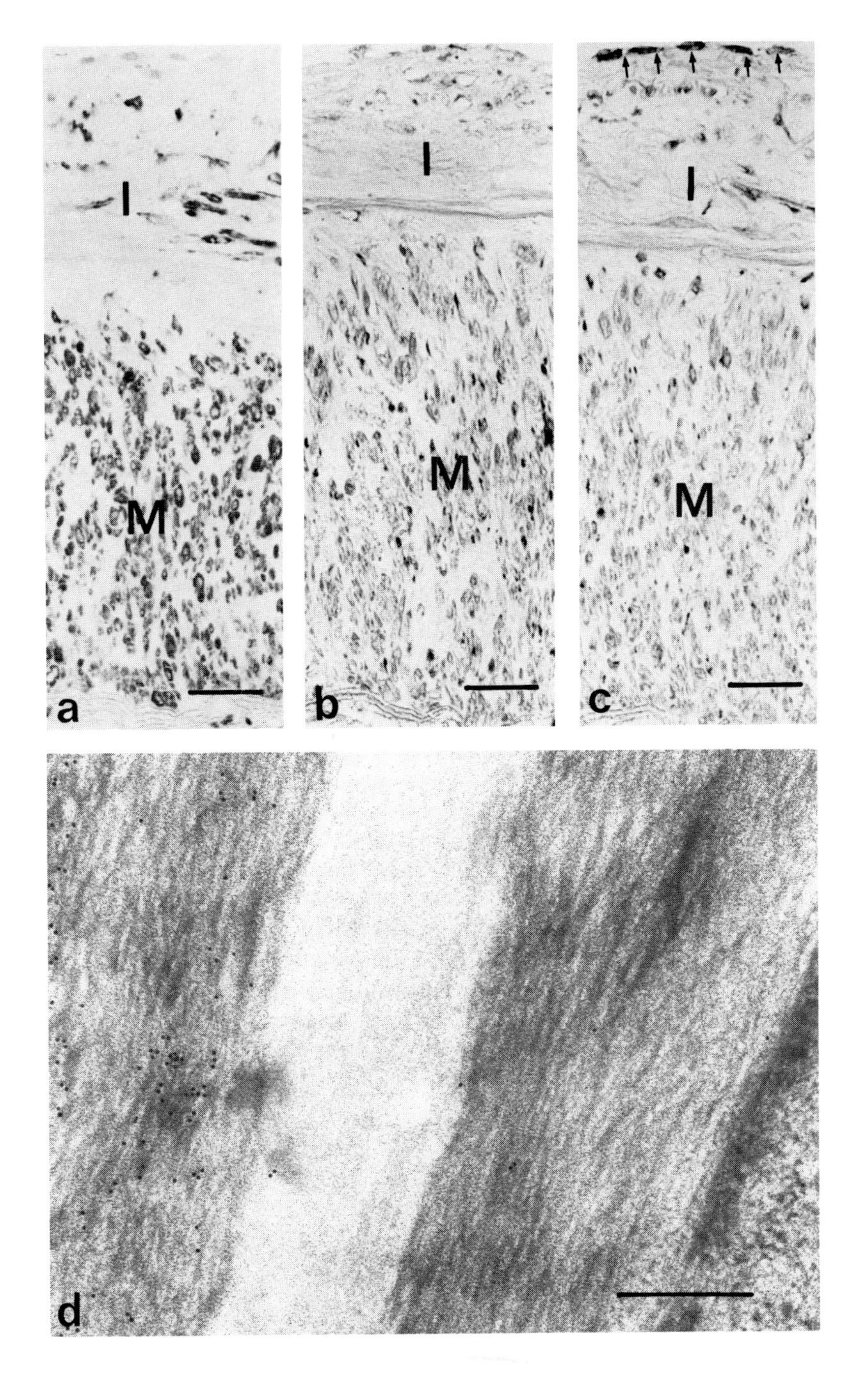

FIGURE 4.

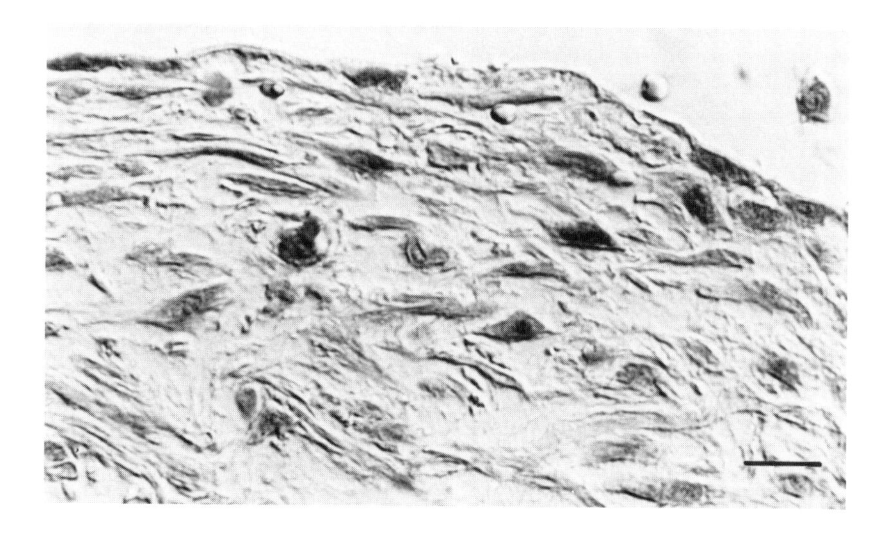

FIGURE 5. Spindle cells expressing proliferating cell nuclear antigen in distinctly fibrotic intima with extensive luminal obliteration. (Bar: 20 μm).

disease.[32-35] Macrophages have been shown to be directly involved in the metabolism of lipoproteins in the vessel wall and, in particular, to incorporate the cholesterol-binding lipoproteins entering the intima. If macrophages have taken up more cholesterol than they can "digest", the excess cholesterol is stored as droplets in their own cytoplasm. While the majority of foam cells in the atherosclerotic plaque appear to be generated in this way, an origin from smooth muscle cells, though not excluded, would appear to play a more minor role.[36,37] The subendothelial monocyte/macrophage represents a physiological component of the subendothelial layer in many large arteries of several species and of man;[38] metabolic activation of these cells carries high atherogenic potential. In renal transplant arteriopathy, too, the central role of macrophages and of growth factors secreted by them, and their subsequent effect on smooth muscle cells and endothelial cells, is interpreted as a crucial pathogenetic factor.[4,39]

Atherosclerosis is of multifactorial genesis, and some aspects of the disease resemble an inflammatory process. Thus, increased attention has been focused on lymphocytes in the pathogenesis of the disease, but no comprehensive concept has so far emerged for the understanding of such immunologic pathomechanisms.[28]

A more detailed interpretation has, however, been proposed for the immunologic pathomechanisms of transplant rejection. After organ transplantation, lymphocytes as well as monocyte/macrophages of the recipient will be circulating through the transplant itself. Among them, T lymphocytes are capable of recognizing any foreign (donor) antigens of the HLA system. Having thus been sensitized, part of the T cell population seems to acquire some cytotoxic potential; other T cells, leaving the transplant via efferent lymph vessels, will accumulate in regional lymph nodes and start to proliferate, eventually returning to settle in the transplant in the role of cytotoxic lymphocytes. Depending on the actual vascularity of the transplant, on its specific or nonspecific previous impairment or damage, and on the release of tissue antigens, the process will result in the delayed sensitization of the B-cell system. As soon as the synthesis of specific antibodies is under way, the active rejection is increasingly complicated by humoral immune reactions that (via complement activation, kinine release, etc.) will result in an antibody-dependent, cellular cytotoxic reaction. This highly complex situation, involving humoral and cell-mediated immune reactions to varying degrees, also includes other immune-competent cells, such as monocyte/macrophages and killer cells. The vascular endothelial cell, seen as the primary target for the initiation of obliterative transplant arteriopathy, also plays an important part in the development and progression of the disease, as do interactions with smooth muscle cells.[4,13,24]

B. Results

Here we present an attempt at reliable differentiation of mononuclear cells of the inflammatory infiltrate, including foam cells, in chronically rejected renal transplants, performed with the aid of highly specific markers.

In chronically rejected renal transplants, at an early stage of vascular alteration, we found, as in atherosclerosis, a mononuclear infiltration arising from the lumen and adhering to the endothelium, penetrating it, and settling in the subendothelial space. These cells, originating from the recipient, belong to the monocyte/macrophage and lymphocyte populations. Immunohistochemically, the monocyte/macrophage populations first show the phenotype of blood monocytes labeled by the specific antibody 27-E-10 (Figure 6a).[40] These cells migrate into the deeper layers of the vascular wall where they transform into mature tissue macrophages labeled by the specific marker 25-F-9,[41] and often appear as foam cells (Figure 6b). Some of the foam cells also express PCNA (Figure 6c). In contrast to 25-F-9, the monocyte-specific antibody 27-E-10 produces no positive labeling in these foam cells.

In parallel with the picture in atherosclerosis, the renal arteries in the hilar region of transplant kidneys (where lesions are usually less severe than in medium-sized parenchymal arteries) reveal subendothelial T lymphocytes, which are identified with Dako-UCHL1 (a pan T-marker directed against CD45RO antigen), and seem to have immigrated from the lumen via altered endothelium (Figure 6d). In the interlobular arteries, which often have extensive luminal obliteration, immunohistochemical labeling reveals many lymphoid T cells in foam cell clusters, identified with the antibody Dako-UCHL1 (Figure 6e). B cells, demonstrable with the antibody Dako L26 against CD22 antigen, were virtually nonexistent in the vascular lesions of chronically rejected renal transplants. Analyzing a series of consecutive sections for a semiquantitative survey, we identified up to 50% of T lymphocytes as T-helper/inducer cells, as recognized by the marker Dako-OPD4 (Figures 6f, g). A specific antibody to identify T suppressor/cytotoxic lymphocytes, suitable for use in paraffin sections of chemically fixed tissue, is not yet available. The preferred site of all T-lymphocyte infiltrations was in the intimal area between the zone of spindle-shaped cells and the foam cell cushions.

VI. APOLIPOPROTEINS

A. General Remarks

The leading role of cholesterol in the initiation and progression of atherosclerosis is now generally accepted. Epidemiologic studies have identified hypercholesterolemia, together with hypertension and cigarette abuse, as the dominant risk factors for ischemic sequelae.[42,43] Many experimental studies, and data from patients with familial hypercholesterolemia, have confirmed that elevated serum cholesterol alone, without other concurrent risk factors, may promote atherosclerosis.

As cholesterol and other plasma lipids are insoluble in water, they are bound to proteins in the liver or intestine and are transported as lipoprotein complexes via the blood to distant organs (see Chapter 10 for detailed discussion). The various lipoproteins are distinguished by their differing contents of apoproteins and lipids and are subdivided into classes according to density. The lipoprotein fractions of particular relevance to cholesterol metabolism and atherogenesis are low-density lipoprotein (LDL) and high-density lipoprotein (HDL). Two thirds of serum cholesterol is contained in LDL, and one third in HDL. Lipoproteins of minor density (chylomicrons and very-low-density lipoproteins (VLDL)) are predominantly concerned with the transport of triglycerides and carry only a minor quantity of cholesterol.

LDL has the highest cholesterol content of all lipoproteins (50%) and is the main supplier of cholesterol to peripheral cells. Cholesterol uptake is initiated by LDL binding to specific receptors at the cell surface.[44] The receptor recognizes LDL by the lipoprotein's main structural component, Apo B.

HDL has a distinctly lower content of cholesterol, but has the highest Apo content of all lipoproteins, enabling it to act as an accepter of cholesterol. HDL

FIGURE 6. 27-E-10-positive monocytes (a) migrating from the lumen through the endothelium into the intima where they mature to 25-F-9-positive tissue macrophages (b) that are subsequently transformed into foam cells. Some of these cells are often found to express proliferating cell nuclear antigen (c). Numerous, predominantly subendothelial T lymphocytes positive for UCHL-1, immigrating from the lumen through altered endothelium (d), are finally deposited between intra-arterial intimal foam cell beds (e). Consecutive sections reveal that among pan-T-cells (f), up to 50% can be identified as T-helper/inducer cells (g); (Paraffin). (Bars: (a, c) 20 μm; (b, d, f, g) 50 μm; (e) 30 μm).

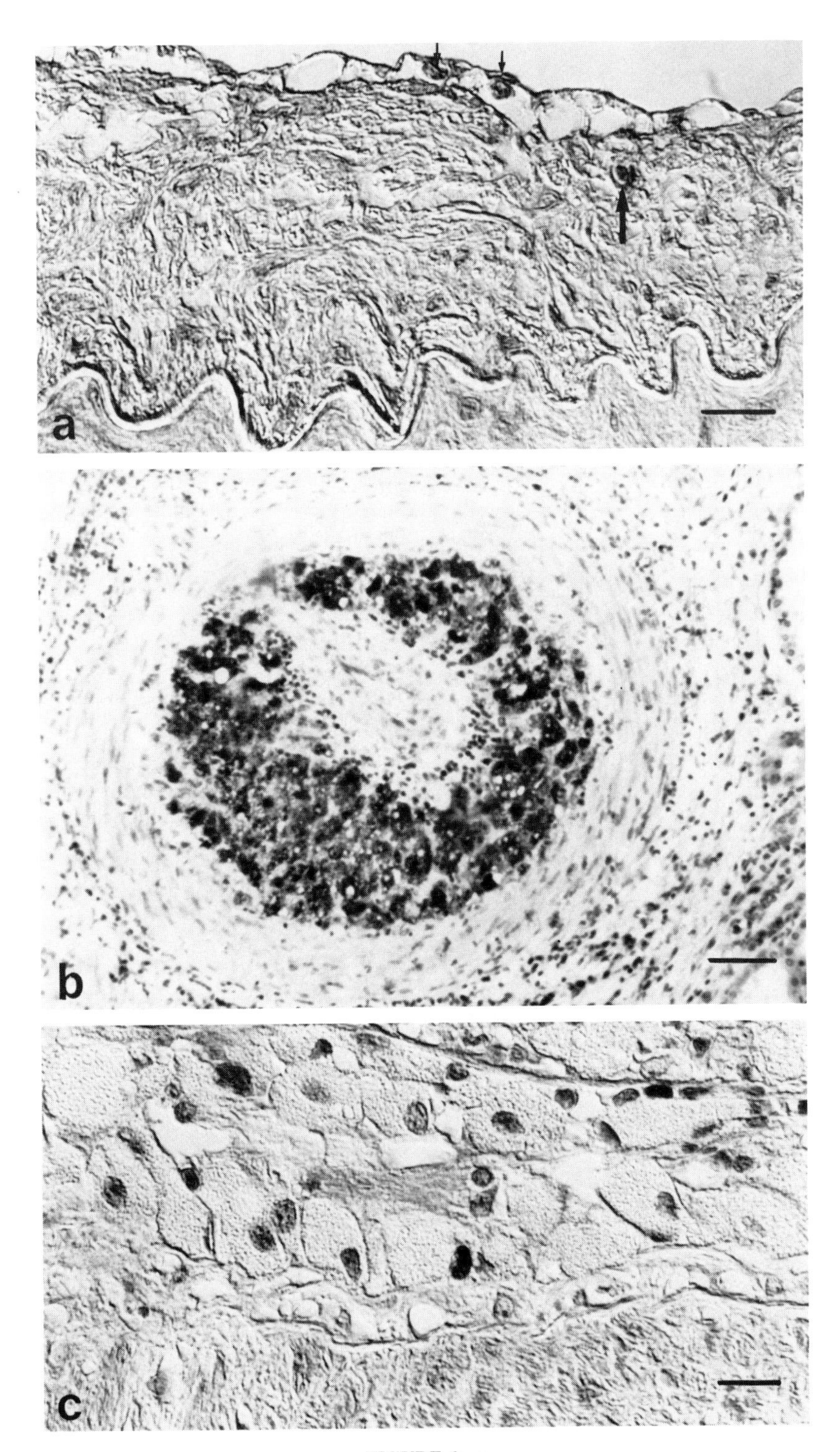

FIGURE 6a–c.

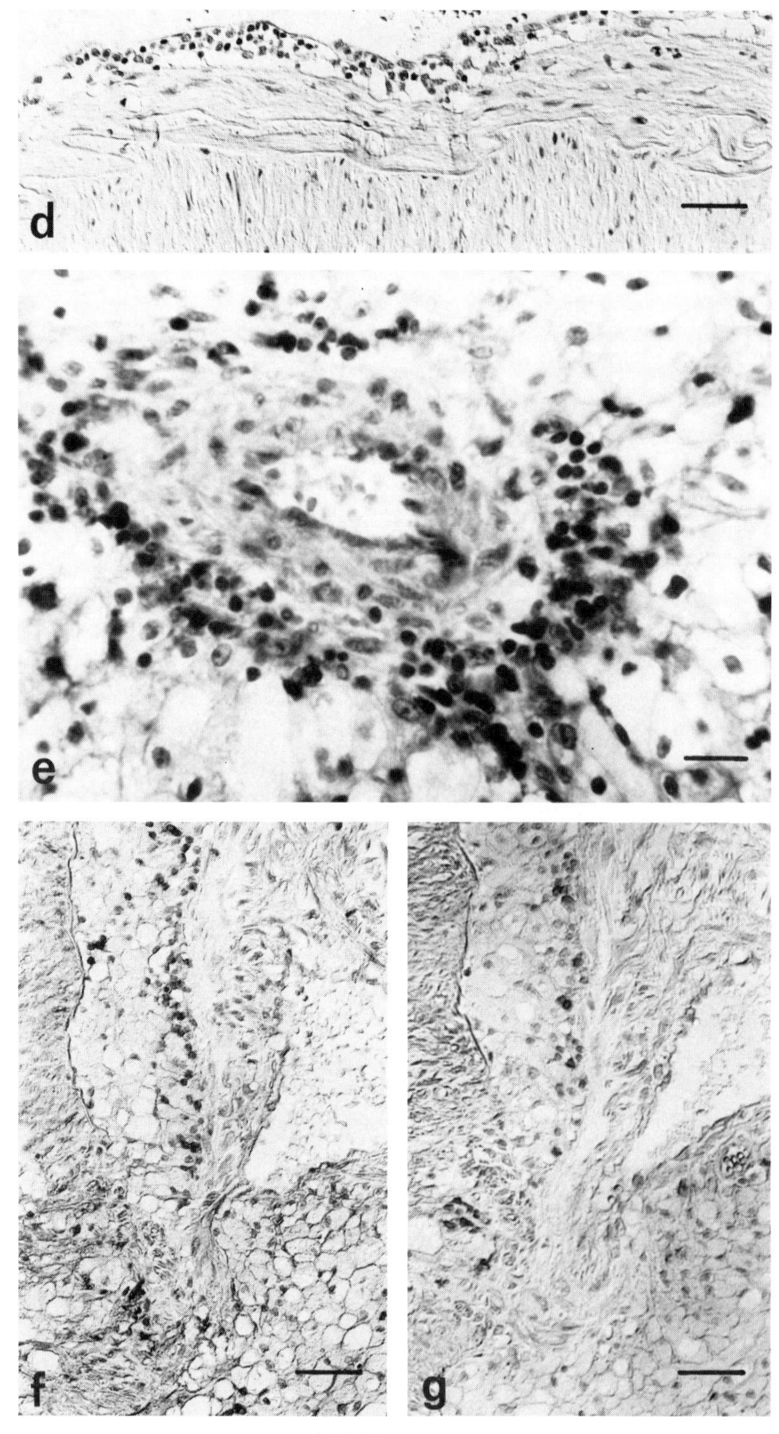

FIGURE 6d–g.

is thought to function in "reverse cholesterol transport," carrying surplus cholesterol from the peripheral tissues back to the liver,[45-47] where it is excreted as bile acids. Cholesterol uptake into HDL is regulated by Apo A_1, which activates LCAT (the enzyme lecithin:cholesterol-acyl-transferase) to esterify free cholesterol and direct it into the interior of HDL particles. It is still unclear precisely when in the process of "reverse cholesterol transport" esterification takes place.[48] Current understanding is that esterification is effected immediately after the return of HDL particles into the bloodstream. Recent investigations, in which two new HDL types (pre-β-I and II) have been identified, suggest that LCAT may be less important in the earlier stage of the "reverse cholesterol transport" than was previously thought. Both these HDL types lack LCAT activity and appear to operate as early accepters of cellular cholesterol. Although they represent a minor proportion of total plasma HDL, they may be the dominant forms of HDL in the extracellular liquid of the vascular bed. If this is so, these minute HDL particles would have ready access to peripheral cells, and "reverse cholesterol transport" could conceivably take place without LCAT activity actually within the vascular intima.

Following this concept and modern lipid theory, the genesis and progression of atherosclerosis is determined not solely by the absolute quantity of cholesterol carrying lipoproteins circulating in the blood, but also by the proportion of the individual subtypes of lipoproteins. HDL (linked to Apo A_1) exerts a suppressive effect, and LDL (linked to Apo B) has a promoting effect on atherosclerosis. Another factor now attracting renewed interest is lipoprotein(a), an antigenic variant of LDL associated with a high risk of atherosclerosis.[49] It is important to bear in mind that plasma levels of LDL and HDL, while undoubtedly important, do not necessarily give a direct reflection of the lipoprotein distributions actually within the atherosclerotic lesion: a factor of obvious importance that demands detailed study.[50]

A key function in lipid metabolism is now attributed to the macrophage. Unlike other cell types of the arterial wall (such as fibroblasts and smooth muscle cells in which LDL cholesterol uptake is suppressed by regulation of specific LDL receptors[51]), macrophages may avidly ingest LDL cholesterol via a "scavenger pathway."[46] In this pathway, however, modified variants of LDL, rather than native LDL, are internalized. According to Steinberg et al.,[52] the capacity of scavenger-type pathways may be considerable, involving several mechanisms, both receptor dependent and less specific in nature. Macrophages themselves have a capacity for Apo synthesis.[53,54] This macrophage-derived Apo, an Apo E, is recognized by the LDL (B/E) receptor and has a potentially important role in lipoprotein metabolism. Apo E, together with cholesterol, may contribute to the formation of HDL particles and, hence, to "reverse cholesterol transport."[46,55,56] Apo A_1 is thought also to be involved in this process.[57] Recent studies suggest that smooth muscle cells may be capable of Apo E secretion,[54,58,59] and Murase et al.[60] have reported immunohistochemical evidence for the presence of Apo E in coronary endothelial cells, raising the possibility of Apo E synthesis within the endothelium. Apo E occurs in various forms that have different influences on serum cholesterol concentration; for a more detailed discussion on the significance of Apo E polymorphism, the reader is referred to Davignon et al.[61]

The frequent iatrogenic disorders of lipid metabolism associated with kidney transplantation have been extensively reported in the literature.[7,11,62-64] These disorders are considered to be major risk factors in relation to obliterative vascular changes and concomitant cardiovascular complications. Clearly, the foam cells of macrophage origin found in the "cushions" typically present in the intima are likely to play a central part in the lipid metabolism of the arterial wall.[4-6] Foam cells of smooth muscle cell origin, if present, may also contribute. However, firm knowledge on the precise role of foam cells in lipid metabolism during renal transplant arteriopathy is still rather limited.

Against this background of known atherogenic effects of high serum LDL concentration, especially when associated with low levels of HDL, we undertook a detailed *in situ* analysis of Apo distribution and localization in the arterial walls, and of the cell types involved in the vessels of chronically rejected renal transplants.

B. Results

Lesion grade was found to correlate with the amount of HDL- or LDL-associated Apo accumulation in the vascular wall. Apo deposits occurred in both intracellular and extracellular locations, predominantly in the intima. Deposits in the media were generally confined to very advanced lesions. Apo A_1 and A_2 were initially localized predominantly in extracellular spaces, but were seen intracellularly with increasing frequency, especially in foam cells, as vascular obliteration advanced (Figures 7a,b, Plate 1a).* Our preliminary studies at the ultrastructural level showed an accumulation of Apo A_1 in electron-dense cytoplasmic vesicles that are most likely of lysosomal origin. The overall extracellular amount of Apo deposits increased with the grade of the lesion, especially in cell-depleted zones of sclerosis. The internal elastic lamina often appeared to act as a barrier, but as the lesion progressed,* Apo A deposits eventually spread to the media where they became increasingly conspicuous within spindle-shaped smooth muscle cells (Plate 1b). Apo A_1 and Apo A_2 showed similar patterns of deposition, with the quantity of the former always exceeding that of the latter.

In all grades of lesion, Apo B appears mainly as extracellular deposits often gathering around foam cells (Plate 2a).* Only very advanced stages show some Apo B in the honeycomb-like cytoplasm of histiocytic foam cells, whereas

---

* Color Plates 1 and 2 follow page 118.

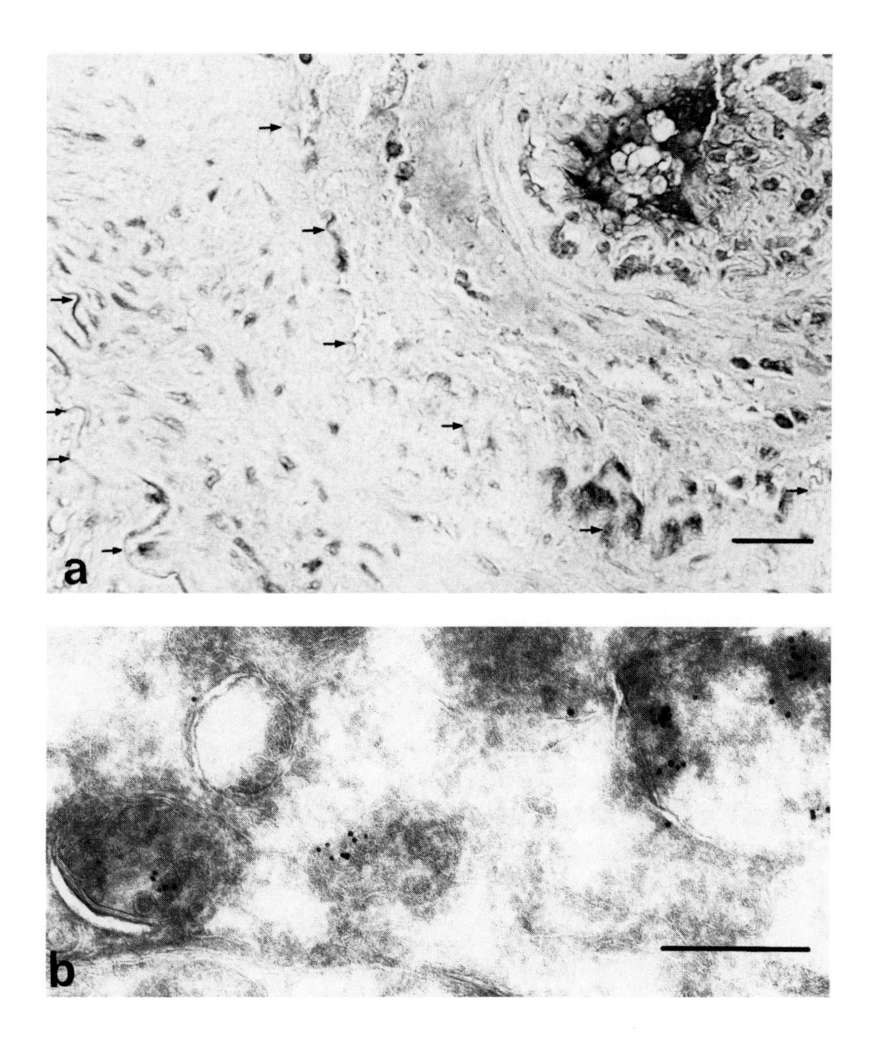

FIGURE 7. Mainly intracellular, but also extracellular, deposits of Apo A_1 are localized predominantly in the intima (a) (arrows indicate internal and external laminas). At the ultrastructural level, intracellular evidence of Apo A_1 in electron-dense cytoplasmic vesicles is observed (b) (a, Paraffin section; b, ultracryosection). (Bars: (a) 300 μm; (b) 0.5 μm).

smooth muscle cells, even those in foamy transformation, will rarely contain Apo B. As in atherosclerosis, renal transplant arteriopathy specimens can display positive Apo immunostaining arising from blood influx from the vasa vasorum torn by preparative techniques,[50] but this is easily distinguished from the disease-related changes described.

On the whole, a clearer picture of Apo deposition and other pathomorphologic changes is found in medium-sized arteries of the renal parenchyma than in those of minor or major size.

VII. DISCUSSION

A. Endothelium

In our study two distinct endothelial cell phenotypes were distinguished by conventional morphological criteria: a flat, spindle-shaped form and a cuboid or bulging form. Endothelial cells were often detached from the basal lamina, a phenomenon associated with the presence of numerous mononuclear cells in contact with the endothelium.

The phenotypic differences between endothelial cells are currently attributed to differences in functional status, reflecting metabolic performance.[25] In atherosclerosis an increase of cuboid cells is supposed to occur in sites that have a predilection for the disease; local hemodynamic conditions seem to be of additional importance.[65] Such areas also show an increased accumulation of macromolecules from the blood.[66]

Detachment of endothelial cells from the basal lamina, a frequent phenomenon in renal transplant arteriopathy, and integration of partly foamy mononuclear cells into the endothelial structure, are also observed in atherosclerosis. The presence of PCNA confirms the high regenerative potential of the endothelium proper in response to endothelial damage. In transplant arteriopathy, endothelial injury may be of immunologic origin, in view of the involvement of immunocompetent cells observed in our investigations, and data from the literature.[4,13] In atherosclerosis, however, the role of immune mechanisms is still controversial.[19,21,67,68]

Subendothelial lipoproteins are known to affect circulating monocytes by exerting chemotactic effects and inducing expression of endothelial adhesive factors. The adhesion of circulating leukocytes to the wall of a blood vessel is an essential component of acute and chronic inflammation in general, and is a key event in atherosclerosis and obliterative transplant arteriopathy, too. Activation of vascular endothelia by certain inflammatory stimuli leads to an enhanced expression of adhesion molecules on the cell surface. Specific cell-surface carbohydrates have been identified as mediators of physiological immune modulation in cell/cell interactions.[69]

Various immune modulators bind specifically to target cells such as macrophages, cytotoxic cells, natural killer cells, and B cells. There is accumulating evidence that the activities of such immune modulators may be specifically suppressed by glycolipids, glycopeptides, and even simple sugars. Apparently, the interaction of certain lymphokines with their target cells involves lectin-like reactions. It may be concluded that some important functions of immunocompetent cells, including proliferation and differentiation, and of effector molecules are induced and mediated by lymphocytes and macrophages by means of lectin-like carbohydrate recognition proteins.

Not much is known, so far, about the physiological function of these endogenous carbohydrate binding activities. Our investigations have shown that leukocyte/endothelial interactions are more pronounced in renal transplant arteriopathy than in atherosclerosis. Labeling by UE1, however, is distinctly less strong compared with that to F VIII, which appears to be a superior endothelial marker in renal transplant arteriopathy. Ultrastructurally, F VIII was localized in the cytoplasm, and in previous studies of endothelial cell cultures, we have demonstrated F VIII in the secretory apparatus, particularly in granular endoplasmic reticulum, and in Weibel-Palade bodies.[26]

Recent studies have identified several endothelial adhesive components that act as recognition structures for leukocytes. One of these, known as ELAM I (endothelial leukocyte adhesion molecule), shows rapid, but transitory, expression on the surface of endothelial cells.[26,70] ELAM I is of functional importance mainly for the selective adherence of neutrophils, but it may also act as a chemotactic agent for the migration of monocyte/macrophages and lymphocytes.

In the recruitment of lymphocytes, which also occurs via adhesion to endothelia, another adhesive molecule, VCAM (vascular cell adhesion molecule) plays a role. Compared to ELAM, VCAM shows a somewhat delayed, but more persistent, expression on stimulated endothelial cells. Other endothelial memory structures for leukocytes are ICAM I and II (intercellular adhesion molecules), which are structurally related to the immunoglobulin superfamily. These proteins bind poly- and mononuclear cells by way of heterodimer accepter structures, including the so-called integrins. In addition to the expression of adhesive properties, the endothelium manifests a range of coordinating activities, both in atherosclerosis and in renal transplant arteriopathy, producing other signal substances of importance in immune defense. Endothelial cells can express receptors for Fc and complement, thus influencing the phagocytotic activity of immunocompetent cells. Another contribution to inflammation control is made by activated endothelial cells secreting various immune-modulating cytokines.

Under pathologic conditions, arterial endothelial cells will also express MHC class II molecules and so may participate with monocytes in the cellular immune response, assuming the part of antigen-presenting cells. The expression of such antigens plays an important role, too, in the transmigration of lymphocytes (see below) that occurs in both obliterative transplant arteriopathy and in atherosclerosis.[21,68]

B. Smooth Muscle Cells

The phenotypic characterization of smooth muscle cells, achieved at the light and electron microscopic level, aided by additional immunocytochemical characterization, yielded similar findings in renal transplant arteriopathy to those established for atherosclerosis. In progressing obliterative changes, the smooth muscle cell phenotype is increasingly transformed from the contractile to the synthesizing form. In the media, usually affected at a later stage, we may still see many smooth muscle cells rich in filaments and poor in organelles, while the intima contains an increased number of synthesizing smooth muscle cells containing many organelles. The antibody against alpha and gamma actin used in this study is a suitable marker of smooth muscle cells, as these variants of actin are specific for these cells. In contrast to the antibody against desmin, the actin marker labels all the smooth muscle cells in the intima and media.

Current concepts suggest that phenotypic modulation from the myofilament-rich, contractile type to the myofilament-depleted, organelle-rich, synthesizing type is the basic prerequisite for proliferation of smooth muscle cells,[28] and this, in turn, forms the basis for understanding the principal changes in atherosclerotic vessels. In our proliferation kinetic studies, proliferating smooth muscle cells were particularly common in renal transplant arteriopathy, especially where intimal fibrosis was marked. In atherosclerosis, however, Gordon et al.[71] recorded only minimal proliferation of smooth muscle cells, and our own (unpublished) observations on atherosclerosis have revealed only occasional smooth muscle cells expressing Ki-67 (a proliferation-associated nuclear antigen) or PCNA. Distinct smooth muscle cell proliferation, especially in the intima, has been reported by other investigators, in transplanted kidneys and hearts.[72] Thus, renal and other forms of transplant arteriopathy are widely envisaged as a kind of accelerated atherosclerosis.

From our observation of reduced desmin staining in proliferating smooth muscle cells, we may deduce that phenotypic alteration may be triggered by atherogenic stimuli that encourage either proliferation in the media or migration into the intima followed by proliferation there. The antibody to the intermediate filament protein vimentin, in contrast to that against desmin, is expressed not only in smooth muscle cells, but also in inflammatory cells and endothelia. Vimentin is the dominating intermediate filament observed at the light and electron microscopic level in all layers of the vascular wall, a finding confirmed by other studies on atherosclerosis.[73,74] A characteristic feature of vimentin-containing intermediate filaments is their frequent perinuclear concentration in smooth muscle cells. This association suggests that vimentin may give mechanical support to the nucleus, maintaining its position within the cells.[75] Newer data ascribe a distinct function to vimentin in the intracellular accumulation of lipid droplets; vimentin filaments appear to build a kind of mesh around these droplets,[76] as confirmed by our own observations on smooth muscle cells. The coexistence of vimentin and desmin in smooth muscle cells was once controversial, but the studies of Fujimoto et al.[77] have helped to close the argument. These authors were able to demonstrate, at the ultrastructural level, the coexistence of vimentin and desmin within the same intermediate filament and attributed the altered expression of intermediate filaments to disassembly, degradation, and reassembly of existing proteins, rather than exclusively *de novo* production.

C. Monocyte/Macrophages and Lymphocytes

Our immunohistochemical investigations demonstrated that mononuclear cells migrate from the blood into the intima. Mononuclear inflammation of the vascular wall also occurs in atherosclerosis, but the monocyte/macrophages and lymphocyte infiltrates in renal transplant arteriopathy differ in that they are derived from the transplant recipient's cell populations. Nevertheless, there are important functional and morphological parallels in both disease processes. The adhesion of monocyte/macrophages to the endothelium generally takes place without any marked damage to endothelial structure; subsequent passage through interendothelial slits by protrusion of pseudopodia has been detailed in ultrastructural studies. Soon after migration into the vascular wall, the monocytes transform to mature tissue macrophages. Evidence of this transformation is found in the phenotypic modulation observed in our antibody labeling studies. The cells first express an antigen to which the antibody 27-E-10, directed exclusively against blood monocytes,[40] binds; at a later stage they stain with the antibody 25-F-9, directed exclusively against mature tissue macrophages without monocytic features.[41] As in atherosclerosis, these macrophages take up lipids. Cholesterol thus has a special role in that it accumulates in the cytoplasm of these cells. The numbers of foam cells of histiocytic origin (stained with 25-F-9) increases progressively as the lesion develops. Such foamy macrophages may even reveal PCNA in cases of severe obliterative transplant arteriopathy. Although histiocytic foam cells are commonly considered to be the fully differentiated final form, we may infer from our observations, in contrast to current notions, that they may still be capable of proliferation. Although these proliferating foam cells have not yet been subjected to immunohistochemical double labeling to exclude a smooth muscle cell origin, such an origin would appear unlikely in view of the very distinct staining with 25-F-9 (a highly specific marker for macrophages). This interpretation is further supported by the study of Müller-Hermelink and Dämmrich[4] on 15 transplanted kidneys, which suggested that the foam cells observed were always derived from macrophages.

The literature on atherosclerosis contains only fragmentary reports on proliferation of foam cell macrophages.[33,78] Some recent investigations of Rosenfeld and Ross,[79] using simultaneous autoradiographic determination of thymidine incorporation and immunohistochemistry in a rabbit model, have shown that up to 30% of proliferating cells (depending on the lesion stage) could be confidently identified as macrophages.

In contrast to the slowly developing process of human atherosclerosis, the proliferation of the foam cells observed in renal transplant arteriopathy appears to be an ad hoc process arising from the need for an additional supply of macrophages to cope with the extra demand for lipid clearance that cannot be met by the existing mature macrophages. This yet again emphasizes the concept of renal transplant arteriopathy as a kind of accelerated atherosclerosis, though it must be borne in mind that in transplant arteriopathy, the immigrating cells are monocyte/macrophages of the recipient, directed explicitly against the heterogenic transplant they invade. The same is true for lymphocytes infiltrating the arterial vessels, most of which belong to the T population. In the usual forms of atherosclerosis, T cells also predominate in the lymphocytic infiltration of vascular walls.[68,80,81] In our material, up to 50% of the cells in the lymphocytic infiltration belonged to the T helper/inducer type. For mainly technical reasons, we are as yet unable to determine exactly the proportion of T suppressor/cytotoxic lymphocytes (CD8 positive); according to the literature, this population predominates, especially in earlier lesions.[4] In weak reject reactions,[82] however, a predominance of T helper/inducer cells is found in the inner layers of the vascular wall. Our present data on transplanted kidneys are consistent with the latest results of Salomon et al.[24] on coronary transplant arteriopathy, in which approximately equal numbers of T helper/inducer and T suppressor/cytotoxic cells were reported.

T lymphocyte (sub)populations are responsible for recognition, initiation, and regulation of the immune response. Both the normal and the pathologic response will be initiated by antigen recognition by T lymphocytes in connection with major histocompatibility complex proteins (MHC) on antigen-presenting cells. Tissues with normal, delayed-type immune responses, as well as lesions of autoimmunologic origin, will always contain a large number of T lymphocytes and antigen-presenting cells expressing MHC class II.

Two classes of human leukocyte antigen (HLA), found on all nucleated cells and platelets of the organism, are distinguished. Antigens of HLA class I are recognized by T suppressor/cytotoxic lymphocytes. Cells of a tissue expressing alien HLA-I may be attacked cytotoxically by these lymphocytes. In contrast to HLA class I, the distribution of HLA class II is much more restricted; HLA class II are found on B lymphocytes, macrophages, dendritic cells, and Langerhans cells of the skin. HLA II are target antigens for T helper/inducer lymphocytes, which, upon activation, secrete cytokines. Such T cell activation is known to occur following transplantation of heterologous tissue, as in the renal transplants studied here. The parallels with atherosclerosis so far considered with respect to T cell infiltration suggest a similar process in atherosclerotic lesions. In atherosclerosis, CD8-positive lymphocytes may be involved in the initiation of the atherosclerotic lesion, as in transplant arteriopathy. Despite the different external causes for initiation of the vascular process, we recognize some striking immunomorphological parallels in both diseases. In the general framework of cell-mediated immune reactions of delayed type, T lymphocytes clearly appear to play a major part, but they do so only in combination with MHC. Although the sequential mechanism is as yet unexplained in many details, a certain pathomechanism common to both lesions may be proposed as follows:

1. T suppressor/cytotoxic lymphocytes detecting altered or alien HLA I will induce HLA II, e.g., on the endothelium; cells expressing that antigen class may be the target of a cytotoxic attack.
2. HLA-II-positive cells are subsequent target antigens for T helper/inducer lymphocytes activated, in turn, by such HLA-II induction.
3. The maintenance of these pathomechanisms is ensured by the specialized functions of each of the various T cell subpopulations.

This concept does not exclude the equally important effector mechanisms of B lymphocytes, such as the production of antibodies and immune complexes, and complement-mediated or antibody-dependent cytotoxicity.

D. Apolipoproteins

Our studies have revealed a correlation between the grade of transplant arteriopathy and the quantity of Apo accumulation in the vascular wall. Apo deposits are found in both intra- and extracellular locations, predominantly in the intima, with relatively few deposits in the media. Apo A_1 and A_2 are mainly intracellular and particularly prominent in foam cells. The extracellular deposition of Apo A rises in line with severity of the lesion. Deposits of Apo A_1 are more numerous than those of A_2. Apo B is predominantly extracellular in all grades and often localized around foam cells. These findings show a close correspondence to the distribution patterns of these Apos reported in "conventional" atherosclerosis.[50,83] Such changes may enhance any imbalance in lipoprotein metabolism in the course of the pathogenesis of renal transplant arteriopathy.

Endothelial function, as we have seen, plays a particularly important role in the pathogenesis of the disease. The endothelium shows distinct alterations in renal transplant arteriopathy as a result of cellular or humoral reaction mechanisms. Such immune-mediated dysfunction may be further promoted by other risk factors (hypertension, hyperlipidemia, and even iatrogenic factors, e.g., cyclosporin cytotoxicity), and leads to increased influx of lipoproteins into the vessel wall, thereby disturbing homeostasis. It is within this framework that we find parallels with pathogenetic concepts of atherosclerosis.

In view of the differing plasma levels and divergent effects of Apo B as compared with Apo A_1 and A_2, the concomitant increase of all three Apos within the vessel wall during lesion progression might appear an unexpected finding. Although a rise in Apo B content might be predicted as lesions advanced, Apo A_1 and A_2 content might be expected to fall, as HDL (of which Apo A_1 and A_2 are the principal constituents) is supposed to exert an inhibitory effect on lesion development, at least in atherosclerosis. Our results on the HDL-associated lipoprotein accord with those of Carter et al.,[84] who suggested that the protective effect of Apo A is not as high within the lesion as in the plasma. Even for the atherogenic proteins Apo B and Apo (a), recent data of Rath et al.[85] suggest no clear correlation between serum levels of these proteins and build up in the vessel wall. These results indicate that accumulation of Apos is primarily determined by morphologic changes of the vessel wall (associated with cellular dysfunction), and only secondarily reflects lipoprotein concentrations in the blood plasma.

Our own preliminary immunoelectron microscopic studies of atherosclerosis[50] are consistent with the idea that after initial insudation of Apos (i.e., lipoproteins) into the intima, they become bound to extracellular matrix components, while the internal elastic lamina apparently acts as a barrier to deeper intrusion. A transitory situation follows in which additional intracellular Apo A_1

and A_2 are formed, and this diminishes only after increased build up of Apo B deposits in the progressing vascular lesion. Current quantitative analyses of vessels in chronically rejected renal transplants suggest a similar trend (Vollmer, unpublished).

That Apo A is detected in macrophages in advanced stages of lipid accumulation suggests that a mechanism to counter against uptake of excess cholesterol may operate involving recruitment of HDL as a cholesterol accepter from preexisting extracellular deposits. This would allow mobilization of the macrophage's cholesterol, with transfer to HDL and subsequent release of the cholesterol-enriched HDL. It is known from *in vitro* studies that lipid accumulation in macrophages is halted by adding HDL to the culture medium.[86]

Another important apolipoprotein in cellular cholesterol homeostasis is Apo E. In culture, monocyte/macrophages are capable of actively secreting Apo E,[87] and smooth muscle cells may have the same potential.[58] In atherosclerosis, Apo E has been demonstrated in the vascular wall, in foam cells of histiocytic, as well as in those of smooth muscle cell, origin.[54] These and the preceding findings are consistent with the idea that HDL-associated Apo A detected in macrophages and smooth muscle cells in renal transplant arteriopathy reflects HDL metabolism. HDL particles may be internalized by receptor-mediated endocytosis, and Apo A is detected in electron-dense organelles presumed to be lysosomes. Another conceivable source of the intracellular Apo A is *de novo* synthesis.

Thus, in macrophages the metabolism of Apo A may act as a kind of compensatory mechanism against cholesterol overloading. Our findings are consistent with the idea that reverse cholesterol transport, as discussed in atherosclerosis, may also be at work, to some degree, in renal transplant arteriopathy.

VIII. SUMMARY AND CONCLUSIONS

The results of our light and electron microscopical studies have traced numerous functional and pathomorphological parallels between atherosclerosis and renal transplant arteriopathy. In both diseases the central role during the initial stages may be attributed to the endothelium. Another protagonist is the T lymphocyte, which, when activated, exerts a modulating effect on various functions of the other cells populating the arterial wall. This leads to phenotypic change of smooth muscle cells, reflected, for example, in intermediate filament and actin expression, and results in their proliferation. Besides their immunologic functions, monocyte/macrophages play a central role in lipid metabolism of the vascular wall.

The effects of HDL- and LDL-associated Apos, as established from their concentration in the plasma, have to be applied with caution to predict the significance of deposited Apos in the arterial wall. In renal transplant arteriopathy, as in atherosclerosis, lipids and their metabolism by various cells, both native and immigrant, are subject to constant reappraisal with regard to their significance in the highly complex pathways of arteriosclerotic pathomechanisms.

We wish to emphasize that despite a different etiology, the natural history of renal transplant arteriopathy takes an accelerated, but in many respects similar, course to that of atherosclerosis. We may thus take transplant arteriopathy as a human model that permits a more detailed and rapid analysis of certain pathogenetic steps and interactive mechanisms crucial in the development and progression of atherosclerosis in general. A morphologic analysis of the sequence of events in renal transplant arteriopathy may help to define the functional steps involved in both diseases and so improve our insight and understanding of the atherosclerotic lesion.

ACKNOWLEDGMENT

This work was supported by the Deutsche Forschungsgemeinschaft, grant Vo 386/1.

REFERENCES

1. **Michon, L., Hamburger, J., Oeconomos, N., Delinotte, P., Richet, G., Vaysse, J., and Antoine, B.,** Une tentative de transplantation rénale chez l'homme — aspects médicaux et biologiques, *Presse méd.*, 61, 1419, 1953.
2. **Zollinger, H.U., Mihatsch, M.J., Thiel, G., Harder, F., Heitz, P., Uebersax, S., and Gudat, E.,** *Renal Pathology in Biopsy*, Springer-Verlag, New York, 1978.
3. **Cerilli, J., Brasile, L., Sosa, J., Kremer, J., Clarke, J., Leather, R., and Shah, D.,** The role of autoantibody to vascular endothelial cell antigens in atherosclerosis and vascular disease, *Transplant. Proc.*, 19 (Suppl. 5), 47, 1987.
4. **Müller-Hermelink, H.K. and Dämmrich, J.R.,** Die obliterative Transplantatvaskulopathie: Pathogenese und Pathomechanismen, *Verh. Dtsch. Ges. Pathol.*, 73, 193, 1989.
5. **Roessner, A., Bögeholz, J., Bosse, A., Vollmer, E., Buchholz, E., Winde, G., and Bründermann, H.,** Immunhistologische Differenzierung der Zellen in der Arterienwand von Transplantatnieren, *Verh. Dtsch. Ges. Pathol.*, 73, 242, 1989.
6. **Vollmer, E., Bosse, A., Bögeholz, J., Roessner, A., Blasius, S., Fahrenkamp, A., Sorg, C., and Böcker, W.,** Apolipoproteins and immunohistochemical differentiation of cells in the arterial wall of kidneys in transplant arteriopathy, *Pathol. Res. Pract.*, 187, 257, 1991.
7. **Kasiske, B.L. and Umen, A.J.,** Persistent hyperlipidemia in renal transplant patients, *Medicine*, 66, 309, 1987.
8. **Bauch, H.J., Vischer, P., Lauen, A., Raidt, H., Graefe, U., and Hauss, W.H.,** Atherogenese bei dialysepflichtiger Niereninsuffizienz: Einfluβ von Serumfaktoren auf die Regulation der Prostaglandinsynthese bei vaskulären Endothelzellen in der Arteriosklerose. *Neue Aspekte aus Zellbiologie und Molekulargenetik, Epidemiologie und Klinik*, Assmann, G., Betz, E., Heinle, H., and Schulte, H., Eds., Friedr. Vieweg u. Sohn, Braunschweig, 1990, 143.
9. **Mihatsch, M.J., Thiel, G., and Ryffel, B.,** Brief review of the morphology of cyclosporin nephropathy, *Contrib. Nephrol.*, 51, 156, 1986.

10. **Hall, B.M., Tiller, D.J., Hardie, I., Mahony, J., Matthew, T., Thatcher, G., Miach, P., Thomson, N., and Sheil, A.G.R.,** Comparison of three immunosuppressive regimens in cadaver renal transplantation: long-term cyclosporine, short-term cyclosporine followed by azathioprine and prednisolone, and azathioprine and prednisolone without cyclosporine, *N. Engl. J. Med.*, 318, 1499, 1988.
11. **Vathsala, A., Weinberg, R.B., Schoenberg, L., Grevel, J., Dunn, J., Goldstein, R.A., van Buren, C.T., Lewis, R.M., and Kahan, B.D.,** Lipid abnormalities in renal transplant recipients treated with cyclosporine, *Transplant. Proc.*, 21, 3670, 1989.
12. **Kasiske, B.L.,** Risk factors for accelerated atherosclerosis in renal transplant recipients, *Am. J. Med.*, 84, 985, 1988.
13. **Hruban, R.H., Beschorner, W.E., Baumgartner, W.A., Augustine, S.M., Ren, H., Reitz, B.A., and Hutchins, G.M.,** Accelerated arteriosclerosis in heart transplant recipients is associated with a T-lymphocyte-mediated endothelialitis, *Am. J. Pathol.*, 137, 871, 1990.
14. **Cordell, J.L., Falini, B., Erber, W.N., Ghosh, A.K., Abdulaziz, Z., MacDonald, S., Pulford, K.A.F., Stein, H., and Mason, D.Y.,** Immunoenzymatic labeling of monoclonal antibodies using immune complexes of alkaline phosphatase and monoclonal anti-alkaline phosphatase (APAAP complexes), *J. Histochem. Cytochem.*, 32, 219, 1984.
15. **Sitte, H., Edelmann, L., and Neumann, K.,** Cryofixation without pretreatment at ambient pressure, in *Cryotechniques in Biological Electron Microscopy*, Steinbrecht, R.A. and Zierold, K., Eds., Springer-Verlag, New York, 1987, 87.
16. **Humbel, B. and Schwarz, H.,** Freeze substitution for immunochemistry, in *Immuno-Gold Labeling in Cell Biology*, Verkleij, A.J. and Leunissen, J.L.M., Eds., CRC Press, Boca Raton, FL, 1989, 115.
17. **Sjöstrand, F.S.,** Common sense in electron microscopy. About cryofixation, freeze-substitution, low temperature embedding, and low denaturation embedding, *J. Struct. Biol.*, 103, 135, 1990.
18. **Tokuyasu, K.T.,** Immuno-cryoultramicrotomy, in *Immunolabeling for Electron Microscopy*, Polak, J.M. and Varndell, J.M., Eds., Elsevier, Amsterdam, 1984, 71.
19. **Munro, J.M. and Cotran, R.S.,** Biology of disease. The pathogenesis of atherosclerosis, atherogenesis and inflammation, *Lab. Invest.*, 58, 249, 1988.
20. **Hort, W. and Bürrig, K.F.,** Endothel und Arteriosklerose, *Z. Kardiol.*, 78 (Suppl. 6), 105, 1989.
21. **Gimbrone, M.A., Bevilacqua, M.P., and Cybulsky, M.I.,** Endothelial-dependent mechanisms of leukocyte adhesion in inflammation and atherosclerosis, *Ann. N.Y. Acad. Sci. U.S.A.*, 598, 77, 1990.
22. **Steinbrecher, U.P., Parthasarathy, S., Leake, D.S., Witztum, J.L., and Steinberg, D.,** Modification of low density lipoprotein by endothelial cells involves lipid peroxidation and degradation of low density lipoprotein phospholipids, *Proc. Natl. Acad. Sci. U.S.A.*, 81, 3883, 1984.
23. **Libby, P., Salomon, R.N., Payne, D.D., Schoen, F.J., and Pober, J.S.,** Functions of vascular wall cells related to development of transplantation-associated coronary arteriosclerosis, *Transplant. Proc.*, 21, 3677, 1989.
24. **Salomon, R.N., Hughes, C.G.W., Schoen, F.J., Payne, D.D., Pober, J.S., and Libby, P.,** Human coronary transplantation-associated arteriosclerosis, *Am. J. Pathol.*, 138, 791, 1991.
25. **Gerrity, R.G.,** Arterial endothelial structure and permeability as it relates to susceptibility to atherogenesis, in *Pathobiology of the Human Atherosclerotic Plaque*, Glagov, S., Newmann, W.P., and Schaffer, S.A., Eds., Springer-Verlag, New York, 1990, 13.
26. **Vollmer, E., Roessner, A., Bosse, A., Voss, B., Goerdt, S., Sorg, C., and Böcker, W.,** Immunoelectron microscopic investigations for the phenotypical characterization of endothelial cells, in *Modern Aspects of the Pathogenesis of Arteriosclerosis*, Hauss, W.H., Wissler, R.W., and Bauch, H.J., Eds., Westdeutscher Verlag, Düsseldorf, 1989, 235.

27. **Benditt, E.P. and Benditt, J.M.,** Evidence for a monoclonal origin of human atherosclerotic plaques, *Proc. Natl. Acad. Sci. U.S.A.*, 70, 1753, 1973.
28. **Ross, R.,** Mechanisms of atherosclerosis — a review, *Adv. Nephrol.*, 19, 79, 1990.
29. **Tachas, G., Rennick, R.E., Ang, A.H., Bateman, J.F., Campbell, J.H., and Campbell, G.R.,** Smooth muscle phenotype: controlling factors and collagen gene expression, in *Modern Aspects of the Pathogenesis of Arteriosclerosis*, Hauss, W.H., Wissler, R.W., and Bauch, H.J., Eds., Westdeutscher Verlag, Düsseldorf, 1989, 37.
30. **Campbell, G.R. and Campbell, J.H.,** The phenotypes of smooth muscle expressed in human atheroma, *Ann. N.Y. Acad. Sci. U.S.A.*, 598, 143, 1990.
31. **Gabbiani, G.,** Cytoskeletal characterization of smooth muscle cells of human and experimental atherosclerotic plaques, in *Pathobiology of the Human Atherosclerotic Plaque*, Glagov, S., Newman, W.P., III, and Schaffer, S.A., Eds., Springer-Verlag, New York, 1990, 63.
32. **Roessner, A., Vollmer, E., Zwadlo, G., Sorg, C., Greve, H., and Grundmann, E.,** Zur Differenzierung von Makrophagen und glatten Muskelzellen mit monoklonalen Antikörpern in der arteriosklerotischen Plaque der menschlichen Aorta, *Verh. Dtsch. Ges. Pathol.*, 70, 365, 1986.
33. **Stary, H.C.,** Macrophages, macrophage foam cells, and eccentric intimal thickening in the coronary arteries of young children, *Atherosclerosis*, 64, 91, 1987.
34. **Stary, H.C.,** Evolution and progression of atherosclerosis in the coronary arteries of children and adults, in *Atherogenesis and Aging*, Bates, S.R. and Gangloff, E.D., Eds., Springer-Verlag, New York, 1987, 20.
35. **Mitchinson, M.J., Carpenter, K.L.H., and Ball, R.Y.,** The role of macrophages in human atherosclerosis, in *Pathobiology of the Human Atherosclerotic Plaque*, Glagov, S., Newman, W.P., and Schaffer, A.S., Eds., Springer-Verlag, New York, 1990, 121.
36. **Masuda, J. and Ross, R.,** Atherogenesis during low level hypercholesterolemia in the nonhuman primate. I. Fatty streak formation, *Arteriosclerosis*, 10, 164, 1990.
37. **Masuda, J. and Ross, R.,** Atherogenesis during low level hypercholesterolemia in the nonhuman primate. II. Fatty streak conversion to fibrous plaque, *Arteriosclerosis*, 10, 178, 1990.
38. **Massmann, J.,** Mononuclear cell infiltration of the aortic intima in domestic swine, *Exp. Pathol.*, 17, 110, 1979.
39. **Alpers, C.E., Gordon, D., and Gown, A.M.,** Immunophenotype of vascular rejection in renal transplants, *Mod. Pathol.*, 3, 198, 1990.
40. **Zwadlo, G., Schlegel, R., and Sorg, C.,** A monoclonal antibody to a subset of human monocytes found only in the peripheral blood and inflammatory tissues, *J. Immunol.*, 137, 512, 1986.
41. **Zwadlo, G., Bröcker, E.B., v. Bassewitz, D.B., Feige, U., and Sorg, C.,** A monoclonal antibody to a differentiation antigen present on mature human macrophages and absent from monocytes, *J. Immunol.*, 134, 1487, 1985.
42. Lipid Research Clinics Program, The lipid research clinics coronary primary prevention trial results. II. The relationship of reduction in incidence of coronary heart disease to cholesterol lowering, *JAMA*, 251, 365, 1984.
43. **Assmann, G.,** Nationale Cholesterin-Initiative. Ein Strategie-Papier zur Erkennung und Behandlung von Hyperlipidämien, *Dtsch. Ärztebl.*, 87, B991, 1990.
44. **Goldstein, J.L. and Brown, M.S.,** The low-density lipoprotein pathway and its relation to atherosclerosis, *Ann. Rev. Biochem.*, 46, 897, 1977.
45. **Mahley, R.W.,** Cellular and molecular biology of lipoprotein metabolism in atherosclerosis, *Diabetes*, 30, 60, 1981.
46. **Brown, M.S. and Goldstein, J.L.,** Lipoprotein metabolism in the macrophage: implications for cholesterol deposition in atherosclerosis, *Ann. Rev. Biochem.*, 52, 223, 1983.

47. **Brown, M.S. and Goldstein, J.L.,** Arteriosklerose und Cholesterin: Die Rolle der LDL-Rezeptoren, *Spektrum d. Wissenschaft*, 1, 96, 1985.
48. **Kovanen, P.T.,** Atheroma formation: defective control in the intimal round-trip of cholesterol, *Eur. Heart J.*, 11 (Suppl. E), 238, 1990.
49. **Hegele, R.A.,** Lipoprotein (a): an emerging risk factor for atherosclerosis, *Can. J. Cardiol.*, 5, 263, 1989.
50. **Vollmer, W., Brust, J., Roessner, A., Bosse, A., Kaesberg, B., Harrach, B., Robenek, H., and Böcker, W.,** Distribution patterns of apolipoproteins A_1, A_2 and B in the wall of atherosclerotic vessels, *Virch. Arch. [A]*, 419, 79, 1991.
51. **Brown, M.S., Ho, Y.K., and Goldstein, J.L.,** The low-density lipoprotein pathway in human fibroblasts, *Ann. N.Y. Acad. Sci. U.S.A.*, 275, 244, 1976.
52. **Steinberg, D., Parthasarathy, S., Carew, T.W., Khoo, J.C., and Witztum, J.L.,** Beyond cholesterol. Modifications of low-density lipoprotein that increase its atherogenicity, *N. Engl. J. Med.*, 320, 915, 1989.
53. **Robenek, H.,** Topography and internalization of cell surface receptors as analysed by affinity and immunolabeling combined with surface replication and ultrathin sectioning techniques, in *Electron Microscopic Analysis of Cell Dynamics*, Plattner, H., Ed., CRC Press, Boca Raton, FL, 1989, 141.
54. **Vollmer, E., Roessner, A., Bosse, A., Böcker, W., Kaesberg, B., Robenek, H., Sorg, C., and Winde, G.,** Immunohistochemical double labeling of macrophages, smooth muscle cells, and apolipoprotein E in the atherosclerotic plaque, *Path. Res. Pract.*, 187, 184, 1991.
55. **Roessner, A., Schmitz, G., and Sorg, C.,** What's new in the pathology of atherosclerosis? *Pathol. Res. Pract.*, 182, 694, 1987.
56. **Roessner, A., Vollmer, E., Harrach, B., Zwadlo, G., Sorg, C., and Grundmann, E.,** Immunelektronenmikroskopische Charakterisierung von Makrophagen in der Zellkultur, *Verh. Dtsch. Ges. Pathol.*, 71, 386, 1987.
57. **Riesen, W.F.,** Klinische Bedeutung der Apolipoproteine, *Diagnose u. Labor*, 4, 147, 1989.
58. **Driscoll, D.M. and Getz, C.S.,** Extrahepatic synthesis of apolipoprotein E, *J. Lipid Res.*, 25, 1368, 1984.
59. **Majack, R.A., Castle, C.K., Goodman, L.V., Weisgraber, K.H., Mahley, R.W., Shooter, E.M., and Gebicke-Haerter, P.J.,** Expression of Apolipoprotein E by cultured vascular smooth muscle cells is controlled by growth state, *J. Cell Biol.*, 107, 1207, 1988.
60. **Murase, T., Oka, T., Yamanda, N., Mori, N., Ishibashi, S., Takuka, F., and Mori, W.,** Immunohistochemical localization of apolipoprotein E in atherosclerotic lesions of the aorta and coronary arteries, *Atherosclerosis*, 60, 1, 1986.
61. **Davignon, J., Gregg, R.E., and Sing, C.F.,** Apolipoprotein E polymorphism and atherosclerosis, *Arteriosclerosis*, 8, 1, 1988.
62. **Kobayashi, N., Okubo, M., Marumo, F., and Uchida, H.,** De novo development of hypercholesterolemia and elevated high-density lipoprotein cholesterol: apoprotein A_1, ratio in patients with chronic renal failure following kidney transplantation, *Nephron*, 35, 239, 1983.
63. **Harris, K.P.G., Russell, G.J., Parvin, S.D., and Veitch, P.S.,** Alterations in lipid and carbohydrate metabolism attributable to cyclosporine A in renal transplant recipients, *Br. Med. J.*, 292, 16, 1986.
64. **Lowry, R.P., Soltys, G., Peters, L., Mangel, R., and Sniderman, A.D.,** Type II hyperlipoproteinemia, hyperapobetalipoproteinemia, and hyperalphalipoproteinemia following renal transplantation: implications for atherogenic risk, *Transplant. Proc.*, 19, 3426, 1987.
65. **Bürrig, K.F. and Hort, W.,** Das koronare Endothelmuster beim Menschen, *Z. Kardiol.*, 78 (Suppl. 6), 100, 1989.

66. **Gerrity, R.G., Richardson, M., Bell, F.P., Somer, J.B., and Schwartz, C.J.,** Endothelial cell morphology in areas of in vivo Evans blue uptake in the young pig aorta. II. Ultrastructure of the intima in areas of differing permeability to proteins, *Am. J. Pathol.*, 89, 313, 1977.
67. **Hansson, G.K., Jonasson, L., Seifert, P.S., and Stemme, St.,** Immune mechanisms in atherosclerosis, *Arteriosclerosis*, 9, 567, 1989.
68. **Vollmer, E., Maurer, Th., Roessner, A., Bosse, A., Winde, G., and Böcker, W.,** Immunhistologische Untersuchungen zum Lymphozyteninfiltrat in unterschiedlichen Stadien der humanen Arteriosklerose, *Verh. Dtsch. Ges. Pathol.*, 72, 600, 1988.
69. **Vierbuchen, M.,** Lectin receptors, in *Cell Receptors*, Seifert, G., Ed., Current Topics in Pathology, Vol. 83, Springer-Verlag, New York, 1991, 271.
70. **Schulze-Osthoff, K., Meinardus-Hager, G., and Sorg, C.,** Das Endothel und seine Schlüsselrolle in der Entzündung, *Jahrb. Dermatologie*, 41, 1989.
71. **Gordon, D., Schwartz, S.M., Benditt, E.P., and Wilcox, J.N.,** Growth factors and cell proliferation in human atherosclerosis, *Transplant. Proc.*, 3692, 1989.
72. **Demetris, A.J., Zerbe, T., and Banner, B.,** Morphology of solid organ allograft arteriopathy. Identification of proliferating intimal cell populations, *Transplant. Proc.*, 21, 3667, 1989.
73. **Gabbiani, G., Schmid, E., Winter, S., Chaponnier, C., de Chastonay, C., Vandekerckhove, J., Weber, K., and Franke, W.W.,** Vascular smooth muscle cells differ from other smooth muscle cells: predominance of vimentin filaments and a specific alpha type actin, *Proc. Natl. Acad. Sci. U.S.A.*, 78, 298, 1981.
74. **Kocher, O. and Gabbiani, G.,** Cytoskeletal features of normal and atheromatous human arterial smooth muscle cells, *Hum. Pathol.*, 17, 875, 1986.
75. **Lazarides, E.,** Intermediate filaments as mechanical integrators of cellular space, *Nature*, 283, 249, 1980.
76. **Franke, W.W., Hergt, M., and Grund, C.,** Rearrangement of the vimentin cytoskeleton during adipose conversion: formation of an intermediate filament cage around lipid globules, *Cell*, 49, 131, 1987.
77. **Fujimoto, T., Tokuyasu, K.T., and Singer, S.J.,** Direct morphological demonstration of the coexistence of vimentin and desmin in the same intermediate filaments of vascular smooth muscle cells, *J. Submicrosc. Cytol. Pathol.*, 19, 1, 1987.
78. **Villaschi, S. and Spagnoli, L.G.,** Autoradiographic and ultrastructural studies on the human fibro-atheromatous plaque, *Atherosclerosis*, 48, 95, 1983.
79. **Rosenfeld, M.E. and Ross, R.,** Macrophage and smooth muscle cell proliferation in atherosclerotic lesions of WHHL and comparable hypercholesterolemic fat-fed rabbits, *Arteriosclerosis*, 10, 680, 1990.
80. **Jonasson, L., Holm, J., Skalli, O., Gabbiani, G., and Hansson, G.K.,** Expression of class II transplantation antigen on vascular smooth muscle cells in human atherosclerosis, *J. Clin. Invest.*, 76, 125, 1985.
81. **Emeson, E.E. and Robertson, A.L.,** T-lymphocytes in aortic and coronary intimas, *Am. J. Pathol.*, 130, 369, 1988.
82. **Bishop, G.A., Hall, B.M., Duggin G.G., Horvath, J.S., Sheil, A.G.R., and Tiller, D.J.,** Immunopathology of renal allograft rejection analysed with monoclonal antibodies to mononuclear cell markers, *Kidney Int.*, 29, 708, 1968.
83. **Vollmer, E., Brust, J., Roessner, A., Bosse, A., Harrach, B., Robenek, H., Herrera, A., and Böcker, W.,** Immunhistochemische Untersuchungen zur Verteilung von Apolipoproteinen in der arteriosklerotischen Gefäßwand menschlicher Arterien, *Verh. Dtsch. Ges. Pathol.*, 73, 445, 1989.
84. **Carter, R.S., Siegel, R.J., Chai, A.U., and Fishbein, M.C.,** Immunohistochemical localization of apolipoproteins A-I and B in human carotid arteries, *J. Pathol.*, 153, 31, 1987.

85. **Rath, M., Niendorf, A., Reblin, T., Dietel, M., Krebber, H.J., and Beisiegel, U.,** Detection and quantification of lipoprotein(a) in the arterial wall of 107 coronary bypass patients, *Arteriosclerosis*, 9, 579, 1989.
86. **Ho, Y.K., Brown, M.S., and Goldstein, J.L.,** Hydrolysis and excretion of cytoplasmic cholesteryl esters by macrophages: stimulation by high density lipoproteins and other agents, *J. Lipid Res.*, 21, 391, 1980.
87. **Basu, S.K., Ho, Y.K., Brown, M.S., Bilheimer, D.W., Anderson, R.G.W., and Goldstein, J.L.,** Biochemical and genetic studies of the apoprotein E secreted by mouse macrophages and human monocytes, *J. Biol. Chem.*, 257, 9788, 1982.

Chapter 4

COLLAGEN AND COLLAGEN SYNTHESIS IN THE ATHEROSCLEROTIC VESSEL WALL

Jürgen Rauterberg and Elisabeth Jaeger

TABLE OF CONTENTS

ISBN 0-8493-5505-2

I. INTRODUCTION

Enhanced deposition of collagen-rich extracellular matrix and changes in the ratio of the different collagen molecules are major features that characterize the atherosclerotic vessel wall. In this article we review the current data on vascular collagen and collagen synthesis, which, in our opinion, are of relevance to atherosclerosis. Within this framework we will describe in detail our studies on the detection of mRNA by *in situ* hybridization for characterizing the activation of collagen synthesis in the atherosclerotic vessel wall.

Extracellular matrix components are essential contributors to the normal function of the arterial wall. Not only are mechanical properties of this tissue, such as tensile strength and elasticity, governed by collagen fibrils and elastic "membranes", but a broad spectrum of other functions is subserved by matrix components. The intravasal and transvasal diffusion of proteins is greatly influenced by basement membranes, for example, and subendothelial matrix is a potent inducer of platelet aggregation.

Comprehensive understanding of the structure-function relationships of vessel wall extracellular matrix components still seems a distant hope. A complete knowledge of the identities and numbers of different proteins contributing to the extracellular matrix has yet to be attained, and each year brings the discovery of further new molecules. Structural components that are defined by mainly morphological criteria, for instance basement membranes, have been characterized as complicated supramolecular structures formed by several chemically distinct components. Even the collagen fibrils with their characteristic electron microscopical appearance, long regarded as polymers solely of type I collagen, have turned out to consist of several collagen types in a number of different tissues.

Table 1 gives a short and necessarily incomplete summary of the morphologically defined structural components of the extracellular matrix of the vessel wall, together with details of their composition and their possible functions. This table emphasizes that most of the structural components are polyfunctional. Collagen fibrils, for example, are not only responsible for giving tensile strength to the tissue, but they also have regulatory effects on a variety of cell activities and, of all the vessel wall components, are probably the most effective inducers of platelet activation.

Vessel wall extracellular matrix molecules may be divided into the following groups, according to their chemical characteristics: (1) collagens, (2) elastin, (3) structural glycoproteins, and (4) proteoglycans. Each group, apart from elastin, consists of a number of individual proteins.[1,2] The extracellular matrix of the vessel wall provides an interesting illustration of how an amazing variety of components is needed to build up the supramolecular structures required for proper tissue function. Atherosclerosis is characterized by a number of dramatic changes of the extracellular matrix, which lead to various dysfunctions, in turn leading to further progress of the disease. Fibrous plaque formation involves extensive deposition of collagen fibers and microfibrillar material, and fragmentation of elastic membranes within the hugely expanded intimal region of the vessel. Our focus on the various roles of the collagens starts with an examination of their diversity, distribution, and functions in the normal vessel wall.

II. COLLAGENS OF THE ARTERIAL WALL

More than 15 members of the collagen family have so far been defined as individual proteins.[3-6] It is still not known with certainty how many of these occur in the vessel wall, and further types may await discovery. Table 2 summarizes collagen types that, according to current knowledge, are present in the arterial wall.

A. Fibril-Forming Collagens

The fibril-forming collagens I, III, and V constitute by far the highest proportion of vessel wall collagens. Quantitative data are only available for these collagen types, and these show that type I is dominant, followed by types III and V. A summary of published results on quantitative ratios of collagen types in the normal and atherosclerotic vessel wall is given in Section IV.A. The morphological counterparts of type III and V collagen are still not completely clear. During the last few years, evidence has accumulated indicating that type III collagen mainly participates in the formation of typical collagen fibrils. Henkel et al.[7] demonstrated the occurrence of crosslink-containing peptides in digests of vessel wall collagen in which type I- and III-derived chain fragments are linked. By double staining methods in electron microscopic immunohistochemistry, the colocalization of type I and type III collagen in the same collagen fibrils has been demonstrated in a number of tissues.[8] Fleischmayer et al.[9] reported that during

TABLE 1
Extracellular Matrix Components of Vessel Walls: Morphologically Defined Structures, Their Constituent Macromolecules, and Functions

Morphologically defined structures	Macromolecules	Function(s)
Collagen fibrils	Type I collagen Type III collagen Type V collagen	Tensile strength; induction of platelet aggregation; regulation of cellular functions
Fibril-associated components	Type XII collagen (?) Type XIV collagen (?) Decorin Fibronectin	Modulation of fibril functions ?
Microfibrils ("beaded fibrils") Linkage structures	Type VI collagen Fibronectin Fibrillin Undulin Tenascin Thrombospondin von Willebrand protein and others	Linkage between different matrix components and between matrix components and cells; regulation of cellular activities
Elastic fibers and membranes	Elastin Fibrillin	Elasticity; (Elastin fragments: chemotaxis)
Basement membranes	Type IV collagen Laminin Nidogen (Entactin) BM 40 (SPARC) Type VIII collagen (?)	Filtration; Support for endothelial cells; anchorage of cells within the matrix
"Structureless ground substance"	Proteoglycans Hyaluronic acid	Water and ion reservoir; regulation of cellular activities; binding of growth factors

different stages of development, the ratio of collagen III to I was inversely proportional to the average diameter of collagen fibrils, and that antibodies specific for type III collagen or for its N-terminal propeptide mainly labeled thin fibrils. These observations suggest a role for type III collagen (or its precursor form) in the regulation of fibril diameters. Recently, Romanic et al.[10] were able

TABLE 2
Quantitative Ratios of Collagen Types in Normal and Atherosclerotic Human Aorta

Collagen Type	Nonatherosclerotic		Atherosclerotic		References
	Intima	Media	Intima	Media	
I	67	65	74	47	d
	67	56	76	56	66
	68[a]	61[a]	66[a]		67
		65[b]	88[c]		68
III	33	35	26	53	d
	33	44	24	44	66
	18[a]	26[a]	13[a]		67
		37[b]	12[c]		68
IV	0.08[a]		0.4[a]		67
V	15.2[a]	13.3[a]	33.5[a]	13.3[a]	66
	15[a]	14[a]	21[a]		67

[a] Values without superscript 'a' are derived from cyanogen bromide peptide patterns and expressed in % of I + III collagens. Values with superscript 'a' are obtained from peptic extracts and expressed in % of I + III + V collagens.
[b] Media + intima from aorta.
[c] Mean value of three plaques from different aortas.
[d] M. Althaus, J. Rauterberg, unpublished results.

to demonstrate *in vitro* copolymerization of collagen type I with pN* collagen III after treating a mixture of pC* collagen I and procollagen III with procollagen C-proteinase. These authors were able to show a direct inverse relation between fibril diameter and their pN collagen-III portion.

Conflicting data have been reported on the size of vessel wall collagen fibrils. Shirashi et al.[11] have described two populations of fibrils both of which show increasing diameters up to the age of approximately 20 years: the smaller ones (70% of fibrils) reaching about 28 nm, and the larger ones (30% of fibrils) reaching 42 nm. The diameters of both these populations are relatively small compared to those of the collagen fibrils of tendon (e.g., 318 ± 12 nm for rat-tail tendon) or skin (89.0 ± 12.9 for 20-year-old men). Other authors have described the existence of only one population of fibrils with a relatively small range of diameters in the aortic media of rats and cattle.[12]

Relating type V collagen to distinct structural components of the matrix has been problematic since its discovery in peptic extracts of fetal membranes and placenta. Two molecular forms of type V collagen that show different chain compositions have been identified. These are $[\alpha1(V)]_2\alpha2(V)$ (the "two-chain

* pN collagen, pC collagen: intermediates in the processing of procollagen to collagen in which the N terminal or the C terminal propeptides, respectively, are retained.

form") and α1(V)α2(V)α3(V) (the "three-chain form").[13] The two-chain form is widespread throughout the organism, and relatively high amounts occur in smooth muscle cell-rich tissues (vessel wall, uterus, chicken gizzard) and in bone and cornea. The occurrence of the three-chain form is more restricted; it accounts for a comparatively large proportion of the type V molecules in endometrial villi of the placenta and in the uterus.[14,15] Apart from a conspicuous presence in these tissues, type V collagen is found in association with basement membranes and in the form of stromal or cell-associated microfibrils ("exocytoskeleton").[16-20] However, from the amino acid sequence of type V collagen, structural features typical of fibril-forming interstitial collagens are suggested,[21] and *in vitro* formation of cross-striated fibrils has been demonstrated.[22] Adachi and Hayashi[23] were able to show *in vitro* formation of hybrid fibrils from type I–V mixtures. Cross-striated collagen fibrils that react positively for type V collagen on gold immunolabeling have been reported in the embryonic chicken cornea.[24] Birk et al.[25] have demonstrated dose-dependent downregulation of fibril diameters by addition of acid-soluble type V collagen prior to *in vitro* fibril formation. This result may point to a regulatory role of type-V collagen in fibril formation in a similar way to that suggested for type III collagen.

Recently, a new family of collagenous molecules has emerged that also participates in collagens fibril formation and for which the name FACIT collagens (fibril-associated collagens with interrupted triple-helices) has been suggested.[26] So far, there is no information about the occurrence of these molecules in vessel walls.

B. Basement Membrane Collagens

Basement membranes, as defined by their characteristic electron microscopical appearance, comprise a cell-associated continuous layer in which an electron-dense layer, the lamina densa, is separated from the adjacent cell by a more transparent zone, the lamina lucida. In vessel walls, basement membranes occur as a continuous layer underlying the endothelium and as an exterior coat around smooth muscle cells of the media. Knowledge on the chemical composition of basement membranes has been derived mainly from work on epithelial basement membranes (for review see Leblond and Inoue[27]), the major components of which are the basement-membrane-specific type IV collagen, the glycoproteins laminin and nidogen (which form a 1:1 molecular complex), and the basement membrane-specific heparan sulfate proteoglycan. That these components also occur in endothelial and smooth muscle cell-associated basement membranes has been demonstrated by immunohistochemistry. Type IV collagen molecules are heterotrimers consisting of two α1(IV) and one α2(IV) chain (for review see Timpl[28]). The recently described α3, α4, and α5 chains seem to be restricted to distinct kidney basement membranes.[29,30]

Vessel-derived cultivated endothelial cells have been shown to produce another type of collagen, type VIII.[31] This is also produced by cultivated corneal endothelial cells and has been localized in the Descemets membrane of the cornea.[32] Whereas type IV collagen is secreted by all endothelial cells so far investigated, the production of type VIII collagen is not a common feature of cultivated endothelial cells (e.g., it is absent from human umbilical vein endothelial cells, and its production may be linked to certain states of cell activation).[33] The presence of type VIII collagen in extracts of fetal bovine aorta has been demonstrated,[34] and Kittelberger et al.[35] reported immunohistochemical evidence for its presence in the subendothelium of fetal and adult sheep aortas. However, Kapoor et al.[36] failed to detect type VIII collagen in fetal bovine vessel walls, using a monoclonal antibody that reacted with bovine Descemets membranes. Thus, the question as to whether type VIII collagen is a regular or an occasional component of the subendothelial basement membrane of blood vessels remains open.

C. Microfibrillar Collagens: Type VI Collagen

Type VI collagen was initially discovered as a vessel wall collagen and was provisionally named "intima collagen" owing to its detection in peptic extracts of the intimal layer of aortas.[37] It is ubiquitous and abundant in the extracellular matrices of various tissues and has a peculiar molecular and genomic structure, which is unique among the molecules of the collagen family (for review see Rauterberg et al.[38] and Engvall et al.[39]). Type VI collagen molecules are composed of three different chains, $\alpha 1$(VI), $\alpha 2$(VI)and $\alpha 3$(VI). The $\alpha 1$(VI) and $\alpha 2$(VI) chains are similar in length (both ca. 1000 amino acid residues) and domain structure. An N-terminal noncollagenous domain (235 residues) is followed by the triple helix (336 residues) and a noncollagenous C-terminal portion (430 residues).[40,41] The $\alpha 3$(VI) chain is of unusual size (2600 residues) largely because of an enormous N-terminal nonhelical domain of 1600 residues.[42,43] The noncollagenous domains on both sides of the triple helix, representing altogether ca. 70% of the molecular mass, impart to the molecule its typical dumbbell-like shape. These molecules aggregate to form dimers and tetramers which represent the secreted form of type-VI collagen and which form fibrillar structures ("beaded fibrils") by head-to-head aggregation.

The noncollagenous domains contain a number of repetitive sequences that are homologous to the A domain of von Willebrand protein,[40-44] and such sequences are characteristic of several collagen-binding proteins. The collagenous domain of all three chains contains several arginine-glycine-aspartate (RGD) sequences that are known to mediate cell binding via integrin-type receptors. Thus, type VI collagen may be specialized for linking collagens to cells and to other matrix components.

III. COLLAGEN SYNTHESIS IN VASCULAR CELLS

Three types of mesenchymal cells are involved in the construction of blood vessels: (1) endothelial cells, which form the blood tissue barrier, (2) smooth muscle cells, the only cell type of the normal media, and (3) fibroblasts, which are present in the adventitia. Smooth muscle cells are the only possible source of the medial extracellular matrix. In the intima of the normal artery, a thin layer of extracellular matrix containing a few smooth muscle cells exists between the subendothelial basement membrane and the internal elastic lamina. This thin layer represents the part of the arterial wall that becomes involved in atherosclerosis. In diseased vessels this layer becomes dramatically thickened by processes that involve cell proliferation, invasion of blood-derived cells, enhanced deposition of extracellular matrix, and intercalation of blood-derived substances. It is generally agreed that synthetic-state smooth muscle cells are the main producers of the extracellular matrix in the thickened intima.[45]

A. Smooth Muscle Cells

The quantitative ratio of collagen types produced by cultivated arterial smooth muscle cells has been characterized by a number of authors.[46-49] For human aortic smooth muscle cells grown from explants of aortic media, the ratio of type I to type III was found to be approximately 3:1, the same as that of most atherosclerotic plaques. An increase of type III and decrease of type I has been demonstrated during infancy; approximately 10% of the total collagen is type III in fetal aortic smooth muscle cells, rising to 20 to 30% in smooth muscle cells from 12-year olds.[46] Beyond this age no further increase in type III occurs. Distinct differences have been found in the ratio of type I to type III produced by smooth muscle cells obtained from different species. Burke et al.[47] found that in monkey aortic smooth muscle cells, far more type III collagen was produced than type I. Ratios similar to those of human smooth muscle cells have been found for pig and guinea pig aortic smooth muscle cells.[48-50] Our own results indicate unusually low levels of type III (approximately 5%) in smooth muscle cells cultivated from rat aortas (Allam and Rauterberg, unpublished results). Steady-state levels of mRNAs of type I and type III polypeptide chains are found to show about the same ratio as those observed for the secreted collagens.[51]

Qualitative detection of collagens type IV and V has been done by biochemical, immunochemical, and immunofluorescence methods.[52-54] There are, however, few quantitative data available. By electrophoresis of pepsin-resistant proteins secreted by cultivated rabbit aortic smooth muscle cells, Okada et al.[54] determined the amounts of collagens type I, III, IV, and V synthesized by these cells as 82.7, 14.9, 0.8, and 1.6% of total collagen, respectively. They found a remarkable shift of this ratio in cells treated with dimethylsulfoxide, which induces change to a more contractile state; type III collagen decreased to 2.9%, and type IV increased to 9.2%, of total collagen. No data are yet available on the amount of type VI collagen produced by smooth muscle cells.

Mayne et al.[50,52] and Majack and Bornstein[55] described a short pepsin-resistant fragment (45 kDa) among the collagenous proteins produced by smooth muscle cells from guinea pig, rhesus monkey, and rat aortas. The nondegraded fragment has a molecular weight of 60 kDa and is deposited mainly into the cell layer.[56] Cultivated smooth muscle cells are thus capable of producing the spectrum of collagens found in the extracellular matrix of the normal vessel wall, the deposition of which is pathologically enhanced during the formation of atherosclerotic lesions.

B. Endothelial Cells

Although smooth muscle cells are undoubtedly responsible for the bulk of matrix synthesis in the normal and diseased vessel wall, the ability of endothelial cells to synthesize matrix components should not be overlooked. Cultivated endothelial cells are active producers not only of basement membrane components, but also of interstitial collagens, fibronectin, other glycoproteins, and elastin (for review see Sage[57]). These findings raise the possibility that endothelial cells may participate in forming extracellular matrix structures even below their basement membrane.

One important result found from cell culture studies is that the pattern of collagens synthesized by endothelial cells varies according to species, culture conditions, and the vessel of their origin.[58] This variability suggests the existence of mechanisms that regulate collagen synthesis *in vivo*. Though presumably only marginal contributors to the overproduction of collagen in atherosclerotic lesions, a dysregulation in endothelial collagen synthesis may be involved in alterations of the endothelium and subendothelium, which may participate in the onset of the atherosclerotic process. Diabetic angiopathies are, for instance, characterized by an overproduction of endothelium-associated basement membranes, and the high risk of diabetic patients for atherosclerosis is well documented.

Recently, Iruela-Arispe et al.[59] described the induction of collagen type I and collagen type VIII synthesis in bovine aortic endothelial cells undergoing the "sprouting" phenomenon thought to be related to angiogenesis. And evidence from our *in situ* hybridization studies suggests that thrombus formation in the aorta leads to expression of collagens I and III and proliferation in the underlying endothelial cells.[60] These findings suggest that induction of the biosynthesis of certain collagen types may be linked to endothelial cell activation. In earlier work we were able to demonstrate by immunocytochemistry that several matrix proteins (type V collagen, laminin, and fibronectin) become more abundant on the luminal surface of the rat aortic endothelium with experimentally induced hypertension.[61] Thus, cell activation may lead to changes in the overall spectrum of synthesized collagens and also to dysfunction of the polarity of collagen secretion. These findings, placed in the context of the influence of collagens on platelet activation, emphasize the relevance of cell activation-linked changes in collagen synthesis to atherosclerosis.

IV. COLLAGENS IN NORMAL AND ATHEROSCLEROTIC VESSELS

A. Quantitative Ratios of Collagens

Owing to the highly crosslinked state of collagens in the arterial wall, exact quantitation of each collagen type from tissue extracts is extremely difficult. This is reflected in the conflicting results that have been reported on arterial collagen composition (for reviews see Barnes,[62,63] Rauterberg et al.,[64] and Wagner[65]). Two main methods of solubilization have been developed: pepsin treatment and cyanogen bromide digestion. Each of these approaches has advantages and drawbacks, depending on which collagen type is to be measured.

1. Solubilization by Pepsin Treatment

Pepsin treatment at low temperature digests terminal non-triple-helical regions of collagen while leaving triple-helical regions intact. Since intermolecular crosslinks involve the terminal peptides of the molecules, collagen is solubilized by digestion of these regions, and after denaturation, the polypeptides can be identified and quantified by gel electrophoresis and densitometry. The most serious disadvantage of this method is the differing rate of solubilization, which varies according to tissue- and age-dependent degree of crosslinking, and to the collagen type.

2. Solubilization by Cyanogen Bromide Digestion

Cyanogen bromide digestion results in complete solubilization in the form of defined peptides. Since only two distinct sites within the triple helix of fibrillar collagens are involved in crosslinking, solubilization and electrophoretic mobility of most peptides is independent of crosslink formation. Most investigations using this method relate to the ratio of type I and III collagens. Recently, analysis of cyanogen bromide peptides has also been described for type V collagen-derived peptides.[66]

Results achieved by various authors using these methods have been reviewed by Barnes[63] and Wagner.[65] Table 2 summarizes a selection of these data, which reveal two tendencies in the atherosclerotic intima. First, the ratio of type I to type III collagen is shifted in favor of type I collagen. Second, the proportion of type V collagen is clearly elevated. A similar increase in type V collagen is found in other pathological conditions characterized by new formation of extracellular matrix, namely hypertrophic scar tissue,[69] Dupuytren's contracture,[70] inflamed gingiva in periodontitis,[71] and experimentally-induced granulation tissue.[72]

B. Immunohistochemical Localization

The availability of monospecific antibodies against different collagen types has allowed immunohistochemical investigation of the local distribution of collagens within the different layers of the vessel wall. A number of reports on the normal human and atherosclerotic vessel wall and bovine aorta have been published,[17,18,73-76] and all are in broad agreement on the distribution of types I, III, and IV collagen. Type I and III are found to be colocalized in the matrix between smooth muscle cells and elastic membranes in the media, in the interstitial matrix of the adventitia, and in the matrix underlying the endothelial basement membrane in the intima. Type IV collagen is found restricted mainly to the subendothelial basement membrane, to the surface area of intimal and medial smooth muscle cells, and to the basement membranes of the vasa vasorum in the outer media and in the adventitia. Major divergences emerge between the reported distributions of type V collagen, however. In earlier reports, strong staining of endothelial and smooth muscle cell-related basement membranes was described.[73] This has not been confirmed by later studies in which a distribution more closely resembling that of types I and III collagen is reported, with more pronounced immunostaining in the intima than in the media, especially in atherosclerotic plaques.[19] Type VI collagen-specific antibodies give diffuse staining of the medial intercellular space, with more marked staining of the plaque matrix, a distribution similar to that of fibronectin.[77]

V. COLLAGEN SYNTHESIS IN THE NORMAL AND ATHEROSCLEROTIC VESSEL WALL

Several approaches have been applied to investigate the mechanisms regulating the synthesis of collagen and other matrix proteins in atherosclerosis. Cell culture model systems are particularly useful; they have, for example, revealed the key importance of cytokines and matrix components as regulators. Cell culture, however, is itself a kind of multifactorial activation process, as demonstrated by the switch in smooth muscle cell differentiation to the synthetic phenotype that it induces.[78] For this reason, several approaches to the study of biosynthetic activity of arterial wall cells have been developed to avoid tissue destruction and preserve the various relationships of cells with their natural surroundings.

A. Studies on Overall Collagen Synthesis in Tissues

Incorporation of radioactively labeled proline has been most frequently used for the determination of collagen synthesis. This procedure is based on the finding that posttranslational modification of proline to hydroxyproline is practically collagen-specific. In the interstitial collagen, hydroxylation is restricted to proline residues in the y position of the collagen characteristic sequence triplet Gly-x-y, and a hydroxylation rate of 50% can be assumed.[79] After administration of radiolabeled proline into animals, or after incubating tissue samples in radiolabeled proline-containing media, the activity of collagen synthesis can be deduced from the measurement of total incorporation of proline into the tissue and by determination of the ratio of labeled proline and hydroxyproline. Where comparison is made between tissue samples, however, care has to be taken to ensure that the specific radioactivity of the intracellular proline pool (which determines the specific radioactivity of newly synthesized collagen) remains comparable.[80]

Only a few whole-animal studies using this approach in vessel walls have been reported. This is in part because of the high quantity of radiolabeled proline required, and in part because of uncertainty concerning the distribution and compartmentation of the injected proline (which may differ in animals with and without experimentally induced atherosclerosis). An early investigation was reported by Hauss et al.,[81] who reported significantly elevated levels of radiolabeled hydroxyproline in aortas of hypertensive rabbits. Nissen et al.[82] examined the collagen turnover rate in hypertensive rats. After injection of the radiolabel, the authors followed the decay of tritiated proline over a period of 100 d. By repeated administration of large quantities of unlabeled proline, they avoided the influence of reutilization of labeled proline derived from degradation of collagen. They found that the half-life of newly synthesized collagen was reduced from 70 d in "normal" animals to 17 d in high-pressure animals. Biosynthesis rates after injection of radiolabeled proline have been determined by Opsahl et al.[80] in a diet-induced model of atherosclerosis in rabbits. By applying a large quantity of unlabeled proline together with labeled proline, they maintained comparable and approximately constant specific activities of serum proline in normal and atherosclerotic animals during the incubation period (5 h). In the intimal-medial layer of animals on an atherogenic diet, the collagen synthesis rate was increased tenfold, and the collagen proportion of newly synthesized protein was doubled, compared with controls, thus indicating specific enhancement of collagen synthesis over and above activation of protein synthesis in general.

More frequently applied methods for studying collagen synthesis in atherosclerosis use either radiolabeled proline incubation of arterial tissue samples from experimental animals in organ culture,[82-84] or measurement of the activity of enzymes involved in posttranslational modification of collagen, such as prolylhydroxylase[85-89] or lysyloxidase.[90,91] Using these approaches, elevated collagen synthesis has been reported in the vessel walls of rabbits with diet-induced atherosclerosis, rabbits subjected to hypoxia,[82-84] hypertensive rats,[88-91] and female breeder rats developing spontaneous atherosclerosis.[92] After separation and isolation of different collagen types from peptic extracts of incubated tissues, specific activities of hydroxyproline in these collagen types have been determined, and ratios of synthesis rates deduced. In this way, Deyl et al.[93] found that synthesis of type I exceeded that of type III in spontaneously hypertensive rats with established hypertension, whereas in prehypertensive rats at the age of 4 weeks, type III collagen showed a higher rate of synthesis. In balloon-catheterized rabbits, Barnes[63] reported a preferential enhancement of type I collagen, which was somewhat reduced by adding a cholesterol-enriched diet. Synthesis of type III and type V collagens was also enhanced, but not to the same extent. A similar marked increase in type I collagen, compared to types III and V, was reported by Ryan et al.[94] in external jugular veins of sheep in which atherosclerosis was induced by anastomosis with the common carotid artery.

B. Steady-State Levels of mRNAs in Vessel Wall Extracts

In addition to the possibilities offered by established techniques of protein chemistry, the development of recombinant DNA technology during the last few years now allows further insights into the mechanisms of biosynthetical activation processes at the cellular level. Using messenger RNAs or genomic DNAs extracted from cells or tissues, complementary DNAs (cDNAs) can be synthesized that are then used to construct labeled DNA or RNA probes complementary to the desired DNA or RNA sequence. With the help of labeled complementary probes, it is possible to identify and quantify sequences, e.g., of messenger RNA (mRNA) specific for a desired target protein after extraction from tissues via dot blot or slot blot hybridization or northern blot analysis. Additionally, the specificity of complementary binding provides the possibility of performing hybridization directly on frozen tissue sections and thus of localizing biosynthetically activated cells (see section V.C.).

A prerequisite for specific hybridization is the use of complementary probes without cross-hybridization activity. Unspecific hybridization to highly homologous or noncomplementary DNA or RNA sequences or to the highly abundant ribosomal cellular RNA should be excluded by southern or northern analysis of the hybridization probe. The labeled probe is hybridized to cellular DNA or RNA digested with restriction enzymes and immobilized on a nitrocellulose filter or nylon membrane after electrophoretic separation. The number and position of additional unexpected bands indicate interaction of the probe with noncomplementary sequences and thus allows assessment of probe specificity.

1. Cell Culture Studies

Initial insights into control mechanisms of matrix protein mRNA levels in vessel wall cells came from cell culture experiments.[95,96] Alterations in collagen synthesis accompanying smooth muscle cell phenotypic change during development and (in the reverse direction) during atherosclerosis[45] have recently been investigated at the protein and mRNA levels in enzymatically isolated rabbit aortic smooth muscle cells.[51] These studies show dramatically elevated collagen biosynthesis in smooth muscle cells of the synthetic phenotype and a shift in the ratio of type I to type III collagen in favor of type I, both at the protein and the mRNA level. The close association of these changes with the synthetic state was demonstrated by their reversal upon return of the cells to the contractile state at confluency.

2. mRNAs Extracted from Tissues

Quantitation of mRNA steady-state levels in RNA extracts from vessel wall tissue offers an additional approach to estimate the rate of *in vivo* protein synthesis. In the investigation of alterations in protein synthesis under experimental conditions, determination of mRNA steady-state levels is particularly useful for providing data on stimulating or inhibiting regulatory mechanisms at the transcriptional level. In contrast to the determination of overall collagen synthesis, e.g., by measurement of proline incorporation, the determination of mRNA levels in vessel wall extracts allows parallel assessment of different collagen mRNAs within one tissue extract, as well as the comparison to other mRNAs of interest.

Using a combination of morphological and immunohistochemical techniques, and quantitation of extracted mRNAs for laminin and type IV collagen, McGuire et al.[97] demonstrated that the cellular subendothelial thickening of aortic tissues of spontaneously hypertensive rats is at least in part due to a significant increase in mRNA transcription and deposition of basement membrane macromolecules. In aortic tissue from hypertensive rats fed a deoxycorticosterone/salt diet, Takasaki et al.[98] found an increase of up to threefold in fibronectin mRNA, which was also inducible by infusion of angiotensin II. Corresponding to the elevation of fibronectin mRNA, an increase in aortic fibronectin was found. Expression of mRNA coding for type I collagen, however, remained unchanged. In pulmonary arteries from calves with hypoxic pulmonary hypertension, Crouch et al.[99] described a significant increase of type I, III, IV, and V collagens, as well as enhanced transcription of mRNAs coding for type I and IV collagens. The authors found these differences between normal and hypertensive calves in organ cultures as well as in cultures of medial smooth muscle cells. Additionally, they presented evidence that cells from hypertensive animals secrete a factor that enhances collagen and elastin synthesis in normal fibroblasts and smooth muscle cells. In elucidating the mechanisms leading to enhanced transcription of genes coding for extracellular matrix proteins in the vessel wall, the participation of endogenous cytokines has to be considered. Determination of cytokine mRNA levels in vessel wall extracts has been conducted in experimental angiopathies. Northern blots of RNA extracted from aortas from normal and hypertensive rats revealed that, of various growth factors and cytokines tested, only transforming growth factor β (TGF-β) was enhanced during hypertension.[100] TGF-β has been reported to stimulate synthesis of extracellular matrix components in a number of mesenchymal cells,[101] including smooth muscle cells (see Chapter 5).

C. Localization of mRNAs in Tissue Sections by *in situ* Hybridization

Although hybridization assays of extracted mRNAs have enabled valuable insights into mechanisms of growth control in the animal and human arterial wall, this method provides information only on the average mRNA levels of the whole extracted tissue sample. Using hybridization of distinct DNA or RNA probes to the appropriate mRNA on tissue slices, it has become possible to study the expression of mRNAs of interest in selected cell types *in situ* (for review see References 103 to 105). The method of *in situ* hybridization was originally introduced in 1969 by Gall and Pardue[106] and has primarily been used in the localization of DNA sequences. During the following years the method became a valuable tool in the localization of viral DNAs and RNAs,[107] in chromosomal gene mapping,[108,109] and in localization of mRNAs coding for proteins of interest in developmental and in pathologic processes.[110-112] Several attempts have been made to improve the sensitivity, the rapidity, and the efficiency of *in situ* hybridization analysis.[113,114] Some methodological aspects of these developments will be reviewed in this section.

1. Methodology of in situ *Hybridization*

The cDNA constructed from the cellular mRNA may either be used as double-stranded DNA probe or as template to construct single-stranded DNA probes. Double-stranded DNA probes can be enzymatically labeled by nick translation[113,115] or by random-primed DNA synthesis using Klenow fragments of DNA polymerase.[116,117] Advantages and disadvantages of these two methods of probe labeling have been reviewed in detail by Chan and McGee.[118] Single-stranded DNA probes can be synthesized with the help of the cloning vector M13.[119] In contrast to double-stranded DNA probes, the use of single-stranded probes does not require a denaturation step to prevent reannealing of the probe and thus results in the formation of a stronger signal.[120] Synthetic oligonucleotides with high specific activity, good stability, and target penetration are also applied as hybridization probes because of their relatively short lengths.[121]

Disadvantages of this approach are the reduced stability of the hybrids in comparison to RNA-mRNA hybrids, and the limited degree of labeling due to the small probe size. Recently, a method of primer-directed synthesis of specific DNA regions (polymerase chain reaction, PCR) has been developed that allows amplification of the template sequence to exponentially increasing copy numbers.[122] Using PCR, the detection of mRNAs of extremely low concentrations in tissue samples has become possible.[123]

A number of advantages is provided by the use of single-stranded RNA probes. They can be transcribed from a cDNA inserted into specially constructed transcription vectors containing a transcription promoter in the presence of labeled and unlabeled ribonucleotides and the corresponding RNA polymerase. Vectors with two different RNA promoters ("gemini vectors") flanking the multiple cloning site provide the possibility of bidirected transcription of the inserted cDNA to create both the coding (sense) and the noncoding (antisense) RNA probe. Sense probes can be used as negative control probes to exclude nonspecific hybridization. RNA-mRNA hybrids are more stable than are DNA-mRNA hybrids,[124] thus providing the possibility of RNase digestion of nonspecific hybrids after the hybridization process. An often used control to show that the probe binds to intracellular RNA is the pretreatment of samples with RNase; this should lead to complete loss of signal. In the case of single-stranded riboprobes, this control should be performed by using micrococcal nuclease; this enzyme is not active under the conditions of hybridization and thus cannot cause digestion of the probe itself.[125] The high stability of RNA-mRNA hybrids allows the application of highly stringent washing conditions without reduction of the specific signal. Riboprobes constructed in an *in vitro* run-off transcription reaction are of defined length and high specific activity, and are free of nonspecific vector sequences. A special difficulty in the use of riboprobes is the need to avoid RNase contamination of the tissue examined. A detailed review on the use of complementary riboprobes in *in situ* hybridization has been published by Gibson and Polak.[126] Labeling of riboprobes can be performed using radioisotopes or nonisotopic substances (for methodology see Dirks et al.[127]). The most widely used isotopes are the following: (1) ^{32}P, which, with a short exposure time to X-ray films or autoradiographic emulsion, produces a high signal density, leading to poor resolution because of high energy emittance; (2) ^{35}S, which provides a good signal resolution and high sensitivity; and (3) ^{3}H, which has the signal resolution of ^{35}S, but requires prolonged exposure times (for review see Brady and Finlan[128]). Disadvantages of radioactive labeling systems relate to problems of safety and waste disposal. However, autoradiographic detection is characterized by high specificity and resolution and provides better opportunities for signal quantification than colorimetric detection systems do, as was recently emphasized in a quantitative study using a human papilloma virus model system with known copy numbers of viral DNA.[129] In recent years a number of approaches to quantitative analysis of *in situ* hybridization results in tissue sections have been reported.[130-132] As nonradioactive labels, biotin and digoxigenin are most widely used. They provide good signal resolution and stability, very short processing times, and increased signal stability compared to radioactively labeled probes. Biotin can be enzymatically incorporated, or chemically introduced, into the probe and detected by standard antibody procedures or by means of the avidin/streptavidin technique.[118] Very recently, localization of mRNAs coding for interleukin-1 α and β in tissue sections of nonhuman primate iliac arteries by digoxigenin-labeled DNA probes has been reported.[133] A wide variety of further nonradioactive labeling systems has recently been developed. The haptens or the modified probe are detected in most cases by affinity cytochemical techniques and can often be amplified by immunocytochemical means.[105,126,134,135]

The procedure of *in situ* hybridization of tissue sections includes pretreatment of slides prior to hybridization in order to increase accessibility of the cellular mRNA to probes and to reduce nonspecific binding of probes to cellular structures. The labeled probes are hybridized to tissue sections for a sufficient time, determined by probe specificity and stringency conditions. After hybridization, several steps of posthybridization washings reduce the amount of nonspecific hybrids. Approaches to varying these conditions in order to achieve optimal hybridization rates and a low background have recently been reviewed by Höfler[105] and Gibson and Polak.[126] In the following section we will describe in detail our approach to *in situ* hybridization using ^{35}S-labeled riboprobes on tissue sections of human vessel walls.

a. Preparation of RNA Probes

^{35}S-labeled riboprobes were transcribed from cDNAs inserted into gemini vectors Gem 3 or Gem 4 (Promega Biotec) and linearized with appropriate restriction endonucleases. Linearized plasmid DNA (1.5 μg) was transcribed in a reaction volume of 20 μl of 40 m*M* Tris-HCl (pH 7.5), 10 m*M* NaCl, 6 m*M* $MgCl_2$, 2 m*M* salmon sperm (Sigma), 13 m*M* DTT (Merck), 20 U RNasin (Promega Biotec), 0.5 m*M* each of ATP, CTP, and GTP, and 12.5 μ*M* UTP, using 10 U of SP6 or T7 polymerase in the presence of 100 μCi ^{35}S-UTP (specific activity: 1,350 mM^{-1}) (New England Nuclear) at 40°C for 1 h according to a protocol described by Melton et al.[141] with modifications suggested by the supplier (Promega Biotec). The DNA pellet was subsequently digested with 1.5-U DNase I (Promega Biotec) for 15 min at 37°C. After addition of 1 A_{260}U yeast tRNA (Sigma) and protein denaturation with 0.2% SDS (Sigma), the reaction solution was extracted several times with phenol (Gibco BRL) and Chloroform in a 0.3 *M* solution of sodium acetate and Tris-EDTA (1xTE),[115] and unincorporated nucleotides were removed by precipitating the labeled transcript twice with a 2.5 volume of ethanol. The labeled riboprobes were dissolved in water, the rate of incorporation of labeled UTP was determined by liquid scintillation counting, and the solution adjusted to a probe concentration of 1×10^6 cpm μl^{-1}.

b. Tissue Processing

Immediately after surgical resection, arterial samples were embedded in Reichert-Jung medium, frozen in liquid nitrogen, and stored at –80°C.Five μm-Cryostat sections were cut from the arteries and mounted on poly-L-lysine-hydrogen-bromide-coated slides (100 mg l^{-1}; Sigma). Sections were fixed for 1 min in 4% paraformaldehyde in phosphate-buffered saline (PBS: 0.15 *M* NaCl; 1.6 m*M* NaH_2PO_4; 8 m*M* Na_2HPO_4) and stored at 4°C in 70% ethanol.

c. In situ *Hybridization*

Following rehydration of tissue sections in 2xSSC (standard saline citrate),[115] they were acetylated for 30 min in 25 μ*M* acetic anhydride/0.1-M triethanolamine (pH 8.0) to avoid electrostatic interaction of probe molecules with free amino residues.[142] After rinsing in 2xSSC and PBS treatment, slides were incubated with 0.1 *M* glycine in 0.1 *M* Tris-HCl (pH 7.0) to inactivate endogenous enzymes and quench free aldehydes.[143] All steps were done at room temperature. Prior to hybridization, slides were rinsed in 2xSSC and treated with 50% formamide/2xSSC at 50°C for 5 min in order to denature endogenous RNA and promote hybridization.[143] Formamide has a helix-destabilizing effect resulting in a decreased dissociation temperature of hybrids and thus allows hybridization at lower temperatures and improved preservation of morphological structure. The labeled RNA transcript (1×10^6 cpm per tissue section) was dissolved in a solution containing 50% formamide (Gibco BRL), 2xSSC, 10 m*M* DTT, 2 mg ml^{-1} bovine serum albumin (Sigma), 1 mg ml^{-1} sonicated salmon sperm (Sigma), and 10 mg ml^{-1} yeast t-RNA in order to block nonspecific binding of probe.[144] For hybridization, the mixture was heated to 90°C for 10 min, adjusted to 50°C, and applied to slides (10 µl of hybridization solution per tissue section). Slides were sealed with parafilm and incubated for 3 h at 50°C. A large quantity of nonhybridized single-stranded probe was removed by rinsing slides twice in 2xSSC/50% formamide at 52°C and repeated washing in 2xSSC at room temperature. To reduce the amount of residual single-stranded probe molecules, tissue sections were incubated in 30 µl of 100 µg ml^{-1} RNase A (Sigma) and 1 µg ml^{-1} RNase T_1 (Boehringer Mannheim) in a humid chamber at 37°C for 30 min. Finally, the slides were treated with 2xSSC/50% formamide at 52°C for 5 min, 2xSSC at room temperature for 10 min, dehydrated in ethanol, and air dried.

d. Autoradiography

Autoradiographic detection was performed by covering slides with NTB-2 nuclear track emulsion (Kodak) and exposing 3 d prior to development with D19 developer (Kodak). The slides were subsequently counterstained with hematoxylin/eosin, dehydrated, and mounted.

e. Controls

For controls, slides were treated with ^{35}S-labeled sense probe and with NTB-2 emulsion only. For hybridization with one RNA probe, five serial sections were mounted on one slide. Only results that were reproducible within complete series were included in the analysis of results.

2. *Parallel Application of Immunocytochemistry and* in situ *Hybridization: Characterization of Cell Types in Regions of Active Collagen Synthesis*

The immunocytochemical identification of cells in which collagen mRNA is detected in tissue sections makes a valuable contribution to the interpretation of results provided by *in situ* hybridization studies. Immunocytochemical staining may be performed on neighboring serial sections or on the same section following *in situ* hybridization.

a. Investigation of Serial Sections

Parallel investigation of serial sections has found broad application in the identification of cells expressing mRNAs for extracellular matrix proteins. In pulmonary interlobar arteries from calves with hypoxic hypertension, Prosser et al. were able to demonstrate differential stimulation of tropoelastin and procollagen mRNA within specific regions and populations of smooth muscle cells, using immunohistochemical detection of actin, desmin, and vimentin.[145] Enhanced levels of mRNAs coding for type I and III collagen have been reported in intimal and medial smooth muscle cells in the hyperplastic prestenotic portion of inborn aortic coarctations.[60] Similarly enhanced levels have been detected in intimal

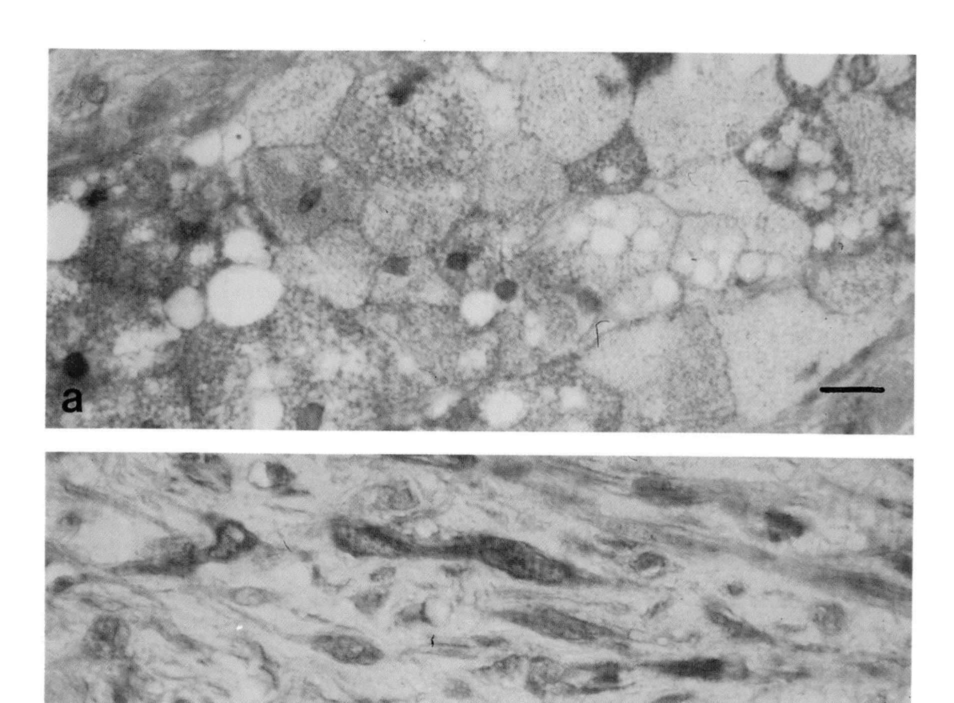

PLATE 1. Chapter 3. Immunohistochemical detection of apolipoprotein A (Apo A) in advancing vascular obliteration. (a) Shows staining for Apo A_1, which is usually intracellular, mainly in foam cells. (b) Shows staining for Apo A_2. This lipoprotein is transferred to the media and is found increasingly in spindle-shaped smooth muscle cells. (Paraffin sections). (Bars: 20 μm).

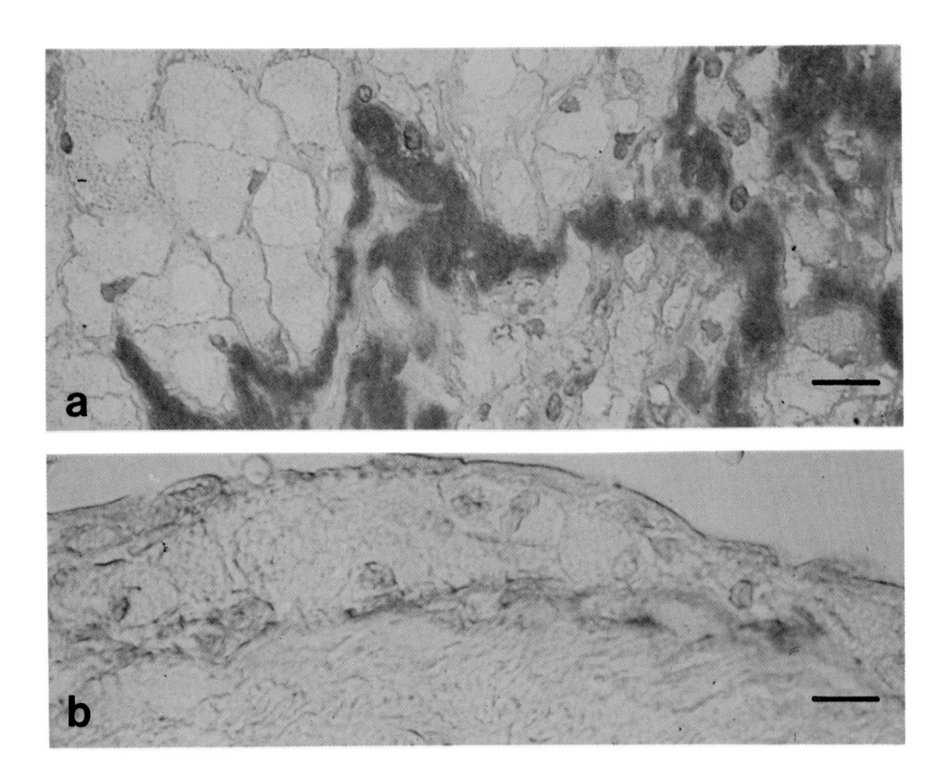

PLATE 2. Chapter 3. (a) Shows immunohistochemical detection of apolipoprotein B (Apo B). In contrast to Apo A (Plate 1), Apo B is mainly extracellular in all grades of lesion, often localized around foam cells that remain immunonegative for Apo B. Panel (b) shows visualization of endothelial splicing with integration of mononuclear cells, some of which are foam cells, between the apical and basal layers of the endothelium which have been stained with antibody against Factor VIII-related antigen. (Paraffin sections). (Bars: 20 μm).

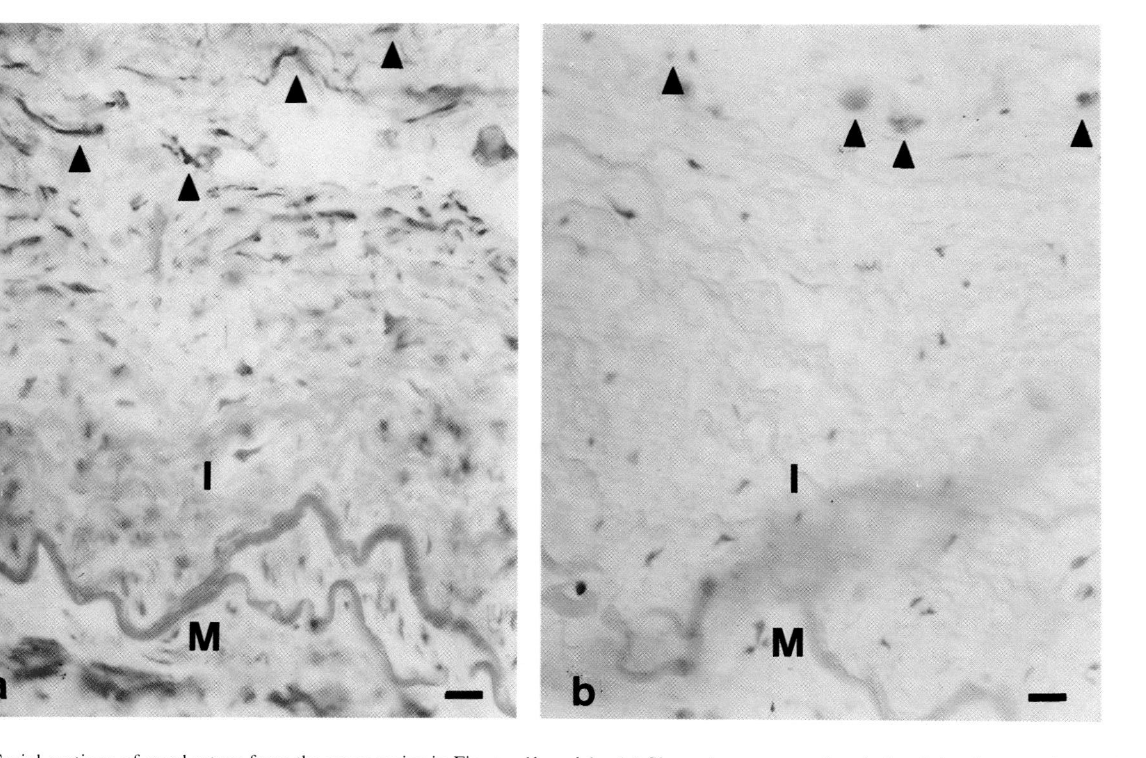

PLATE 3. Chapter 4. Serial sections of renal artery from the same series in Figures 1b and 1c. (a) Shows immunocytochemical staining for smooth muscle cell-specific α-actin; (b) for cytotoxic T lymphocytes (anti-CD8). Areas of closely associated immunostained smooth muscle cells and T lymphocytes are indicated by arrowheads. (I = intima; M = media). (Bars: 10 μm).

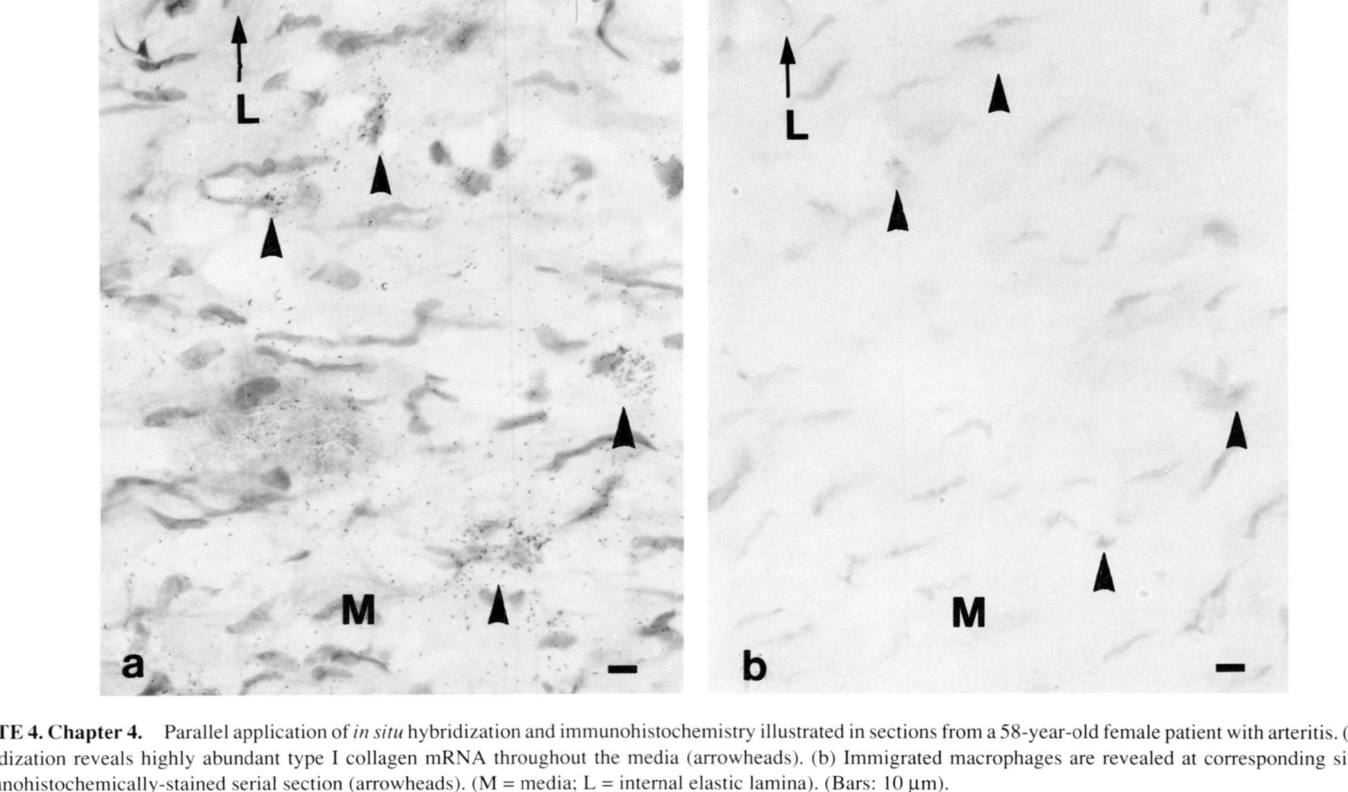

PLATE 4. Chapter 4. Parallel application of *in situ* hybridization and immunohistochemistry illustrated in sections from a 58-year-old female patient with arteritis. (a) *In situ* hybridization reveals highly abundant type I collagen mRNA throughout the media (arrowheads). (b) Immigrated macrophages are revealed at corresponding sites in an immunohistochemically-stained serial section (arrowheads). (M = media; L = internal elastic lamina). (Bars: 10 µm).

smooth muscle cells of atherosclerotic vessels, in close vicinity to immigrated monocyte/macrophages, and identified immunohistochemically with the aid of specific markers.[146] The smooth muscle cells in this system were identified using a monoclonal antibody against smooth muscle cell-specific α-actin[147,148] (HHF35; Enzo Biochem), combined with the alkaline phosphatase-antialkaline phosphatase method.[149]

In order to investigate cytokine expression in vessel walls, detection of mRNA and immunoreactive protein within the same cell can be used to demonstrate that (1) mRNA is translated in a given cell and (2) the corresponding intracellular protein detected in this same cell has not gained entry to it by endocytosis or phagocytosis. Following this reasoning, Moyer et al.[133] were able to show that not only do intimal foam cells and adherent leukocytes express interleukin-1(IL-1), but so too do intimal smooth muscle cells. Investigation of serial sections has also been applied to compare both mRNA expression and protein synthesis of tissue factor, a glycoprotein with functions in the extrinsic coagulation pathway.[150] Wilcox et al. demonstrated tissue factor staining deposited in the extracellular matrix surrounding mRNA-positive cells adjacent to cholesterol clefts within necrotic cores, and in foam cell-rich regions of atherosclerotic plaques. Coincident occurrence of tissue factor mRNA and protein revealed activation of synthesis at both the transcriptional and translational level. In macrophage-rich areas of rabbit and human atherosclerotic lesions, the expression of monocyte chemoattractant protein 1 (MCP-1) at the level of mRNA expression and protein synthesis has also been demonstrated with the serial section approach.[151] The presence of MCP-1 mRNA in macrophage-derived foam cells raises the possibility that a subpopulation of macrophages could be the source of MCP-1 expression. A combination of *in situ* hybridization and immunocytochemical methods has also proved to be a valuable tool for investigating the role of lipid peroxidation in atherogenesis. Using riboprobes and antibodies against 15-lipoxygenase in serial sections of aortic tissue obtained from spontaneously atherosclerotic Watanabe heritable hyperlipidemic rabbits, a strong increase in the abundance of RNA message and protein of 15-lipoxygenase was found in macrophage-rich areas within atherosclerotic lesions. This enzyme plays a central role in lipid peroxidation and was also demonstrated to colocalize with epitopes of oxidized low-density lipoprotein.[152] Macrophage-rich areas and fatty streaks in human atherosclerotic lesions have shown similar results.[153]

b. Combined Application of in situ *Hybridization and Immunohistochemistry in the Same Tissue Section*

Combined *in situ* hybridization and immunohistochemical staining of the same tissue section represents a further technical refinement. It allows the identification of cells expressing enhanced amounts of mRNA, and the simultaneous use of mRNA probes and antibodies specific for the encoded protein helps to gain further insight into the regulation of gene expression at both the transcriptional and the translational levels. An additional important advantage is provided by the possibility of testing specificity of probes for cell-type-specific mRNAs. The method of combined investigation has already been successfully applied in studies on cell cultures, brain tissues, and virus-infected tissues.[154] Simultaneous application has also previously been reported in organ culture.[155]

Recently, human keloid tissue has been used as a model system to establish a protocol for combined utilization of *in situ* hybridization and peroxidase-antiperoxidase immunohistochemistry (PAP) within one tissue section.[154] After localization of type I collagen mRNA, endothelial cells were detected by antibodies specific for von Willebrand protein. The authors found enhanced expression of type I collagen mRNA in endothelial cells of capillaries and in cells in close proximity to medium-sized vessel walls.

Application of combined *in situ* hybridization and immunocytochemistry in studies of human atherosclerotic arteries has been described by Wilcox et al.[156] and by Barath et al.[157] The former authors were able to demonstrate the presence of PDGF A and B chain mRNA in von Willebrand-protein-positive endothelial cells. The latter authors showed that only so-called "mesenchymal-appearing" intimal smooth muscle cells (76% positive for smooth muscle cell-specific marker HHF35) with detectable tumor necrosis factor mRNA also contain tumor necrosis factor protein in their cytoplasm; medial smooth muscle cells with immunoreactive tumor necrosis factor (TNF) on their surfaces appeared devoid of message.

3. Localization of Collagen mRNAs in Normal and Diseased Vessel Walls

In order to elucidate participation of monocytic cells in the activation of collagen expression in the atherosclerotic plaque, we investigated samples of human renal and common iliac arteries by *in situ* hybridization and immunocytochemical staining using monoclonal antibodies specific for blood-derived monocytes (27E10) and for mature macrophages (25F9).[146] In normal renal and common iliac arteries studied by *in situ* hybridization and immunocytochemistry, virtually no collagen mRNA-containing cells were detected within the intima and media, and serial sections were free of cells reacting with monocyte-specific or macrophage-specific antibodies. In atherosclerotic samples of these arteries, however, we found enhanced expression of genes coding for types I and III collagen in intimal smooth muscle cells situated close to immigrated monocytic cells in atherosclerotic renal (Figure 1) and common iliac arteries (Figure 2). Immunocytochemical staining of serial sections for smooth muscle cell-specific α-actin (Plate 3a)* revealed an abundance of smooth muscle cells throughout the whole intima, including regions of enhanced collagen message, confirming that intimal smooth muscle cells are the main source of enhanced collagen gene expression within the atherosclerotic arterial wall. Capillary

* Color Plate 3 follows page 118.

formation in human atherosclerotic plaques has been proposed to be involved in locally restricted cell activation.[156] Immunohistochemical staining of serial sections of atherosclerotic arteries with a von Willebrand-protein-specific antibody revealed a lack of endothelial cells and capillaries within the plaques examined.

In further serial sections, we studied the occurrence of human T helper and cytotoxic T lymphocytes within the same plaques by using CD4-specific and CD8-specific antibodies, respectively. The presence of T lymphocytes in atherosclerotic lesions has recently been described.[158,159] In close proximity to collagen mRNA-containing cells and to invaded monocytic cells, we were able to demonstrate the presence of invading CD8-bearing cytotoxic T lymphocytes (Plate 3). In a corresponding localization, we also found anti-CD4-detectable T-helper cells (not shown) (I. Grimm, unpublished results).

Association between macrophages and collagen mRNA-containing cells was also found in a renal artery with arteritis, where monocytic cell infiltrates were detected throughout the media (Plate 4b).* In several advanced fibrous plaques, monocytes or macrophages were not detectable. These samples were also found to be free of collagen mRNA.

In order to investigate high blood pressure-induced human vessel wall thickening by *in situ* hybridization and immunocytochemistry, we studied inborn aortic coarctations as a model system for comparing collagen mRNA levels in the thickened prestenotic and in unaffected poststenotic areas.[60] In contrast to atherosclerotic lesions, we found that activation of collagen biosynthesis was not dependent on the presence of invaded monocytes or macrophages. Areas of prestenotic intimal and medial hyperplasia containing collagen-mRNA-expressing cells (Figure 3a, b) were devoid of monocytic cells (Figure 3c). The presence of smooth muscle cells throughout the vessel wall and the hypertrophic areas was confirmed by immunocytochemical staining (data not shown).

VI. DISCUSSION AND CONCLUSIONS

A selective thickening of the intimal layer is generally assumed to result from accelerated proliferation of intimal smooth muscle cells combined with excessive synthesis of extracellular matrix components. The fundamental changes intimal smooth muscle cells undergo during atherosclerotic plaque formation have given rise to different hypotheses based on (1) migration of medial smooth muscle cells into the intima,[160] (2) transformation and clonal proliferation of intimal smooth muscle cells,[161,162] and (3) a change of phenotype and synthetic activity to a more "dedifferentiated" form.[163] Phenotypic change to the synthetic type is associated with increased cell volume, enlargement of rough endoplasmic reticulum, and a decline in actomyosin filaments, leading to the "mesenchymal-type" cell typical of atherosclerotic lesions.

---

* Color Plate 4 follows page 118.

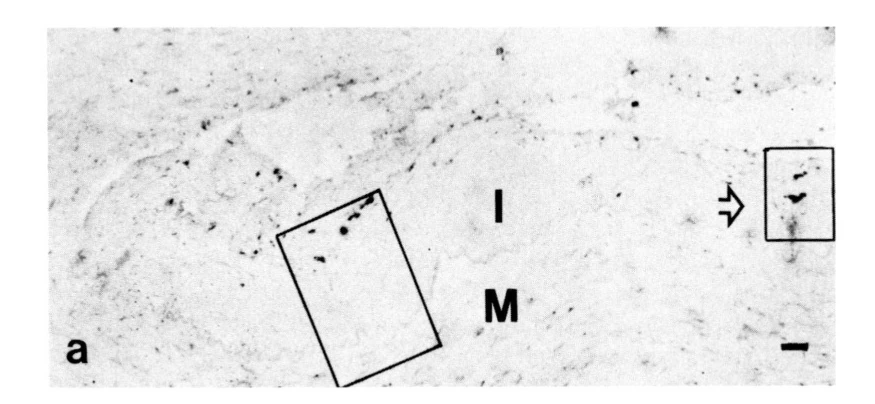

FIGURE 1. (a) Section of a renal artery of a 67-year-old male patient with severe intimal fibrosis immunostained for macrophages (antibody 25F9). Higher magnification of serial sections hybridized with type I collagen probes (b, d) and immunostained for macrophages (c, e, corresponding to b, d) reveals close colocalization of collagen mRNA and macrophages within the subendothelial intima. (I = intima; M = media). (Bars: (a) 50 μm; (b–e) 10 μm).

Detection of mRNAs coding for type I and III collagen in atherosclerotic lesions by *in situ* hybridization reveals remarkable differences in the activity of collagen gene expression between samples of different atherosclerotic vessels, and between plaques of the same vessel. Even within fibrous plaques having abundant matrix deposition, synthetic activity of smooth muscle cells is not necessarily enhanced. Active collagen gene expression seems to be restricted to cells in the vicinity of monocytic cell infiltrates.

Cell culture studies have shown that when smooth muscle cells are grown in a contracted collagen lattice, collagen synthesis is strongly inhibited, though cells retain the phenotype of synthetic cells[164] (see Chapter 5 by Michael Thie). Smooth muscle cells of atherosclerotic plaques that are embedded in a collagenous matrix may similarly be inactive in synthesizing collagen. This quiescence may be suspended by factors produced by invading monocytes.

It is not yet known precisely which cytokines induce collagen synthesis in the plaque. It is not even clear whether macrophage-derived factors are directly responsible or whether macrophage factors operate by activating expression of cytokines in other vessel wall cells that then induce collagen synthesis. Recent immunohistochemical and *in situ* hybridization studies have demonstrated the expression of several cytokines in smooth muscle cells of atherosclerotic vessels.[133,165-167] Thus, activation of collagen synthesis may be induced by the combined influence of secretory products of cell infiltrates and resident smooth muscle cells. The effect of several cytokines on collagen synthesis in smooth muscle cells has been investigated *in vitro*,[168] and TGF-β has been found to be especially effective, as is also the case in fibroblasts.[169] Even in smooth muscle

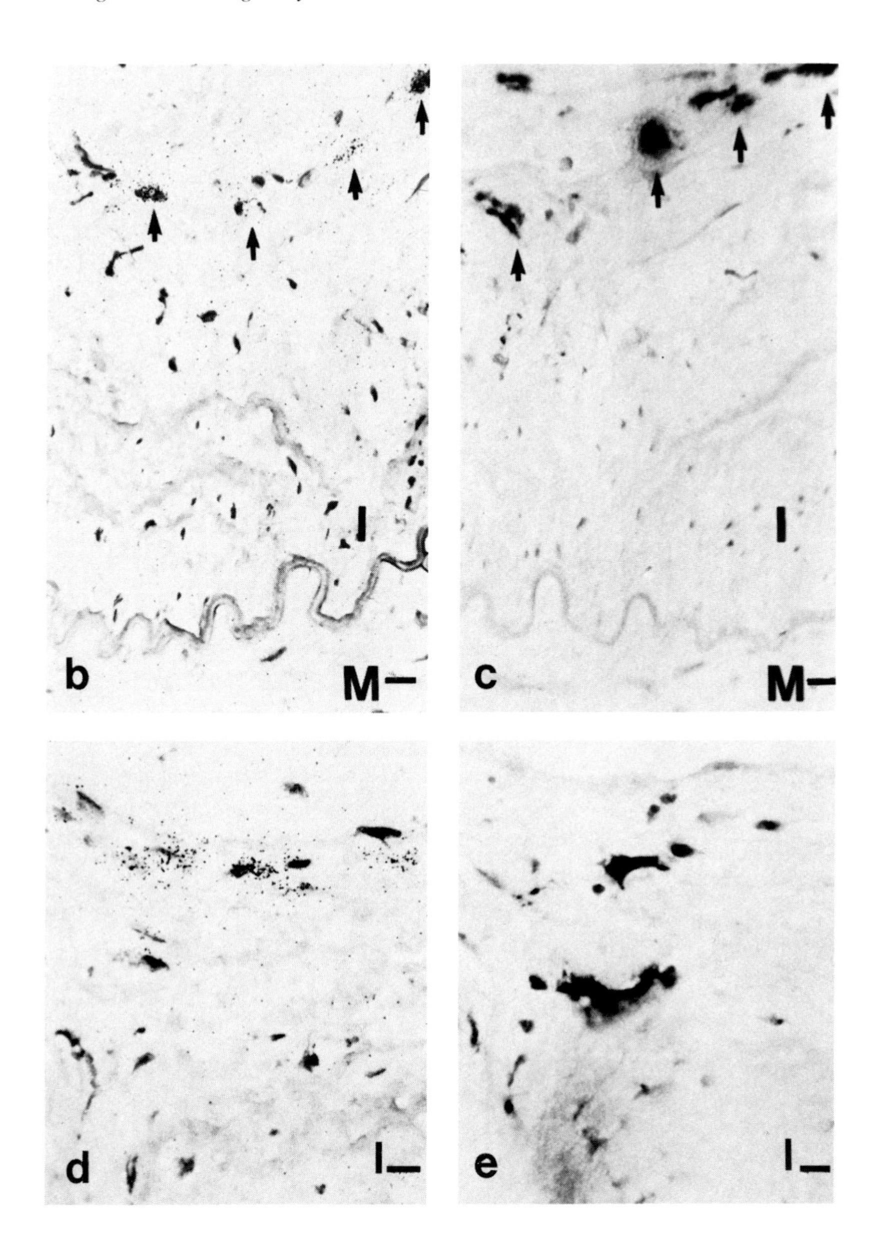

FIGURE 1b–e.

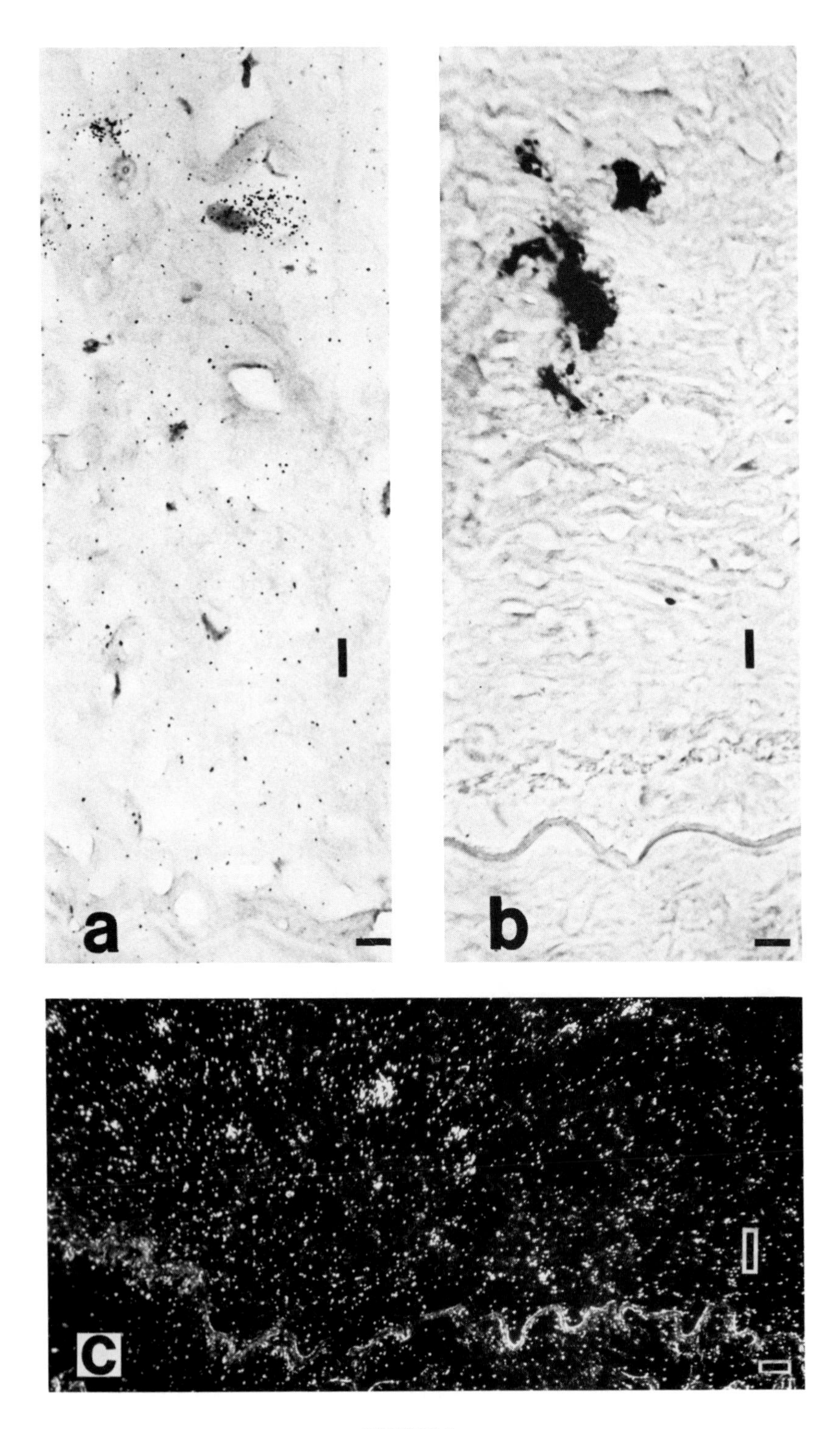

FIGURE 2a–c.

cells cultured in a collagen lattice, collagen synthesis is markedly stimulated by TGF-β, whereas IL-1 and TNF-α have no comparable effects.[168] Wilcox et al.[156] demonstrated that in atherosclerotic plaques of human carotid arteries, the producers of PDGF A and PDGF B chain mRNA were "mesenchymal-appearing" intimal cells, perhaps altered smooth muscle cells, and endothelial cells. They detected penetration of plaques by small blood vessels, the endothelium of which was actively synthesizing PDGF A and B chains. In this case, the endothelium may participate in the activation of intimal smooth muscle cells. In our studies, however, we found no evidence for the presence of capillaries within plaques of renal and iliac arteries. Instead, transcriptional activation of intimal smooth muscle cells occurred in the local vicinity of monocytic cells in plaque regions that were devoid of capillaries.

In contrast to atherosclerotic lesions, transcriptional activation in hyperplastic areas of hypertension-induced arterial wall thickening is clearly not mediated by invading cells of the monocyte/macrophage system. In the thickened prestenotic areas of human aortic coarctations, enhanced transcription of collagen mRNA in intimal and medial smooth muscle cells is independent of the presence of monocytic cells. Activation of smooth muscle cells in hypertension is probably due to the influence of cytokines secreted from neighboring vascular cells, bearing in mind the capacity of smooth muscle cells and endothelial cells to produce several cytokines.

Further studies on cytokines and growth factors at the level of protein and mRNA expression during human atherogenesis and in experimental animals, together with investigations on the effects of matrix components and cytokines in cell culture, are fundamental to further advances in our understanding of atherogenic mechanisms. Equipped with such improved knowledge, we will be better placed to devise new ways of preventing development of atherosclerosis and, it is hoped, of inducing regression of life-threatening lesions.

ACKNOWLEDGMENTS

Experimental work presented in this article was supported by the Deutsche Forschungsgemeinschaft. We thank Drs. E. Vollmer, A. Roessner, and K. Vollmer for their helpful cooperation with immunocytochemistry. We appreciate the photographical assistance of M. Opalka. We thank A. Hövener for preparing the manuscript.

FIGURE 2. Common iliac artery with intimal fibrous plaque in a 37-year-old male patient. (A) *In situ* localization of type I collagen mRNA demonstrates message-containing intimal cells in close vicinity to macrophages visualized by immunocytochemical staining of a serial section (b). (a, b: bar: 10 μm.). (c) Collagen mRNA expressing cells in the intimal atherosclerotic plaque of the same iliac artery photographed by dark field imaging. (I = intima). (c: bar: 20 μm).

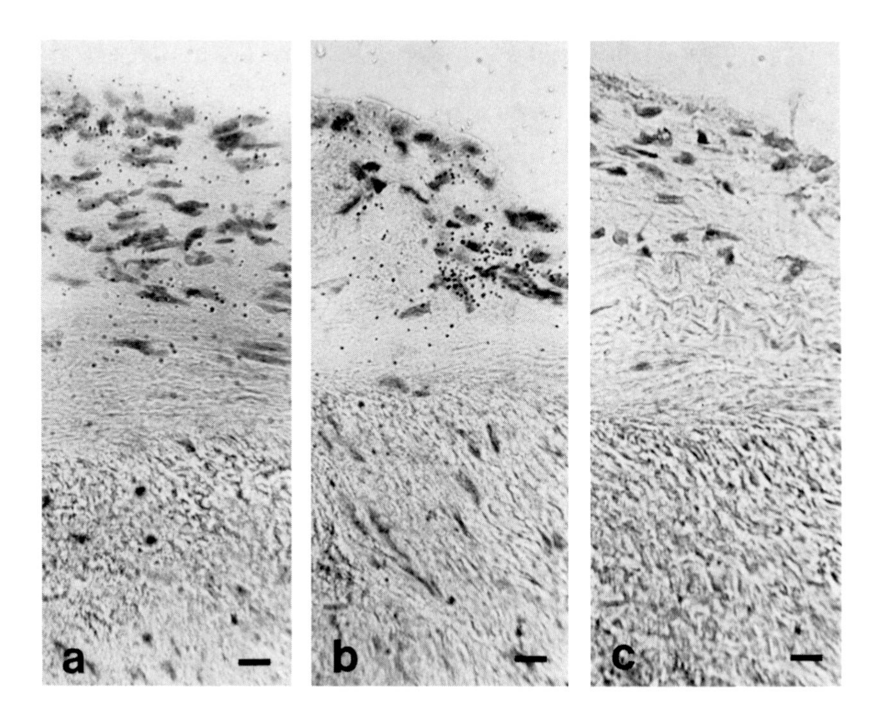

FIGURE 3. Stenotic part of an aortic coarctation in a 16-year-old male patient; *in situ* hybridization for type I (a) and type III (b) collagen mRNA reveals enhanced expression of both collagen types within the hypertrophic intima; (c) immunocytochemical staining shows the absence of monocytic cells within the stenotic area. (Bars: 10 μm).

REFERENCES

1. **Kühn, K.,** The collagen family — variation in the molecular and supermolecular structure, *Rheumatology*, 10, 29, 1986.
2. **Gallagher, J.T.,** The extended family of proteoglycans, *Curr. Top. Cell Biol.*, 1, 1201, 1989.
3. **Kühn, K.,** The Collagens — molecular and macromolecular structures, in *Proteinases in Inflammation and Tumor Invasion*, Tschesche, H., Ed., Walter de Gruyter, Berlin, 1986, 107.
4. **Miller, E.J.,** Recent information on the chemistry of the collagens, in *Chemistry and Biology of Mineralized Tissues*, Butler, W.T., Ed., EBSCO Media, Birmingham, AL, 1985, 80.
5. **Van der Rest, M. and Garrone, R.,** Collagens as multidomain proteins, *Biochimie*, 72, 473, 1990.
6. **Kielty, C.M., Hopkinson, I., and Grant, M.E.,** The collagen family: structure, assembly and organization in the extracellular matrix, in *Extracellular Matrix in Inherited Diseases of Connective Tissue*, Royce, P.M. and Steinmann, B., Eds., Allan R. Liss, New York, in press.
7. **Henkel, W. and Glanville, R.W.,** Covalent crosslinking between molecules of type I and type III collagen. The involvement of the N-terminal, non-helical regions of the α1(I) and α1(III) chains in the formation of intermolecular crosslinks, *Eur. J. Biochem.*, 122, 205, 1982.

8. **Keene, D.R., Sakai, L.Y., Bächinger, H.P., and Burgeson, R.E.,** Type III collagen can be present on banded collagen fibrils regardless of fibril diameter, *J. Cell Biol.*, 115, 2393, 1987.
9. **Fleischmayer, R., Olsen, B.R., Timpl, R., Perlish, J.S., and Lovelace, O.,** Collagen fibril formation during embryogenesis, *Proc. Natl. Acad. Sci. U.S.A.*, 80, 3354, 1983.
10. **Romanic, A.M., Adachi, E., Kadler, K.E., Hojima, Y., and Prockop, D.J.,** Co-polymerization of pNcollagen III and collagen I-pNcollagen III decreases the rate of incorporation of collagen I into fibrils, the amount of collagen I incorporated, and the diameter of the fibrils formed, *J. Biol. Chem.*, 266, 12703, 1991.
11. **Shirashi, M., Segawa, M., and Takebayashi, S.,** An electron-microscopic study on aging of collagenous fibrils in the larger blood vessels of humans, *J. Electron. Microsc.*, 29, 139, 1980.
12. **Raspanti, M., Ottani, V., and Ruggeri, A.,** Different architectures of the collagen fibril: morphological aspects and functional implications, *Int. J. Macromol.*, 11, 367, 1989.
13. **Niyibizi, C., Fietzek, P.P., and van der Rest, M.,** Human placenta type V collagens. Evidence for the existence of an $\alpha1(V)\alpha2(V)\alpha3(V)$ collagen molecule, *J. Biol. Chem.*, 259, 14,170, 1984.
14. **Rauterberg, J. and Troyer, D.,** Type V collagen: heterogeneity in different tissues, in *Marker Proteins in Inflammation*, Vol. 3, Laurent, P., Grimaud, J.A., and Bienvenue, J., Eds., Walter de Gruyter, Berlin, 1986, 343.
15. **Abedin, M.Z., Ayad, S., and Weiss, J.B.,** Type V collagen: the presence of appreciable amounts of $\alpha3(V)$ chain in uterus, *Biochem. Biophys. Res. Commun.*, 102, 1237, 1981.
16. **Madri, J.A., Dreyer, B., Pitlick, F.A., and Furthmayr, H.,** The collagenous components of the subendothelium: correlation of structure and function, *Lab. Invest.*, 43, 303, 1980.
17. **Voss, B. and Rauterberg, J.,** Localization of collagen types I, III, IV and V, fibronectin and laminin in human arteries by the indirect immunfluorescence method, *Pathol. Res. Pract.*, 181, 568, 1986.
18. **Shekhonin, B.V., Domogatsky, A.P., Muzykantov, V.R., Idelson, G.L., and Rukosuev, V.S.,** Distribution of type-1, type-3, type-4 and type-5 collagen in normal and atherosclerotic human arterial wall — immunological characteristics, *Coll. Relat. Res.*, 5, 355, 1985.
19. **Schuppan, D., Becker, J., Boehm, H., and Hahn, E.G.,** Immunofluorescent localization of type-V collagen as a fibrillar component of the interstitial connective tissue of human oral mucosa, artery and liver, *Cell Tissue Res.*, 243, 535, 1986.
20. **Gay, S., Martineshernandes, A., Rhodes, R.K., and Miller, E.J.,** The collagenous exocytoskeleton of smooth muscle cells, *Coll. Relat. Res.*, 1, 377, 1981.
21. **Weil, D., Bernard, M., Gargano, S., and Ramirez, F.,** The pro $\alpha2(V)$ gene is evolutionarily related to the major fibrillar-forming collagens, *Nucleic Acids Res.*, 15, 181, 1987.
22. **Adachi, E. and Hayashi, T.,** Comparison of axial banding patterns in fibrils of type-V collagen and type-I collagen, *Coll. Relat. Res.*, 7, 27, 1987.
23. **Adachi, E. and Hayashi, T.,** *In vitro* formation of hybrid fibrils of type V collagen and type I collagen — limited growth of type I collagen into thick fibrils by type V collagen, *Connect. Tissue Res.*, 14, 257, 1986.
24. **Birk, D.E., Fitch, J.M., Babiarz, J.P., and Linsenmayer, T.F.,** Collagen type I and type V are present in the same fibril in avian corneal stroma, *J. Cell Biol.*, 106, 999, 1988.
25. **Birk, D.E., Fitch, J.M., Babiarz, J.P., Doane, K.J., and Linsenmayer, T.F.,** Collagen fibrillogenesis *in vitro*-Interaction of type-I and type-V collagen regulates fibril diameter, *Microbiol. Cell Biol.*, 95, 649, 1990.
26. **Gordon, M.L and Olsen, B.R.,** The contribution of collagenous proteins to tissue-specific matrix assemblies, *Curr. Opinion Cell Biol.*, 2, 833, 1990.
27. **Leblond, C.P. and Inoue, S.,** Structure, composition, and assembly of basement membrane, *Am. J. Anat.*, 185, 367, 1989.
28. **Timpl, R.,** Structure and biological activity of basement membrane proteins, *Eur. J. Biochem.*, 180, 487, 1989.

29. **Gunwar, S., Noelken, M.E., and Hudson, B.G.,** Properties of the collagenous domain of the α3(IV) chain, the goodpasture antigen, of lens basement membrane collagen — selective cleavage of α(IV) chains with retention of their triple helical structure and noncollagenous domain, *J. Biol. Chem.*, 266, 14,088, 1991.
30. **Fagg, W.R., Timoneda, J., Schwartz, C.E., Langeveld, J.P.M., Noelken, M.E., and Hudson, B.G.,** Glomular basement membrane — evidence for collagenous domain of the α3-chain and α4-chain of collagen-IV, *Biochem. Biophys. Res. Commun.*, 170, 322, 1990.
31. **Sage, H., Pritzl, P., and Bornstein, P.,** Endothelial cells secrete a novel collagen type *in vitro* independently of prolyl hydroxylation, *Coll. Relat. Res.*, 2, 465, 1982.
32. **Labermeyer, U. and Kenney, M.C.,** The presence of EC collagen and type IV collagen in bovine Descemet's membranes, *Biochem. Biophys. Res. Commun.*, 116, 619, 1983.
33. **Iruela-Arispe, M.L., Diglio, C.A., and Sage, E.H.,** Modulation of extracellular matrix proteins by endothelial cells undergoing angiogenesis *in vitro*, *Arterioscler. Thromb.*, 11, 805, 1991.
34. **Jander, R., Korsching, E., and Rauterberg, J.,** Characteristics and *in vivo* occurrence of type-VIII collagen, *Eur. J. Biochem.*, 189, 601, 1990.
35. **Kittelberger, R., Davis, P.F., and Greenhill, N.S.,** Immuno-localization of type VIII collagen in vascular tissue, *Biochem. Biophys. Res. Commun.*, 159, 414, 1989.
36. **Kapoor, R., Sakai, L.Y., Funk, S., Roux, E., Bornstein, P., and Sage, E.H.,** Type VIII collagen has a restricted distribution in specialized extracellular matrices, *J. Cell Biol.*, 107, 721, 1988.
37. **Chung, E., Rhodes, R.K., and Miller, E.J.,** Isolation of three collagenous components of probable basement membrane origin from several tissues, *Biochem. Biophys. Res. Commun.*, 71, 1167, 1976.
38. **Rauterberg, J., Jander, R., and Troyer, D.,** Type VI collagen — a structural glycoprotein with a collagenous domain, *Front. Matrix Biol.*, 11, 90, 1986.
39. **Engvall, E., Hessle, H., and Klier, G.,** Molecular assembly, secretion, and matrix deposition of type VI collagen, *J. Cell Biol.*, 162, 703, 1986.
40. **Bonaldo, P., Russo, V., Bucciotti, F., Bressan, G.M., and Colombatti, A.,** α1 chain of chick type VI collagen. The complete cDNA-sequence reveals a hybrid molecule made of one short collagen and three von Willebrand factor type A-like domains, *J. Biol. Chem.*, 264, 5575, 1989.
41. **Chu, M.L., Pan, T.C., Conway, D., Kuo, H.J., Glanville, R.W., Timpl, R., Mann, K., and Deutzmann, R.,** Sequence analysis of α1(VI) and α2(VI) chains of human type-VI collagen reveals internal triplication of globular domains similar to the A-domains of von Willebrand factor and 2 α2(VI) chain variants that differ in the carboxy terminus, *EMBO J.*, 8, 1939, 1989.
42. **Chu, M.L., Zhang, R.Z., Pan, T.C., Stokes, D., Conway, D., Kuo, J.H., Glanville, R., Mayer, U., Mann, K., Deutzmann, R., and Timpl, R.,** Mosaic structure of globular domains in the human type-VI collagen α3 chain — Similarity to von Willebrand factor, fibronectin, actin, salivary proteins and aprotinin type protease inhibitors, *EMBO J.*, 9, 385, 1990.
43. **Bonaldo, P., Russo, V., Bucciotti, F., Doliana, R., and Colombatti, A.,** Structural and functional features of the α3 chain indicate a bridging role for chicken collagen-VI in connective tissues, *Biochemistry*, 29, 1245, 1990.
44. **Koller, E., Winterhalter, K.H., and Trüeb, B,** The globular domains of type VI collagen are related to the collagen-binding domains of cartilage matrix protein and von Willebrand factor, *EMBO J.*, 8, 1073, 1980.
45. **Campbell, G.R. and Campbell, J.H.,** Smooth muscle phenotypic changes in arterial wall homeostasis: implications for the pathogenesis of atherosclerosis, *Exp. Mol. Pathol.*, 42, 139, 1985.

46. **Rauterberg, J., Allam, S., Brehmer, U., Wirth, W., and Hauss, W.H.,** Characterization of the collagen synthesized by cultured human smooth muscle cells from fetal and adult aorta, *Hoppe-Seylers Z. Physiol. Chem.*, 358, 401, 1977.
47. **Burke, J.M., Balian, G., Ross, R., and Bornstein, P.,** Synthesis of types I and III procollagen and collagen by monkey aortic smooth muscle cells *in vitro*, *Biochemistry*, 16, 3243, 1977.
48. **Barnes, M.J., Morton, L.F., and Levene, C.I.,** Synthesis of collagen types I and III by pig medial smooth muscle cells in culture, *Biochem. Biophys. Res. Commun.*, 70, 339, 1976.
49. **Holderbaum, D. and Ehrhart, L.A.,** Modulation of types I and III pro-collagen synthesis at various stages of arterial smooth muscle cell growth *in vitro*, *Exp. Cell Res.*, 153, 16, 1984.
50. **Mayne, R., Vail, M.S., Miller, E.J., Blose, St. H., and Chacko, S.,** Collagen polymorphism in cell cultures derived from guinea pig aortic smooth muscle: comparison with three populations of fibroblasts, *Arch. Biochem. Biophys.*, 181, 462, 1977.
51. **Ang, A.H., Tachas, G., Campbell, J.H., Bateman, J.F., and Campbell, G.R.,** Collagen synthesis by cultured rabbit aortic smooth muscle cells — alteration with phenotype, *Biochem. J.*, 265, 461, 1990.
52. **Mayne, R., Vail, M.S., and Miller, E.J.,** Characterization of the collagen chains synthesized by cultured smooth muscle cells derived from Rhesus monkey thoracic aorta, *Biochemistry*, 17, 446, 1978.
53. **Heickendorff, L. and Ledet, T.,** Arterial basement-membrane-like material isolated and characterized from rabbit aortic myomedial cells in culture, *Biochem. J.*, 211, 397, 1983.
54. **Okada, Y., Katsuda, S., Matsui, Y., Minamoto, T., and Nakanishi, I.,** Altered synthesis of collagen types in cultured arterial smooth muscle cells during phenotypic modulation by dimethyl sulfoxide, *Acta Pathol. Jpn.*, 39, 15, 1989.
55. **Majack, R.A. and Bornstein, P.,** Heparin and related glycosaminoglycans modulate the secretory phenotype of vascular smooth muscle cells, *J. Cell Biol.*, 99, 1688, 1984.
56. **Majack, R.A. and Bornstein, P.,** Heparin regulates the collagen phenotype of vascular smooth muscle cells: induced synthesis of an M_r 60.000 collagen, *J. Cell Biol.*, 100, 613, 1985.
57. **Sage, H.,** Collagen synthesis by endothelial cells in culture, in *Biology of Endothelial Cells*, E.A. Jaffe, Ed., Martinus Nijhoff, Boston, 1984, 161.
58. **Sage, H., Pritzl, P., and Bornstein, P.,** Secretory phenotype of endothelial cells in culture: comparison of aortic, venous, capillary, and corneal endothelium, *Arteriosclerosis*, 1, 427, 1981.
59. **Iruela-Arispe, M.L., Hasselaar, P., and Sage, H.,** Differential expression of extracellular proteins is correlated with angiogenesis *in vitro*, *Lab. Invest.*, 64, 174, 1991.
60. **Jaeger, E., Rust, S., Scharffetter, K., Roessner, A., Winter, J., Buchholz, A., Althaus, M., and Rauterberg, J.,** Localization of cytoplasmic collagen mRNA in human aortic coarctations: mRNA enhancement in high blood pressure induced intimal and medial thickening, *J. Histochem. Cytochem.*, 38, 1365, 1990.
61. **Kerenyi, T., Voss, B., Rauterberg, J., Fromme, H.G., Jellinek, H., and Hauss, W.H.,** Connective tissue proteins on the injured endothelium of the rat aorta, *Exp. Mol. Pathol.*, 43, 151, 1985.
62. **Barnes, M.J.,** Collagen polymorphism in the normal and diseased blood vessel wall, *Atherosclerosis*, 46, 249, 1983.
63. **Barnes, M.J.,** Collagens in atherosclerosis, *Coll. Relat. Res.*, 5, 65, 1985.
64. **Rauterberg, J., Jander, R., Jaeger, E., Troyer, D., Thie, M., Tümmers, M., Fredrich, M., and Robenek, H.,** Arterial collagens and arteriosclerosis, in *Proceedings of the 8th International Symposium on Atherosclerosis*, Elsevier, Rome, 1989, 43.
65. **Wagner, W.D.,** Modification of collagen and elastin in the human atherosclerotic plaque, in *Pathobiology of the Human Atherosclerotic Plaque*, Glasgov, S., Neumann, W.P., and Schaffer, S.A., Eds., Springer-Verlag, New York, 1990, 167.

66. **Morton, L.F. and Barnes, M.J.,** Collagen polymorphism in the normal and diseased blood vessel wall — investigation of collagen types I, III and V, *Atherosclerosis*, 42, 41, 1982.
67. **Murata, K., Matoyama, T., and Kotake, C.,** Collagen types in various layers of the human aorta and their changes with the atherosclerotic process, *Atherosclerosis*, 60, 251, 1986.
68. **Hanson, A.N. and Bentley, J.P.,** Quantitation of type I to type III collagen ratios in small samples of human tendon, blood vessels and atherosclerotic plaque, *Anal. Biochemistry*, 130, 32, 1983.
69. **Ehrlich, H.P. and White, B.S.,** The identification of αA and αB collagen chains in hypertrophic scar, *Exp. Mol. Pathol.*, 34, 1, 1981.
70. **Ehrlich, H.P., Brown, H., and White, B.S.,** Evidence for type V and I trimer collagens in Dupuytren's contracture palmar fascia, *Biochem. Med.*, 28, 273, 1982.
71. **Narayanan, A.S., Engel, L.D., and Page, R.C.,** The effect of chronic inflammation on the composition of collagen types in human connective tissue, *Coll. Relat. Res.*, 3, 323, 1983.
72. **Hering, T.M., Marchant, R.E., and Anderson, J.M.,** Type V collagen during granulation tissue development, *Exp. Mol. Pathol.*, 39, 219, 1983.
73. **McCullagh, K.G., Duance, V.C., and Bishop, K.A.,** The distribution of collagen types I, III and V (AB) in normal and atherosclerotic aorta, *J. Pathol.*, 130, 45, 1980.
74. **Labat-Robert, J., Szendroi, M., Godeau, G., and Robert, L.,** Comparative distribution patterns of type I and type III collagens and fibronectin in human arteriosclerotic aorta, *Pathol. Biol.*, 33, 261, 1985.
75. **Bartholomew, J.S. and Anderson, J.C.,** Investigation of the relationship between collagens, elastin and proteoglycans in bovine thoracic aorta by immunofluorescence techniques, *Histochem. J.*, 15, 1177, 1983.
76. **Howard, P.S. and Macarak, E.J.,** Localization of collagen types in regional segments of the fetal bovine aorta, *Lab. Invest.*, 61, 548, 1989.
77. **Rauterberg, J., Jander, R., Voss, B., Furthmayr, H., and Glanville, R.,** Characteristics and occurrence of short chain collagen, in *New Trends in Basement Membrane Research*, Kühn, K., Schoene, H., and Timpl, R., Eds., Raven Press, New York, 1982, 107.
78. **Chamley-Campbell, J.H., Campbell, G.R., and Ross, R.,** Phenotype-dependent response of cultured aortic smooth muscle to serum mitogens, *J. Cell Biol.*, 8, 379, 1981.
79. **Krieg, T., Hörlein, M., Wiestner, M., and Müller, P.K.,** Aminoterminal extension peptides from type I procollagen normalize excessive collagen synthesis of scleroderma fibroblasts, *Arch. Dermatol. Res.*, 263, 171, 1978.
80. **Opsahl, W.P., DeLuca, D.J., and Ehrhart, L.A.,** Accelerated rates of collagen synthesis in atherosclerotic arteries quantified *in vivo*, *Arteriosclerosis*, 7, 470, 1987.
81. **Hauss, W.H., Junge-Hülsing, G., and Gerlach, U.,** *Die unspezifische Mesenchymreaktion*, Thieme Verlag, Stuttgart, 1968.
82. **Nissen, R., Cardinale, G.J., and Udenfriend, S.,** Increased turnover of arterial collagen in hypertensive rats, *Proc. Natl. Acad Sci. U.S.A.*, 75, 451, 1978.
83. **Newman, R.A. and Langner, R.O.,** Age-related changes in the vascular collagen metabolism of the spontaneously hypertensive rat, *Exp. Gerontol.*, 13, 83, 1978.
84. **Fischer, G.M., Swain, M.L., and Cherian, K.,** Increased vascular collagen and elastin synthesis in experimental atherosclerosis in rabbit. Variation in synthesis among major vessels, *Atherosclerosis*, 35, 11, 1980.
85. **Crossley, H.L., Johnson, A.R., Mauger, K.K., Wood, N.L., and Fuller, G.C.,** Aortic proline hydroxylase in hypoxia induced arteriosclerosis in rabbits, *Life Sci.*, 11, 869, 1972.
86. **Langner, R.O. and Modrak, J.B.,** Aortic collagen synthesis in rabbits following removal of atherogenic diet, *Exp. Molec. Pathol.*, 26, 310, 1977.
87. **St. Clair, R.W., Toma, J.J., and Lofland, H.B.,** Proline hydroxylase activity and collagen content of pigeon aortas with naturally-occurring and cholesterol-aggravated atherosclerosis, *Atherosclerosis*, 21, 155, 1975.

88. **Ooshima, A., Fuller, G., Cardinale, G., Spector, S., and Udenfriend, S.,** Collagen biosynthesis in blood vessels of brain and other tissues of the hypertensive rat, *Science*, 190, 898, 1975.
89. **Ooshima, A., Fuller, G.C., Cardinale, G.J., Spector, S., and Udenfriend, S.,** Reduction of collagen biosynthesis in blood vessels and other tissues by reserpine and hypophysectomy, *Proc. Natl. Acad. Sci. U.S.A.*, 74, 777, 1977.
90. **Sheridan, P.J., Kozar, L.G., and Benson, S.C.,** Increased lysyl oxidase activity in aortas of hypertensive rats and effect of β-aminopropionitrile, *Exp. Mol. Pathol.*, 30, 315, 1979.
91. **Ooshima, A. and Midorikawa, O.,** Increased lysyl oxidase activity in blood vessels of hypertensive rats and effect of β-aminopropionitrile on atherosclerosis, *Jpn. Circ. J.*, 41, 1337, 1978.
92. **Judd, J.T. and Wexler, B.C.,** Aortic prolyl hydroxylase and the development of arteriosclerosis in the female breeder rat, *Arterial Wall*, 3, 159, 1976.
93. **Deyl, Z., Jelinek, J., Macek, K., Chaldakov, G., and Vankov, V.N.,** Collagen and elastin synthesis in the aorta of spontaneously hypertensive rats, *Blood Vessels*, 24, 313, 1987.
94. **Ryan, P.A., Davis, P.F., and Stehbens, W.E.,** The biochemical composition of haemodynamically stressed vascular tissue. III. The collagen composition of experimental arteriovenous fistulae, *Atherosclerosis*, 71, 157, 1988.
95. **Stepp, M.A., Kindy, M.S., Franzblau, C., and Sonenshein, G.E.,** Complex regulation of collagen gene expression in cultured bovine aortic smooth muscle cells, *J. Biol. Chem.*, 261, 6542, 1986.
96. **Liau, G. and Chan, L.M.,** Regulation of extracellular matrix RNA levels in cultured smooth muscle cells, *J. Cell Biol.*, 264, 10,315, 1989.
97. **McGuire, P.G., Brocks, D.G., Killen, P.D., and Orkin, R.W.,** Increased deposition of basement membrane macromolecules in specific vessels of the spontaneously hypertensive rat, *Am. J. Pathol.*, 135, 291, 1989.
98. **Takasaki, I., Chobanian, A.V., Sarzani, R., and Brecher, P.,** Effect of hypertension on fibronectin expression in the rat aorta, *J. Biol. Chem.*, 265, 21,935, 1990.
99. **Crouch, E.C., Parks, W.C., Rosenbaum, J.L., Chang, D., Whitehouse, L., Wu, L.J., Stenmark, K.R., Orton, E.C., and Mecham, R.P.,** Regulation of collagen production by medial smooth muscle cells in hypoxic pulmonary hypertension, *Am. Rev. Respir. Dis.*, 140, 1045, 1989.
100. **Sarzani, R., Brecher, P., and Chobanian A.V.,** Growth factor expression in aorta of normotensive and hypertensive rats, *J. Clin. Invest.*, 83, 1404, 1989.
101. **Roberts, A.B., Flanders, K.C., Kondaiah, P., Thompson, N.L., van Obberghen-Schilling, E., Wakefield, L., Rossie, P., de Combrugghe, B., Heine, U., and Sporn, M.B.,** Transforming growth factor β: biochemistry and roles in embryogenesis, tissue repair and remodeling, and carcinogenesis, *Recent Prog. Horm. Res.*, 44, 157, 1988.
102. **Majesky, M.W., Daemen, M.J.A.P., and Schwartz, S.M.,** α_1-adrenergic stimulation of platelet-derived growth factor A-chain gene expression in rat aorta, *J. Biol. Chem.*, 265, 1082, 1990.
103. **Coghlan, J.P., Aldred, P., Haralambidis, J., Niall, H.D., Penschow, J.D., and Tregear, G.W.,** Hybridization histochemistry, *Anal. Biochemistry*, 143, 1, 1985.
104. **Höfler, H.,** What's new in "*in situ* hybridization," *Pathol. Res. Pract.*, 182, 421, 1987.
105. **Höfler, H.,** Principles of *in situ* hybridization, in *In Situ Hybridization, Principles and Practice*, Polak, J.M. and McGee, J.O'D., Eds., Oxford University Press, New York, 1990, 15.
106. **Gall, J.G. and Pardue, M.L.,** Nucleic acid hybridization in cytological preparations, *Methods Enzymol.*, 21, 470, 1971.
107. **Teo, C.G.,** *In situ* hybridization in virology, in *In Situ Hybridization, Principles and Practice*, Polak, J.M. and McGee, J.O'D., Eds., Oxford University Press, New York, 1990, 125.

108. **Pardue, M.L.,** *In situ* hybridization to study chromosome organization and gene activity, *J. Histochem. Cytochem.*, 34, 1, 1986.
109. **Bhatt, B. and McGee, J.O'D.,** Chromosomal assignment of genes, in *In Situ Hybridization, Principles and Practice*, Polak, J.M. and McGee, J.O'D., Eds., Oxford University Press, New York, 1990, 149.
110. **Wilkinson, D.G.,** mRNA *in situ* hybridization and the study of development, in *In Situ Hybridization, Principles and Practice*, Polak, J.M. and McGee, J.O'D., Eds., Oxford University Press, New York, 1990, 113.
111. **Nakamura, R.M.,** Overview and principles of *in situ* hybridization, *Clin. Biochemistry*, 23, 255, 1990.
112. **Marles, P.J., Hoyland, J.A., Parkinson, R., and Freemont, A.J.,** Demonstration of variation in chondrocyte activity in different zones of articular cartilage: an assessment of the value of *in situ* hybridization, *Int. J. Exp. Pathol.*, 72, 171, 1991.
113. **Lawrence, J.B. and Singer, R.H.,** Quantitative analysis of *in situ* hybridization methods for the detection of actin gene expression, *Nucleic Acids Res.*, 13, 1777, 1985.
114. **Baldino, F., Jr., Chesselet, M.-F., and Lewis, M.E.,** High-resolution *in situ* hybridization histochemistry, *Methods Enzymol.*, 168, 761, 1989.
115. **Maniatis, T., Fritsch, E.F., and Sambrock, J.,** *Molecular Cloning, A Laboratory Manual*, Cold Spring Harbor Laboratory, Cold Spring Harbor, N.Y., 1982.
116. **Feinberg, A.P. and Vogelstein, B.,** A technique for radiolabeling DNA restriction endonuclease fragments to high specific activity, *Anal. Bioch.em*, 132, 6, 1983.
117. **Thomas-Cavallin, M. and Ait-Ahmed, O.,** The random primer labeling technique applied to *in situ* hybridization on tissue sections, *J. Histochem. Cytochem.*, 36, 1335, 1988.
118. **Chan, V.T.-W. and McGee, J.O'D.,** Non-radioactive probes: preparation, characterization, and detection, in *In Situ Hybridization, Principles and Practice*, Polak, J.M. and McGee, J.O'D., Eds., Oxford University Press, New York, 1990, 59.
119. **Varndell, I.M., Polak, J.M., Sikri, K.L., Minth, C.D., Bloom, S.R., and Dixon, J.E.,** Visualization of messenger RNA directing peptide synthesis by *in situ* hybridisation using a novel single-stranded cDNA probe, *Histochemistry*, 81, 597, 1984.
120. **Cox, K.H., DeLeon, D.V., Angerer, L.M., and Angerer, R.C.,** Detection of mRNAs in sea urchin embryos by *in situ* hybridization using asymmetric RNA probes, *Dev. Biol.*, 101, 485, 1984.
121. **Lathe, R.,** Oligonucleotide probes for *in situ* hybridization, in *In Situ Hybridization, Principles and Practice*, Polak, J.M. and McGee, J.O'D., Eds., Oxford University Press, New York, 1990, 71.
122. **Saiki, R.K., Gelfland, D.H., Stoffel, S., Scharf, S.J., Higuchi, R., Horn, G.T., Mullis, K.B., and Erlich, H.A.,** Primer-directed enzymatic amplification of DNA with a thermostabile DNA polymerase, *Science*, 239, 487, 1988.
123. **Herrington, C.S., Flannery, D.M.J., and McGee, J.O'D.,** Single and simultaneous nucleic acid detection in archival human biopsies: application of non-isotopic *in situ* hybridization and the polymerase chain reaction to the analysis of human and viral genes, in *In Situ Hybridization, Principles and Practice*, Polak, J.M. and McGee, J.O'D., Eds., Oxford University Press, New York, 1990, 187.
124. **Wetmur, J.C., Ruyechan, W.T., and Douthart, R.J.,** Denaturation and renaturation of Penicillium chrysogenum mycophage double-stranded ribonucleic acid in tetraalkylammonium salt solutions, *Biochemistry*, 20, 2999, 1981.
125. **Williamson, D.J.,** Specificity of riboprobes for intracellular RNA in hybridization histochemistry, *J. Histochem. Cytochem.*, 36, 811, 1988.
126. **Gibson, S.J. and Polak, J.M.,** Principles and applications of complementary RNA probes, in *In Situ Hybridization, Principles and Practice*, Polak, J.M. and McGee, J.O'D., Eds., Oxford University Press, New York, 1990, 81.

127. **Dirks, R.W., Raap, A.K., Van Minnen, J., Vreugdenhil, E., Smit, A.B., and Van der Ploeg, M.,** Detection of mRNA molecules coding for neuropeptide hormones of the pond snail *Lymaea stagnalis* by radioactive and non-radioactive *in situ* hybridization: a model study for mRNA detection, *J. Histochem. Cytochem.*, 37, 7, 1989.
128. **Brady, M.A.W. and Finlan, M.F.,** Radioactive labels: autoradiography and choice of emulsions for *in situ* hybridization, in *In Situ Hybridization, Principles and Practice*, Polak, J.M. and McGee, J.O'D., Eds., Oxford University Press, New York, 1990, 31.
129. **Unger, E.R., Hammer, M.L., and Chenggis, M.L.,** Comparison of ^{35}S and biotin as labels for *in situ* hybridization: use of an HPV model system, *J. Histochem. Cytochem.*, 39, 145, 1991.
130. **Stolz, W., Scharffetter, K., Abmayr, W., Koditz, W., and Krieg, T.,** An automatic analysis method for *in situ* hybridization using high-resolution image analysis, *Arch. Dermatol. Res.*, 281, 336, 1989.
131. **Uhl, G.R.,** *In situ* hybridization: quantitation using radiolabeled hybridization probes, *Methods Enzymol.*, 168, 741, 1989.
132. **Davenport, A.P. and Nunez, D.J.,** Quantification of radioactive mRNA *in situ* hybridization signals, in *In Situ Hybridization, Principles and Practice*, Polak, J.M. and McGee, J.O'D., Eds., Oxford University Press, New York, 1990, 95.
133. **Moyer, C.F., Sajuthi, D., Tulli, H., and Williams, J.K.,** Synthesis of IL-1 α and IL-1 β by arterial cells in atherosclerosis, *Am. J. Pathol.*, 138, 951, 1991.
134. **Cremers, A.F.M., Jansen in de Wal, N., Wiegant, J., Dirks, R.W., Weisbeek, P., van der Ploeg, M., and Landegent, J.E.,** Non-radioactive *in situ* hybridization. A comparison of several immunocytochemical detection systems using reflection-contrast and electron microscopy, *Histochem.*, 86, 619, 1987.
135. **Hopman, A.H.N., Ramaekers, F.C.S., and Vooijs, G.P.,** Interphase cytogenetics of solid tumors, in *In Situ Hybridization, Principles and Practice*, Polak, J.M. and McGee, J.O'D., Eds., Oxford University Press, New York, 1991, 165.
136. **Chu, M.-L., Myers, J.C., Bernard, M.P., Ding, J.-F., and Ramirez, F.,** Cloning and characterization of five overlapping cDNAs specific for the human pro $\alpha 1$(I) collagen chain, *Nucleic Acids Res.*, 10, 5925, 1982.
137. **Myers, J.C., Chu, M.-L., Faro, S.H., Clark, W.J., Prockop, D.J., and Ramirez, F.,** Cloning a cDNA for the pro $\alpha 2$ chain of human type I collagen, *Proc. Natl. Acad. Sci. U.S.A.*, 78, 3516, 1981.
138. **Bernard, M.P., Myers, J.C., Chu, M.-L., Ramirez, F., Eikenberry, E.F., and Prockop, D.J.,** Structure of a cDNA for pro $\alpha 2$(I) identifies structurally conserved features of the protein and the gene, *Biochemistry*, 22, 1139, 1983.
139. **Misculin, M., Raymond, D., Kluve-Beckerman, B., Rennard, S.I., Tolstoshev, P., Brantly, M., and Crystal, R.G.,** Human type III collagen gene expression is coordinately modulated with the type I collagen genes during fibroblast growth, *Biochemistry*, 25, 1408, 1986.
140. **Scharffetter, K., Lankat-Buttgereit, B., and Krieg, T.,** Localization of collagen mRNA in normal scleroderma skin by *in-situ* hybridization, *Eur. J. Clin. Invest.*, 18, 9, 1989.
141. **Melton, D.A., Krieg, P.A., Rebagliati, M.R., Maniatis, T., Zinn, K., and Green, M.R.,** Efficient *in vitro* synthesis of biologically active RNA and RNA hybridization probes from plasmids containing a bacteriophage SP6 promoter, *Nucleic Acids Res.*, 12, 7035, 1984.
142. **Angerer, L.M. and Angerer, R.C.,** Detection of poly A^+RNA in sea urchin eggs and embryos by quantitative *in situ* hybridization, *Nucleic Acids Res.*, 9, 2819, 1981.
143. **Dix, D.J. and Eisenberg, R.B.,** *In situ* hybridization and immunocytochemistry in serial sections of rabbit skeletal muscle to detect myosin expression, *J. Histochem. Cytochem.*, 36, 1519, 1988.

144. **Singer, R.H., Lawrence, J.B., and Villnave, C.,** Optimization of *in situ* hybridization using isotopic and non-isotopic detection methods, *Biotechniques*, 4, 230, 1986.
145. **Prosser, I.W., Stenmark, K.R., Manish, S., Crouch, E.C., Mecham. R.P., and Parks, W.C.,** Regional heterogeneity of elastin and collagen gene expression in intralobar arteries in response to hypoxic pulmonary hypertension as demonstrated by *in situ* hybridization, *Am. J. Pathol.*, 135, 1073, 1989.
146. **Jaeger, E., Rust, S., Roessner, A., Kleinhans, G., Buchholz, B., Althaus, M., Rauterberg, J., and Gerlach, U.,** Joint occurrence of collagen mRNA containing cells and macrophages in human atherosclerotic vessels, *Atherosclerosis*, 31, 55, 1991.
147. **Tsukada, T., Tippens, D., Gordon, D., Ross, R., and Gown, A.M.,** HHF35, a muscle-actin-specific monoclonal antibody. I. Immunocytochemical and biochemical characterization, *Am. J. Pathol.*, 126, 51, 1987.
148. **Tsukada, T., McNutt, M.A., Ross, R., and Gown, M.A.,** HHF35, a muscle-actin-specific monoclonal antibody. II. Reactivity in normal, reactive, and neoplastic tissues, *Am. J. Pathol.*, 127, 389, 1987.
149. **Cordell, J.L., Falin, B., Erber, W.N., Ghosh, A.K., Abdulaziz, Z., Macdonald, S., Pulford, K.A.F., Stein, H., and Mason, D.Y.,** Immunoenzymatic labeling of monoclonal antibodies using immune complexes of alkaline phosphatase (APAAP complexes), *J. Histochem. Cytochem.*, 32, 219, 1984.
150. **Wilcox, J.N., Smith, K.M., Schwartz, S.M., and Gordon, D.,** Localization of tissue factor in the normal vessel wall and in the atherosclerotic plaque, *Proc. Natl. Acad. Sci. U.S.A.*, 86, 2839, 1989.
151. **Ylä-Herttuala, S., Lipton, B.A., Rosenfeld, M.E., Särkioja, T., Yoshimura, T., Leonard, E.J., Witztum, J.L., and Steinberg, D.,** Expression of monocyte chemoattractant protein 1 in macrophage-rich areas of human and rabbit atherosclerotic lesions, *Proc. Natl. Acad. Sci. U.S.A.*, 88, 5252, 1991.
152. **Ylä-Herttuala, S., Rosenfeld, M.E., Parthasarathy, S., Glass, C.K., Sigal, E., Witztum, J.L., and Steinberg, D.,** Colocalization of 15-lipoxygenase mRNA and protein with epitopes of oxidized low density lipoprotein in macrophage-rich areas of atherosclerotic lesions, *Proc. Natl. Acad. Sci. U.S.A.*, 87, 6959, 1990.
153. **Ylä-Herttuala, S., Rosenfeld, M.E., Parthasarathy, S., Sigal, E., Särkioja, T., Witztum, J.L., and Steinberg, D.,** Gene expression in macrophage-rich human atherosclerotic lesions, 15-lipoxygenase and acetyl low density lipoprotein receptor messenger RNA colocalize with oxidation specific lipid-adducts, *J. Clin. Invest.*, 87, 1146, 1991.
154. **Sollberg, S., Peltonen, J., and Uitto, J.,** Methods in laboratory investigation. Combined use of *in situ* hybridization and unlabeled antibody peroxidase anti-peroxidase methods: simultaneous detection of type I procollagen mRNAs and factor VIII-related antigen epitopes in keloid tissue, *Lab. Invest.*, 64, 125, 1991.
155. **Trimmer, P.A., Phillips, L.L., and Steward, O.,** Combination of *in situ* hybridization and immunocytochemistry to detect messenger RNAs in identified CNS neurons and glia in tissue culture, *J. Histochem. Cytochem.*, 39, 891, 1991.
156. **Wilcox, J.N., Smith, K.M., Williams, L.T., Schwartz, S.M., and Gordon, D.,** Platelet-derived growth factor mRNA detection in human atherosclerotic plaques by *in situ* hybridization, *J. Clin. Invest.*, 82, 1134, 1988.
157. **Barath, P., Fishbein, M.C., Cao, J., Berenson, J. Helfant, R.H., and Forrester, J.S.,** Tumor necrosis factor gene expression in human vascular intimal smooth muscle cells detected by *in situ* hybridization, *Am. J. Pathol.*, 137, 513, 1990.
158. **Hansson, G.K., Jonasson, L., Seifert, P.S., and Stemme, S.,** Immune mechanisms in atherosclerosis, *Arteriosclerosis*, 9, 567, 1989.
159. **Hansson, G.K., Seifert, P.S., Olsson, G., and Bondjers, G.,** Immunohistochemical detection of macrophages and T lymphocytes in atherosclerotic lesions of cholesterol-fed rabbits, *Arterioscler. Thromb.*, 11, 745, 1991.

160. **Ross, R.,** The pathogenesis of atherosclerosis, *N. Engl. J. Med.*, 314, 488, 1986.
161. **Benditt, E.P. and Benditt, J.M.,** Evidence for a monoclonal origin of human atherosclerotic plaques, *Proc. Natl. Acad. Sci. U.S.A.*, 70, 1753, 1973.
162. **Benditt, E.P. and Gown, A.M.,** Atheroma: the arterial wall and the environment, *Int. Rev. Exp. Pathol.*, 21, 55, 1980.
163. **Schwartz, S.M., Campbell, G.R., and Campbell, J.H.,** Replication of smooth muscle cells in vascular disease, *Circ. Res.*, 58, 427, 1986.
164. **Thie, M., Schlumberger, W., Semich, R., Rauterberg, J., and Robenek, H.,** Aortic smooth muscle cells in collagen lattice culture: effects on ultrastructure, proliferation and collagen synthesis, *Eur. J. Cell Biol.*, 55, 295, 1991.
165. **Klagsbrun, M. and Edelman, E.R.,** Biological and biochemical properties of fibroblast growth factor, *Arteriosclerosis*, 9, 269, 1989.
166. **Libby, P., Ordovas, J.M., Birinyi, L.K., Auger, K.R., and Dinarello, C.A.,** Inducible interleukin-1 gene expression in human vascular smooth muscle cells, *J. Clin. Invest.*, 78, 1432, 1986.
167. **Libby, P., Warner, S.J.C., Solomon, R.N., and Birinyi, L.K.,** Production of platelet-derived growth factor-like mitogen by smooth muscle cells from human atheroma, *N. Engl. J. Med.*, 318, 1493, 1988.
168. **Schlumberger, W., Thie, M., Rauterberg. J., Kresse, H., and Robenek, H.,** Deposition and ultrastructural organization of collagen and proteoglycans in the extracellular matrix of gel-cultured fibroblasts, *Eur. J. Cell Biol.*, 50, 100, 1989.
169. **Kähäri, V.M., Chen, Y.Q., Su, M.W., Ramirez, F., and Uitto, J.,** Tumor necrosis factor-α and interferon-gamma suppress the activation of human type I collagen gene expression by transforming growth factor-β 1. Evidence for two distinct mechanisms of inhibition at the transcriptional and posttranscriptional levels, *J. Clin. Invest.*, 86, 1489, 1990.

Chapter 5

REGULATION OF BIOSYNTHETIC ACTIVITY IN AORTIC SMOOTH MUSCLE CELLS BY EXTRACELLULAR MATRIX COMPONENTS

Michael Thie

TABLE OF CONTENTS

ISBN 0-8493-5505-2

I. INTRODUCTION

The concept of mutual interaction between the extracellular matrix and the tissue resident smooth muscle cell has a central part to play in understanding function in the normal vessel wall and in atherosclerosis. Smooth muscle cells are the only cell type in the tunica media of the arterial vessel wall. These cells are surrounded by a complex network of large organic molecules, called the extracellular matrix. The main components of the extracellular matrix are the fibrillar collagens,[1-4] elastin,[5,6] and the non fibrillar, but fibril-associated glycosaminoglycans/proteoglycans[7-10] and glycoproteins.[11] All these macromolecular components of the extracellular matrix are synthesized and secreted by smooth muscle cells.

The smooth muscle cells derive from mesenchymal cells in the embryo and have initially a fibroblast-like ultrastructure. During early fetal life and postnatal development, the main function of smooth muscle cells is to proliferate and to synthesize extracellular matrix components that constitute the vessel wall. As the vessels approach their final size, these activities cease and the smooth muscle cells modulate to highly specialized cells with a cytoplasm largely filled with myofilament bundles, enabling the function of maintaining vascular tone to be fulfilled.[12,13]

For a long time, the extracellular matrix was thought to serve mainly as an inert scaffold for the cells. It is now evident, however, that the matrix plays an active and complex role in regulating the behavior of the cells by influencing the development, shape, migration, proliferation, and metabolic functions of the cells that are in contact with it. Such influence of the extracellular matrix on cell behavior has been demonstrated in various cell culture experiments on isolated smooth muscle cells. Cultured smooth muscle cells stripped free of matrix components during isolation undergo a radical change of their ultrastructural and functional character.[14] They modulate from the growth-arrested cell responsible for contraction and relaxation of the vessel wall to the proliferating cell specialized in synthesis and secretion of extracellular matrix proteins.

Modulation in the ultrastructure of adult smooth muscle cells can also occur under pathological conditions. In the pathogenesis of atherosclerosis, the smooth muscle cells exhibit a structural modulation from the contractile state characteristic of the healthy vessel to the synthetic state of the atherosclerotic vessel. Whereas smooth muscle cells in the contractile phenotype show a cytoplasm filled with myofilament bundles, those in the synthetic phenotype are characterized by abundant rough endoplasmic reticulum, Golgi apparatus and free ribosomes, and a scarcity of myofilament bundles.[15] These morphological changes of the cells are accompanied by specific modifications of the extracellular matrix composition. Most important among these, the collagen content of the vessel wall increases to 60% of the total protein content of the tissue; collagen thus represents the major extracellular product of the advanced atherosclerotic plaque.[16-21] The major types of collagen present both in the normal and

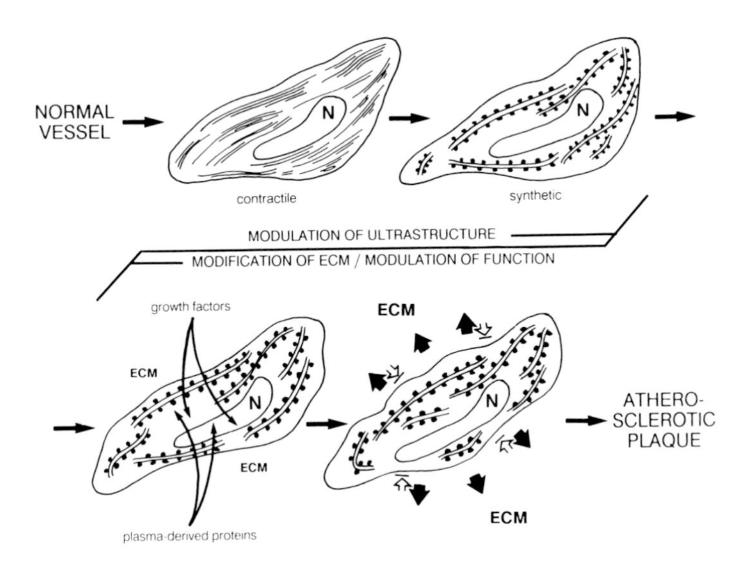

FIGURE 1. Atherogenesis may be envisaged as a multistep process in which the following factors, among others, may be involved. Contractile-state smooth muscle cells convert to the synthetic state and subsequently express a fibroblast-like program. Growth factors and/or plasma-derived proteins in the extracellular space surrounding the cells stimulate or inhibit biosynthetic activity of smooth muscle cells. Feedback loops between synthetic-state smooth muscle cells and their specific extracellular matrix fail, and matrix proteins are overexpressed. Subsequently, the original organization of the healthy extracellular matrix breaks down, and structure and function of the vessel wall alters (n = nucleus; ECM = extracellular matrix; ⇨| = interrupted feedback loop; ➡ = secretion of ECM components).

atherosclerotic vessel wall are type I and type III, which together constitute approximately 80 to 90% of the total collagen.[22] Type I collagen is more abundant than type III collagen, and in atherosclerosis, an alteration in the relative amounts of these collagens takes place, with type I collagen increasing in abundance, and type III collagen diminishing.[23-25] With progression of atherosclerosis, a reduction in the total mass of elastin occurs.[26] Altered production and modulation of proteoglycans and glycoproteins also takes place in atherosclerosis; however, this topic is beyond the scope of the present chapter.

The mechanisms underlying the changes in the ultrastructure of contractile smooth muscle cells to the synthetic state are unknown. Consequences of modified extracellular matrix composition on functional properties of the tissue resident smooth muscle cells themselves are also unknown. Nevertheless, structural modulation of smooth muscle cells,[15] modification of extracellular matrix,[27] and modulation of smooth muscle cell function by altered extracellular matrix[28] are all implicated as important steps in the pathogenesis of atherosclerosis (Figure 1). The aim of the studies presented in this chapter was to investigate the mutual interaction between the extracellular matrix and the functional properties of the smooth muscle cells. We studied the effects of a given, well-characterized matrix on the ultrastructural and biochemical properties of vascular smooth muscle cells, whose phenotype had been changed from the contractile form of the normal vessel to the synthetic form, which is typical of the atherosclerotic vessel. The model used for our studies on the metabolic properties of such cells comprises isolated smooth muscle cells embedded in a network of collagen fibrils.

II. STRUCTURAL AND BIOCHEMICAL BEHAVIOR OF VASCULAR SMOOTH MUSCLE CELLS IN COLLAGEN LATTICE CULTURES

Studies on the metabolic properties of smooth muscle cells have mostly been done in two-dimensional cultures, i.e., monolayer cultures on plastic. However, a main disadvantage of monolayer cultures is the loss of a three-dimensional tissue-like architecture. Therefore, results obtained with monolayers cannot be directly correlated with the *in vivo* situation. Three-dimensional culture systems employing substrates to which all surfaces of the cell are exposed should, in theory, largely overcome this problem by simulating natural conditions more closely than the conventional two-dimensional systems.[29,30]

When mesenchymal cells are embedded into a matrix of collagen fibrils, the cells reorganize the randomly oriented fibrils, causing the matrix to shrink. Three-dimensional cell-collagen systems — "collagen lattices" — are formed that represent living tissue equivalents *in vitro*. Using this culture system, cells can be investigated with respect to synthesis, deposition, supramolecular organization, and metabolism of extracellular material, while in close contact with collagen fibrils.[31-34]

A. Isolation of Smooth Muscle Cells and Preparation of Collagen Lattice Cultures

Smooth muscle cells used for the experiments presented here were prepared from swine. Immediately after exsanguination, the thoracic aorta was dissected out. The vessel was opened longitudinally, and the tunica intima was mechanically scraped off. The tunica media thus exposed was gently peeled off and cut into pieces. Smooth muscle cells were released from these media pieces by collagenase digestion.[18] The digest was passed through a filter, and the freed cells were recovered by gentle centrifugation. Cells were resuspended, seeded out in plastic flasks, and kept in a moist atmosphere of 5% CO_2 and 95% air at 37°C. At confluency the cells were removed from the flasks by trypsinization and used for lattice culture.

For preparation of a hydrated collagen lattice, a defined number of smooth muscle cells in 150 μl of culture medium is added to 1350 μl of a neutralized collagen solution, i.e., 690 μl of a 1.76-fold concentrated medium containing antibiotics, 450 μl of collagen solution (3.3 mg/ml in 0.1% acetic acid), 75 μl of

0.1 N NaOH, and 135 μl of fetal calf serum.[35] This cell-collagen mixture is poured into a 35-mm tissue culture dish. The collagen forms a lattice within some minutes when the culture dish is placed at 37°C. Immediately after its formation the collagen lattice is lifted off the bottom of the dish so that it floats in the medium. The lattice is cultivated with culture medium supplied with 10% fetal calf serum and kept in a moist atmosphere of 5% CO_2 and 95% air at 37°C.

B. Phenotypic Appearance of Lattice-Cultured Smooth Muscle Cells

An important characteristic of collagen lattice cultures is the close contact between the smooth muscle cells and the exogenous collagen fibrils (Figures 2A, 3A). We first studied the effect of membrane-bound collagen fibrils on the phenotypic features of cultured smooth muscle cells. The ultrastructure of the lattice-cultured cells (Figure 2A) was broadly comparable to that of the monolayer-cultured cells (Figure 2B), i.e., they exhibited the synthetic state characterized by a prominent Golgi apparatus and rough endoplasmic reticulum, abundant free ribosomes, and sparse intracytoplasmic filaments. The lattice-cultured cells nevertheless revealed a number of distinctive features. When the cells were examined directly after the collagen lattice had been formed, they were spherical in appearance. With prolonged culture time, the cells spread out and grew cytoplasmic extensions. Smooth muscle cells cultured within the lattice for a longer period appeared mostly spindle shaped. The nucleus was rounded up or elongated, lacking invaginations typical of tissue resident cells. Small and large electron-dense, oval granules (Figure 3B), as well as bodies containing whorled osmiophilic membranes (Figure 3C), were present. A basal lamina and depositions of extracellular material other than collagen fibrils were not observed. Thus, collagen lattice culture did not alter the phenotype of smooth muscle cells, i.e., cells did not revert from the synthetic state to the contractile state.

C. Growth Activity of Lattice-Cultured Smooth Muscle Cells

The loss of the vessel wall environment appears to be a stimulus not only for phenotypic modulation to the synthetic state, but also for inducing smooth muscle cell proliferation. Therefore, we studied the effect of a collagen lattice on the growth activity of synthetic-state smooth muscle cells. Compared to smooth muscle cells in monolayer cultures, proliferation of cells in collagen lattice cultures was strongly suppressed (Figure 4). Two days following lattice formation, the cell number increased, reached a plateau at 2.3-fold the original seeding density at around day 3 of culture, and decreased slowly with prolonged culture time. In contrast, smooth muscle cells cultured as monolayers on plastic demonstrated the conventional growth characteristics.

Smooth muscle cells that had been grown in collagen lattices over a longer period did not lose their capacity to proliferate. This was shown by the following experiment. Cells were cultivated within lattices for 3, 6, and 10 d and suspended by dissolving the lattice in collagenase solution. Subsequently, they were plated

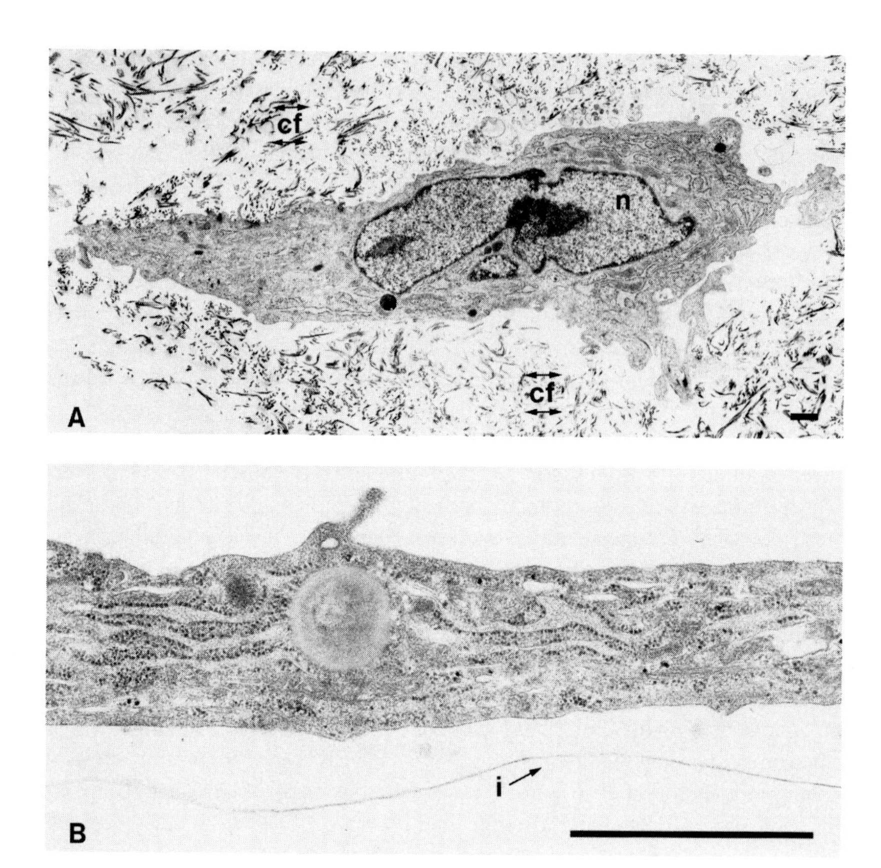

FIGURE 2. Vascular smooth muscle cells cultured within type I collagen lattices (A) and as monolayers on plastic (B) in medium supplemented with 10% fetal calf serum. In lattice cultures, smooth muscle cells are surrounded by collagen fibrils (A). Smooth muscle cells exhibit the synthetic phenotype in lattice cultures (A) as well as in monolayer cultures (B)(n = nucleus; cf = collagen fibrils; i = culture dish-medium interface). (Bars: 1 μm).

on tissue culture dishes. Approximately 70 to 90% of the cells attached within 1 d, flattened on the surface, and proliferated in the same manner as control cells.

Cell proliferation was examined in relation to the DNA content of the cells. To measure DNA content, nucleic acids were stained with propidium iodide and fluorescence intensity was measured with a flow cytometer.[36] Representative data on nuclear DNA content of lattice-cultured smooth muscle cells are shown in Figure 5. With respect to cell cycle distribution, smooth muscle cells cultured within a collagen lattice had significantly fewer cells in the S phase of the cell cycle than did cells cultured as monolayers. With prolonged culture time, the cell cycle distribution of smooth muscle cells in collagen lattices became progres-

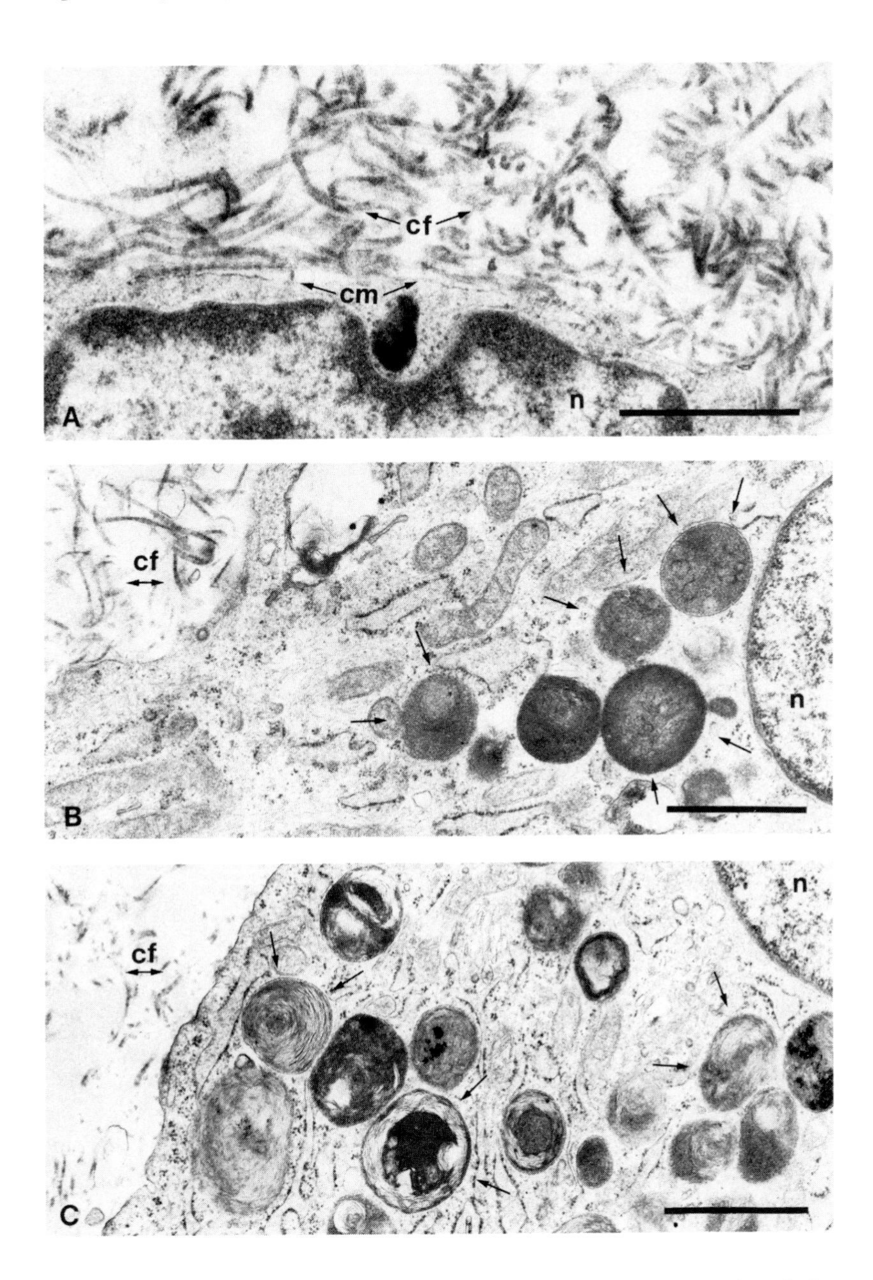

FIGURE 3. Ultrastructure of vascular smooth muscle cells cultured within type I collagen lattices in medium supplemented with 10% fetal calf serum. The cell membrane is in close contact with collagen fibrils (A). Electron-dense, oval granules (arrows, B) as well as bodies containing whorled osmiophilic membranes (arrows, C) are present in the cytoplasm (n = nucleus; cf = collagen fibrils; cm = cell membrane). (Bars: 1 μm).

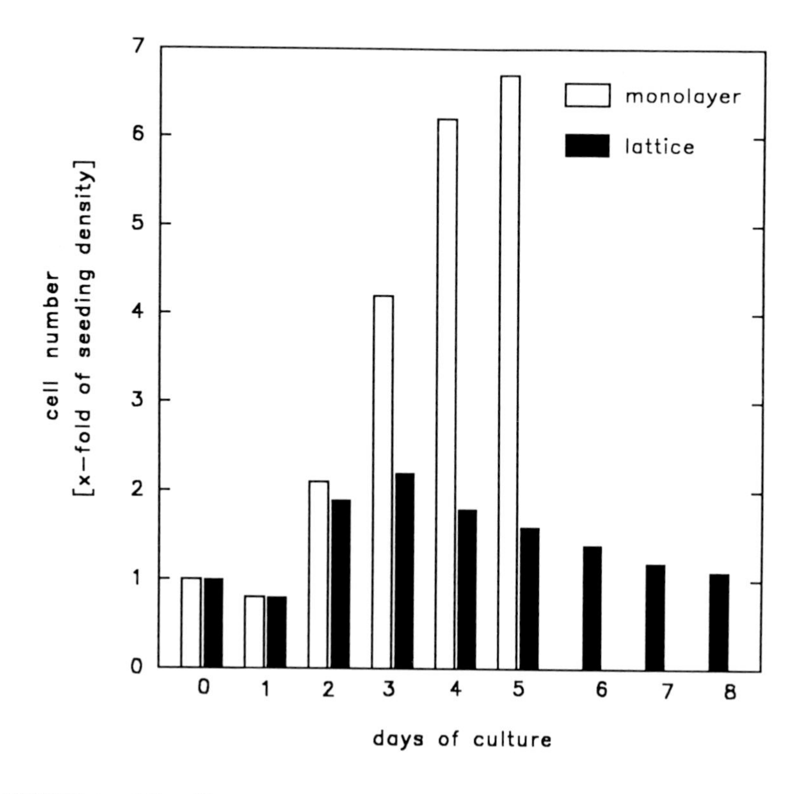

FIGURE 4. Cell proliferation of vascular smooth muscle cells cultured within type I collagen lattices (lattice) in medium supplemented with 10% fetal calf serum. Smooth muscle cells were initially seeded at 250,000 cells per lattice on day 0 and cultured for 8 d. Medium of floating lattices was changed every 2 d. For comparison, growth curves of smooth muscle cells cultured as monolayer on plastic (monolayer) are shown.

sively more similar to that of contractile-state cells of the healthy vessel wall. When dividing smooth muscle cells were embedded within a collagen lattice, they left the proliferating state and stopped to cycle, becoming arrested in the G_0/G^1 phase.

The ability of collagen to suppress cell proliferation of synthetic-state smooth muscle cells is not restricted to type I collagen fibrils. Collagen lattices composed of type III collagen demonstrated the same effect; indeed, suppression of cell proliferation appeared to be more marked with type III collagen than with type I collagen lattices (data not shown).

Thus, although smooth muscle cells did not revert to the contractile phenotype upon collagen lattice culture, they did revert in terms of their cell proliferation program. Fibrils of type I collagen and of type III collagen both have the capacity to inhibit proliferation of synthetic-state smooth muscle cells.

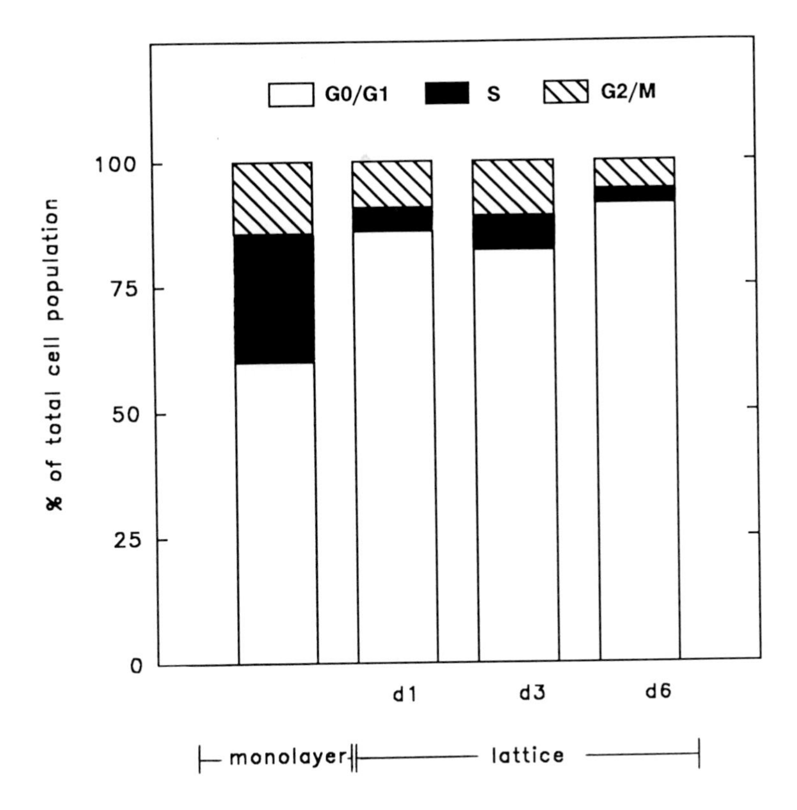

FIGURE 5. Cell cycle distribution in collagen lattice-cultured (lattice) vascular smooth muscle cells. Cells were cultivated in medium supplemented with 10% fetal calf serum for 1 d (d1), 3 d (d3), and 6 d (d6). For comparison, cell cycle distribution of preconfluent monolayer cultures (monolayer) are shown. Data are expressed as a percentage of total population, as determined by propidium iodide fluorescence. Differences between monolayer cultures and collagen lattice cultures are statistically significant ($p < 0.05$).

D. Synthesis of Extracellular Matrix Proteins

To study the effect of a collagenous matrix on the expression of extracellular matrix proteins by synthetic-state smooth muscle cells, total protein synthesis and collagen synthesis were measured. Cultures were prepared according to the following procedure.[37] The standard growth medium used with collagen lattices was replaced with labeling medium, i.e., 1 ml of fresh medium, containing 50 μg/ml of ascorbic acid and 0.37 MBq/ml of ^{14}C-proline. After 24 h in a moist atmosphere of 5% $C0_2$ and 95% air at 37°C, the medium and collagen lattices were collected. Samples were dialyzed against acetic acid, hydrolyzed in HCl, and subjected to ion-exchange chromatography for separation of proline and hydroxyproline. Total protein synthesis was calculated by determining the amount of labeled proline and hydroxyproline. Collagen synthesis was calculated according to the following formula:[38]

$$\frac{2 \times \text{OH–Pro} \times 100}{5 \times (\text{Pro–OH–Pro}) + 2 \times \text{OH–Pro}}$$

In general, smooth muscle cells cultured within collagen lattices synthesized less total protein than did cells cultivated as monolayers (Figure 6). When smooth muscle cells were cultivated within collagen lattices for a longer period, the level of total protein synthesis dropped progressively. Smooth muscle cells cultivated within collagen lattices showed lower values for collagen synthesis than did smooth muscle cells cultured as monolayers (Figure 7). Moreover, the level of synthesized collagen by lattice-cultured smooth muscle cells remained constant during the entire culture period.

Thus, although smooth muscle cells show the synthetic state, they do not necessarily exhibit a high metabolic activity. Smooth muscle cells in the synthetic state appear to be able to sense the macromolecular composition of the collagen lattice and to modify their production of matrix components accordingly. Cell volume also differs between monolayer-cultured and collagen lattice-cultured smooth muscle cells. Monolayer-cultured smooth muscle cells have a diameter of 15 ± 0.8 μm, whereas collagen lattice-cultured cells have a diameter of 13 ± 0.6 μm.

E. Synthesis of Matrix-Degrading Enzymes

Enzyme activity of collagenases within the culture system was measured to study the effect of collagen lattice culture on the production of enzymes capable of degrading the extracellular collagen fibrils. It is known that fibroblasts synthesize and secrete large amounts of collagenase when cultured in collagen lattices. We therefore compared the enzyme activities of smooth muscle cells with those of human skin fibroblasts cultured in both collagen lattices and as monolayers on plastic.

After cultivation for 24 h in serum-free medium, enzyme activity was measured in the medium, in collagen lattices, and in monolayer cells. For this purpose, samples were dialyzed against water, lyophilized, and resuspended in Tris-HCl buffer, pH 7.6 . Latent enzyme was activated by trypsin. Collagenase activity was assayed by degradation of native type I collagen,[39] which was labeled with ^{14}C-formaldehyde.[40] Compared to fibroblasts, smooth muscle cells produced very small amounts of collagenase. Smooth muscle cells in monolayer culture showed only 6% of the enzyme activity secreted by fibroblasts in monolayer culture (Figure 8). While collagen lattice culture stimulated collagenase production by fibroblasts, cultivation of smooth muscle cells within a collagen lattice failed to enhance the production of the enzyme.

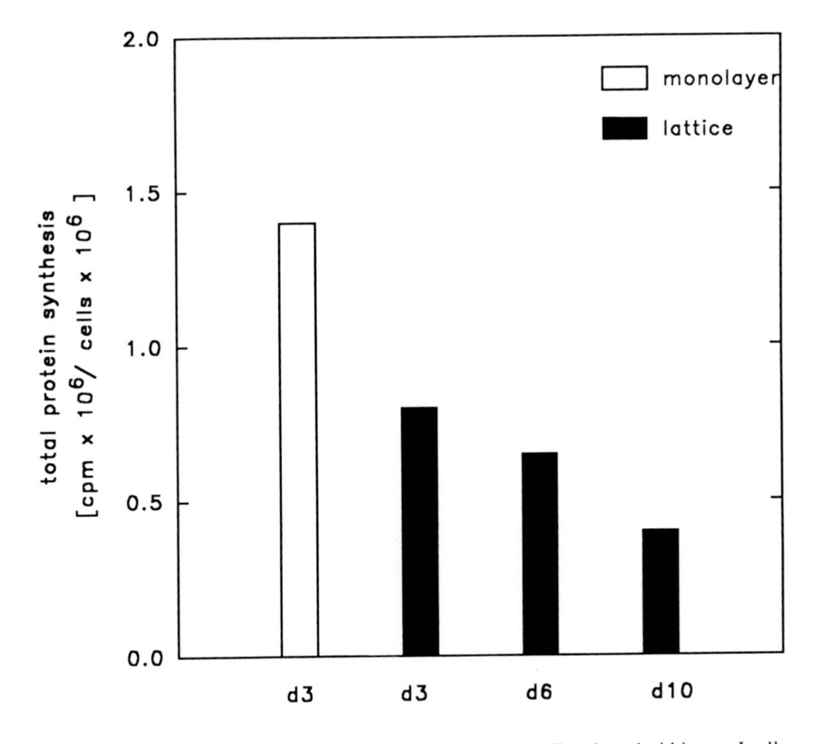

FIGURE 6. Total protein synthesis of vascular smooth muscle cells cultured within type I collagen lattices (lattice) and as monolayers on plastic (monolayer) in medium supplemented with 10% fetal calf serum. Protein synthesis was measured on day 3 (d3), day 6 (d6), and day 10 (d10) of culture. Differences between monolayer cultures and collagen lattice cultures are statistically significant ($p < 0.05$).

III. MODULATION OF BIOSYNTHETIC ACTIVITY OF COLLAGEN LATTICE-CULTURED SMOOTH MUSCLE CELLS BY FETAL CALF SERUM

The experiments so far show that elimination of the physiological extracellular matrix and cultivation of the isolated smooth muscle cells as a monolayer in the presence of fetal calf serum leads to proliferation and increased expression of extracellular matrix proteins by cultured cells. However, reintegration of these cells into a collagenous matrix results in slowed proliferation and a reduction in cell growth and in protein synthesis. These effects may be due to absorbance of growth-stimulating components present in fetal calf serum, or retardation of their effects, by the collagen lattice. In a sequence of experiments, we therefore

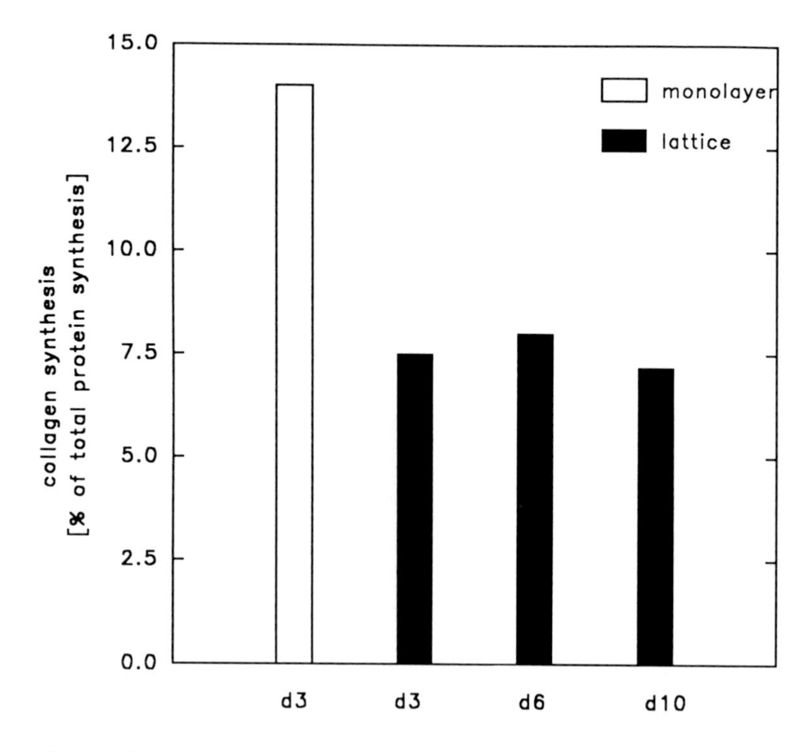

FIGURE 7. Collagen synthesis of vascular smooth muscle cells cultured within type I collagen lattices (lattice) and as monolayers on plastic (monolayer) in medium supplemented with 10% fetal calf serum. Collagen production is expressed as a percentage of total protein synthesized. Production was measured on day 3 (d3), day 6 (d6), and day 10 (d10) of culture. Differences between monolayer cultures and collagen lattice cultures are statistically significant ($p < 0.05$).

studied the effects of increasing concentrations of serum (0.5%, 10%, 20%) on the behavior of synthetic-state smooth muscle cells cultured within collagen lattices.

The amount of labeled proline and hydroxyproline served as an index for total protein synthesis. Cultivation in 0.5% serum showed no significant differences in values for synthesized protein between lattice-cultured cells and monolayer-cultured cells (Figure 9). Using 10 and 20% serum, monolayer-cultured cells showed clear stimulation of protein synthesis, whereas lattice-cultured cells did not. In the lattice cultures the values of total protein synthesis remained low. The amount of collagen synthesized was estimated from the radioactivity associated with nondialyzable hydroxyproline. In both culture systems, increase in serum concentration resulted in a significant decrease in the proportion of collagen synthesis (Figure 10). In each case the proportion of collagen synthesis in collagen lattice-cultured smooth muscle cells was significantly lower than that of the monolayer-cultured cells. The largest difference between the two culture systems was observed with 0.5% serum, and the smallest difference was observed with 20% serum.

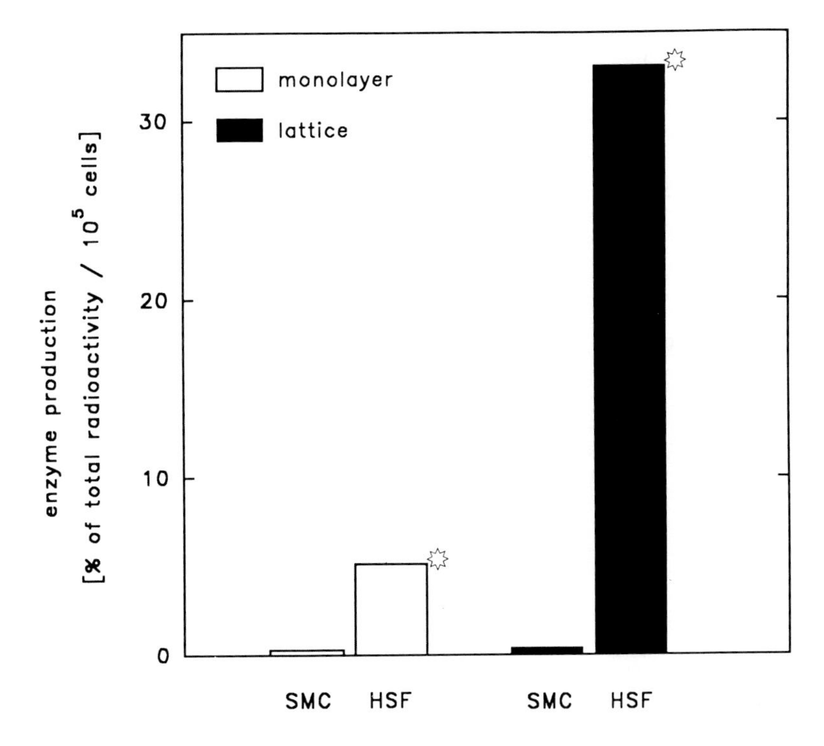

FIGURE 8. Collagenase production in vascular smooth muscle cells (SMC) and human skin fibroblasts (HSF) cultured within type I collagen lattices (lattice) and as monolayers on plastic (monolayer). Collagenase activities are given as a percentage of total radioactivity liberated from ^{14}C-labeled type I collagen. Significant differences ($p < 0.05$) between monolayer cultures and collagen lattice cultures are marked by asterisks.

Stimulation of total protein synthesis of monolayer-cultured smooth muscle cells by 10 and 20% fetal calf serum coincided with stimulation into the growth cycle. As shown in Figure 11, 5.9% of the smooth muscle cell population were in the S phase of the cell cycle with 0.5% serum, whereas 27.4% of the cells were in the S phase with 10% serum. At serum concentrations of 20%, the percentage of cells in the S phase was slightly higher still (29.7%). Similar treatment of lattice-cultured smooth muscle cells, however, failed to stimulate cells to enter the cell cycle and proliferate. When smooth muscle cells were cultivated in 0.5%, 10%, and 20% serum, a constant low percentage of cells in the S phase of the cell cycle was observed, indicating that the cells were arrested in the G_0/G_1 phase.

In collagen lattice-cultured smooth muscle cells, fetal calf serum neither affected the entry into S phase of the cell cycle nor stimulated protein synthesis. However, the volume of lattice-cultured cells rose as the serum concentration was increased. The cell diameters of smooth muscle cells cultivated with 0.5, 10, and 20% serum were 11.8, 12.1, and 13.1 μm, respectively. Conversely, cell volume decreased in monolayer-cultured smooth muscle cells as serum concentration was increased (data not shown).

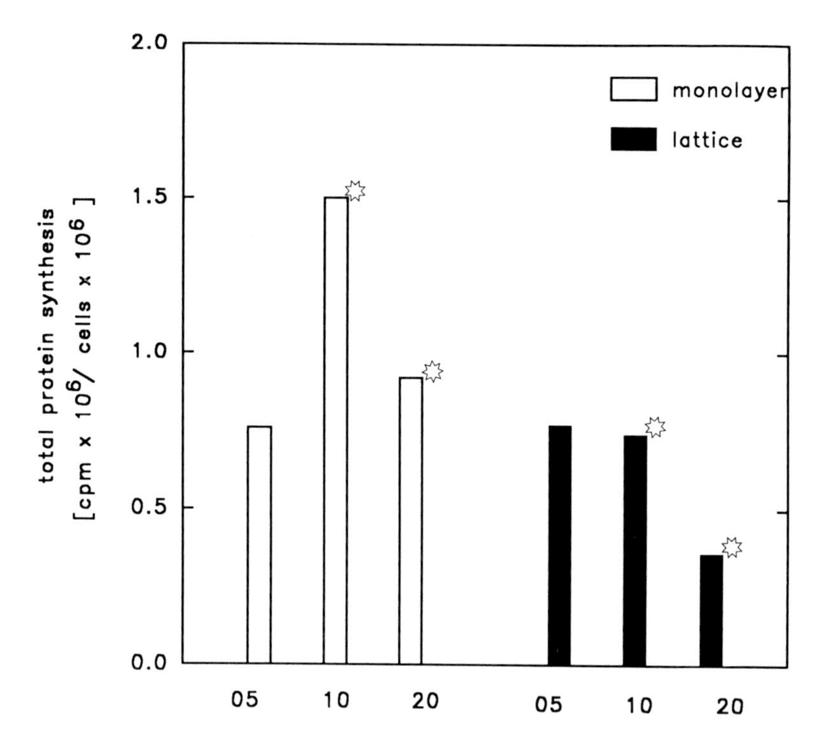

FIGURE 9. Total protein synthesis of vascular smooth muscle cells cultured within type I collagen lattices (lattice) and as monolayers on plastic (monolayer) in medium supplemented with 0.5% (05), 10% (10), and 20% (20) fetal calf serum. Protein synthesis was measured on day 4 of culture. Significant differences ($p < 0.05$) between monolayer cultures and collagen lattice cultures are marked by asterisks.

In conclusion, fetal calf serum stimulates growth and biosynthetic activity in smooth muscle cells cultured as a monolayer. Specifically, increasing concentrations of serum result in an activation of the cell cycle, an acceleration of protein synthesis, and a concomitant decrease in the proportion of collagen synthesis. However, the capacity of fetal calf serum for this stimulatory effect is abolished when smooth muscle cells are cultured within collagen lattices. This suggests an inactivation of serum growth-stimulating substances by the collagen lattice.

IV. MODULATION OF BIOSYNTHETIC ACTIVITY OF COLLAGEN LATTICE-CULTURED SMOOTH MUSCLE CELLS BY GROWTH FACTORS

Fetal calf serum is known to contain a complex mixture of growth-stimulating substances that might act as agonists and antagonists of smooth muscle cell proliferation and protein synthesis. In the atherosclerotic process, monocyte/macrophages and lymphocytes infiltrate the vessel wall[41] and there secrete

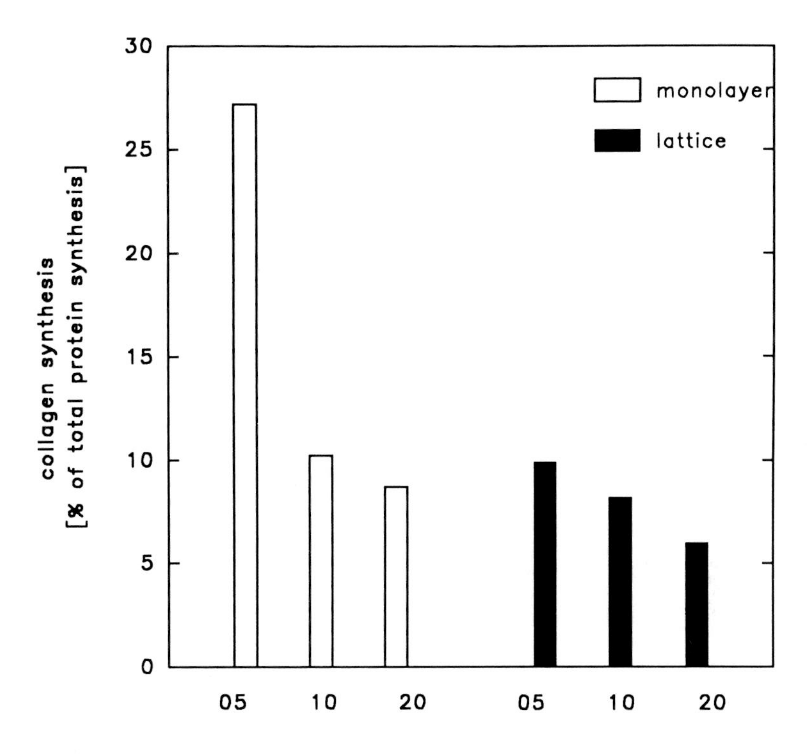

FIGURE 10. Collagen synthesis of vascular smooth muscle cells cultured within type I collagen lattices (lattice) and as monolayers on plastic (monolayer) in medium supplemented with 0.5% (05), 10% (10), and 20% (20) fetal calf serum. Collagen production is expressed as a percentage of total protein synthesized. Production was measured on day 4 of culture. Differences between monolayer cultures and collagen lattice cultures are significant ($p < 0.05$).

molecules that also have the potential to modulate smooth muscle cell functions. Among these various factors we selected transforming growth factor (TGF-β1) and epidermal growth factor (EGF) for more detailed study of growth factor effects on smooth muscle cell activities. TGF-β1 is a homodimeric polypeptide ubiquitously expressed by mammalian cells and recognized by specific receptors on almost every cell type studied. It is thought to be an important mediator in embryogenesis, wound healing, bone remodeling, and vascular diseases.[42,43] Its biological activity on mesenchymal cells *in vitro* includes enhancement of the synthesis of collagen, fibronectin, and proteoglycans.[44-46] For the induction of a number of biological effects by TGF-β1 *in vitro*, the simultaneous presence of EGF seems to be necessary. For example, a synergism of both factors is described for the induction of anchorage-independent growth of NRK fibroblasts[47] and of proliferation of quiescent dermal fibroblasts.[45] The aims of the experiments presented in this section were to investigate the effects of TGF-β1 and EGF on total protein synthesis and collagen synthesis in smooth muscle cells and to determine whether a collagen lattice influenced the responses.

Confluent monolayer cultures and collagen lattice cultures cultivated for 5 d

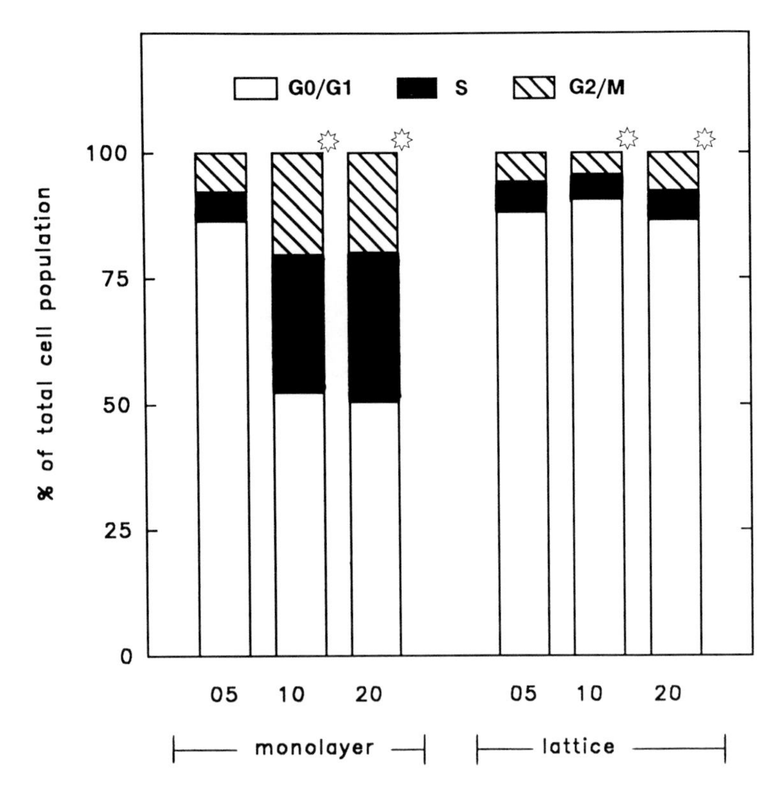

FIGURE 11. Cell cycle distribution in vascular smooth muscle cells cultured within type I collagen lattices (lattice) and as monolayers on plastic (monolayer) in medium supplemented with 0.5% (05), 10% (10), and 20% (20) fetal calf serum. Data are expressed as a percentage of total population as determined by propidium iodide fluorescence. Significant differences ($p < 0.05$) between monolayer cultures and collagen lattice cultures are marked by asterisks.

were preincubated for 24 h with serum-deficient medium (1% fetal calf serum) in the presence of 200 p*M* of TGF-β1 (5 ng/ml) and/or 800 pM of EGF (5 ng/ml) or solvent (4 m*M* HCl in PBS, pH 6.4) as control.[34] To measure protein synthesis and collagen synthesis, 0.37 *M*Bq/ml of ^{14}C-proline and 50 μg of ascorbic acid per milliliter were added to the preincubation medium, and cells were further incubated for 24 h.

When used alone, TGF-β1 induced protein synthesis in smooth muscle cells (Figure 12). Significant stimulation of synthesis was observed both in monolayer cultures and in collagen lattice cultures. In each case synthesis reached approximately double the value of the corresponding controls. By contrast, the culture systems differed significantly in their response to EGF. In monolayer cultures, EGF reduced the protein synthesis by approximately 40% compared to the control, whereas in collagen lattice cultures an increase of approximately 30% was observed. When the cells were exposed to both EGF and TGF-β1, signifi-

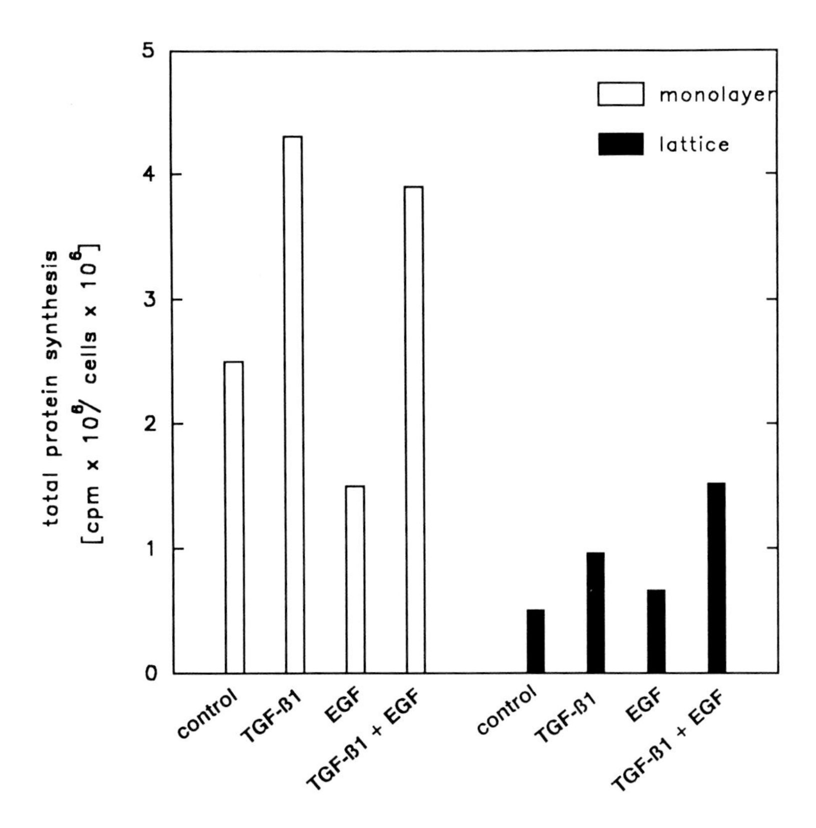

FIGURE 12. Total protein synthesis of vascular smooth muscle cells cultured within type I collagen lattices (lattice) and as monolayers on plastic (monolayer) in medium supplemented with 1% fetal calf serum and TGF-β1, EGF, TGF-β1 + EGF (5 ng/ml each), or solvent alone (control). Differences between growth factors and controls are significant ($p < 0.05$).

cantly enhanced protein synthesis was observed both in monolayer cultures and in collagen lattice cultures compared with the corresponding controls. In collagen lattice cultures, incubation of cells with the combination of growth factors resulted in a 1.8-fold increase in protein synthesis as compared to TGF-β1 alone.

The altered levels of protein synthesis induced by the growth factors were not due to different cell numbers in the culture systems. Incubation with the growth factors for 2 d under the described culture conditions did not result in a significant increase or decrease of the cell numbers.

After incubation of cells with TGF-β1, a specific increase in the proportion of collagen synthesis was observed in both lattice-cultured and monolayer-cultured cells (Figure 13), although this increase was more marked in the latter than in the former. A specific effect of EGF on collagen synthesis could only be observed in monolayer cultures, where a slight increase was observed. EGF plus

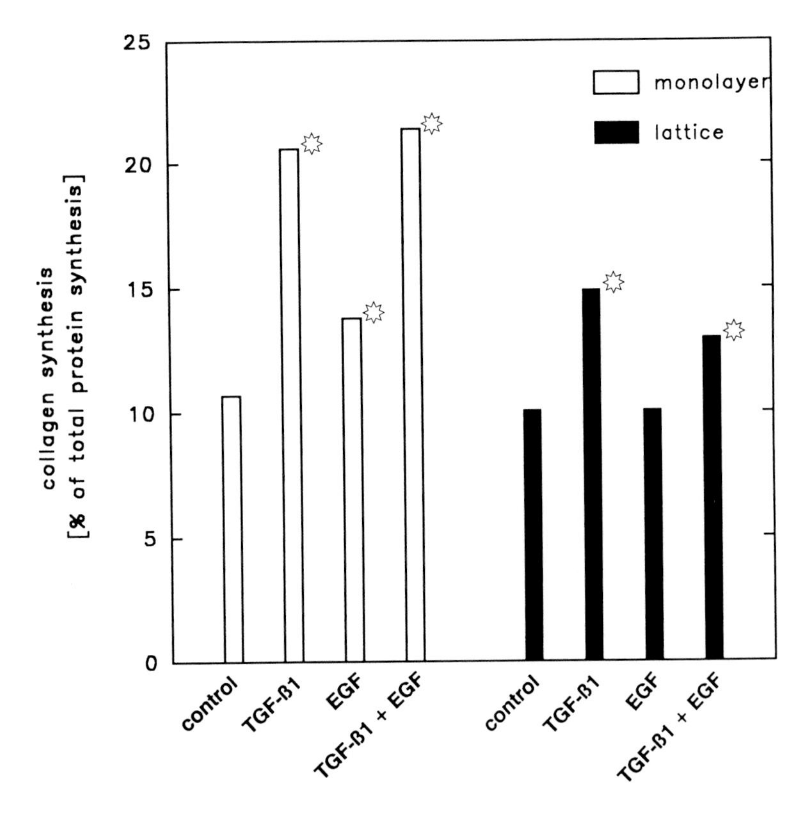

FIGURE 13. Collagen synthesis of vascular smooth muscle cells cultured within type I collagen lattices (lattice) and as monolayer on plastic (monolayer) in medium supplemented with 1% fetal calf serum and TGF-β1, EGF, TGF-β1 + EGF (5 ng/ml each), or solvent alone (control). Collagen production is expressed as a percentage of total protein synthesized. Significant differences ($p < 0.05$) from controls are marked by asterisks.

TGF-β1 enhanced the proportion of collagen synthesis in both monolayer and collagen lattice cultures as compared to controls. These results suggest that smooth muscle cell responses to growth factors are cooperatively regulated by the presence of the collagen lattice.

V. DISCUSSION AND CONCLUSIONS

The foregoing results provide substantial evidence that the extracellular matrix influences the behavior of vascular smooth muscle cells. Electron microscopic studies on cells removed from their physiological matrix environment and subsequently cultured as monolayers have demonstrated loss of myofilament bundles, appearance of extensive rough endoplasmic reticulum, and a large Golgi complex.[19,21,48-50] Biochemical studies indicate enhanced synthesis of extracellular matrix proteins, in particular of both type I and type III collagen, with a preponderance of type I collagen.[1,16,51-54] The work presented here shows that when these cells are reintegrated into an environment similar to that of tissue, i.e., a three-dimensional collagen lattice, these activities are again influenced toward those of the normal vessel. This change mainly affects the biosynthetic activities of cells, while the structural features of smooth muscle cells remain largely unaltered.

The morphological data presented here show that the ultrastructure of collagen-entrapped smooth muscle cells was comparable to that of monolayer-cultured cells. The cytoplasm was extensively filled with organelles characteristic of the synthesis of secretory proteins, confirming that the cells remained in the synthetic state. Reintegration into type I collagen thus failed to cause synthetic-state smooth muscle cells to revert to the contractile state. The synthetic state is typical of metabolically highly active cells as found in early fetal life and postnatal development,[12] in atherosclerosis,[48,55,56] and in monolayer culture.[19] Reversibility of phenotypic state can, however, occur *in vitro* and *in vivo*. When smooth muscle cells are grown in primary culture reaching confluency within not more than five cumulative population doublings, spontaneous reversion to the contractile state can be observed.[17,19] In addition, reversible changes of ultrastructure have been reported in the rabbit carotid after endothelial denudation by balloon catheter injury.[57]

The biochemical data show that smooth muscle cells reduce total protein synthesis and collagen synthesis when reintegrated into a three-dimensional collagen matrix. Thus, smooth muscle cells in the synthetic state sense the macromolecular composition of the extracellular environment. Synthetic-state smooth muscle cells appear to be able to respond to altered extracellular matrix by modifying their production of matrix components accordingly. There is evidence that fibrillar collagen also has the capacity to influence the synthesis and metabolism of other extracellular matrix proteins, such as proteoglycans and glycoproteins. For example, in a collagen lattice, fibroblasts downregulate the synthesis of small dermatan sulfate proteoglycan II to the same level of synthesis as that of large chondroitin sulfate/dermatan sulfate and of heparan sulfate proteoglycans, the synthesis of which remains unaffected.[58] Monkey smooth muscle cells grown as a monolayer on bovine dermal type I collagen appear to modify metabolism of proteoglycans. Here, an analysis of isolated radiolabeled proteoglycans indicated that the cells grown on collagen accumulate less chondroitin sulfate proteoglycan and heparan sulfate proteoglycan, but over twice as much iduronic-acid-rich dermatan sulfate proteoglycan, as do monolayer cultures on plastic. This accumulation seemed to be due in part to a decrease in the degradation and/or release of this proteoglycan.[59] In further support of our conclusion that expression of extracellular matrix components is regulated by the substratum is the observation that fibronectin synthesis is lower in cells grown on fibronectin than in those grown on laminin.[60]

The possibility that interactions of smooth muscle cells with collagen fibrils might influence the production of collagenase was examined with the collagen lattice culture. In the series of experiments presented here, we have obtained clear evidence that synthetic-state smooth muscle cells produce only small amounts of enzyme when cultured as monolayer on plastic, and that cultivation within a collagen lattice neither stimulates nor inhibits production of collagenase. This behavior of smooth muscle cells is in contrast to that of fibroblasts, where collagen lattice culture is a potent stimulator for increase of collagenase production.[61]

Even though we have demonstrated that three-dimensional lattices of preexisting type I collagen fibrils caused smooth muscle cells to reduce collagen synthesis, the precise nature of the mechanisms involved remain unclear. As demonstrated for fibroblasts, there might be a correlation between the extent of contact between cell and matrix material and the level of collagen synthesis.[62] Changes in fibril arrangement have also been reported to affect biosynthetic functions of cells,[63] and there is evidence that the mechanical properties of matrices in which cells are grown themselves represent a regulatory factor for collagen synthesis.[37] Moreover, we cannot exclude the possibility that the propeptides of procollagen molecules are involved in the feedback regulation of collagen synthesis.[64,65]

In the case of monolayer-cultured smooth muscle cells, it has recently been demonstrated that the expression of the genes for type I and type III collagens might be influenced by the proliferative state of the cells. Proliferating cells in monolayer culture were shown to contain low levels of collagen mRNA, whereas quiescent, growth-arrested cells contain high levels.[66] However, as there is generally a correlation between collagen mRNA levels and the extent of protein synthesis, our data obtained from smooth muscle cells cultured within collagen lattices are in contrast with the mRNA data. Changes in growth activity of smooth muscle cells observed during prolonged culture time do not influence synthesis of collagen.

Examining the growth activity of smooth muscle cells by determining cell number and by measurement of nuclear DNA, we found that proliferation was strongly suppressed when cells were reintegrated into an extracellular matrix environment, as compared to monolayer culture. Moreover, it is interesting to note that lattice-cultured smooth muscle cells did not respond to mitogenic stimulation with fetal calf serum. This is in contrast to monolayer-cultured smooth muscle cells.[67] The mechanisms involved in the inhibition of proliferation and the arrest at the G0/G1 phase of the cell cycle in cells integrated into a collagen lattice are poorly understood. Regulatory factors for cell proliferation may be the cell density within the lattice,[68] the number of contact sites between a cell and collagen fibrils,[69] and/or type I collagen fibrils themselves.[70] In the case of fibroblasts it is speculated that differences in extracellular matrix organization and cell shape might, independently of extracellular matrix density, play a role in growth control.[31,71] As demonstrated for MRC5 fibroblasts cultured on collagen gels, other types of collagen also seem to have regulatory effects on proliferation.[72] To what degree other extracellular matrix molecules, such as fibronectin, organized in the type I collagen lattice, might trigger cell growth has still to be elucidated.[73] The significance of growth factor binding to extracellular matrix, the release of sequestered factors from extracellular matrix and the subsequent inhibition or stimulation of cell proliferation remain to be explored.[74-76]

Investigations of the influence of growth factors alone and in association with extracellular matrix proteins on smooth muscle cell proliferation and protein synthesis are crucial to the understanding of atherogenesis. The presence of TGF-β1 in platelets,[77] macrophages,[78] and T lymphocytes,[79] the major cell infiltrates of atherosclerotic plaques, and the well-described stimulating effects of TGF-β1 on matrix synthesis in mesenchymal cells other than smooth muscle cells, has led to the hypothesis that TGF-β1 stimulates extracellular matrix synthesis in smooth muscle cells. When smooth muscle cells were cultured within a collagen lattice, we found that protein synthesis was strongly reduced compared to that observed in monolayer cultures, but incubation with TGF-β1 resulted in a stimulation of protein synthesis, with a specific increase of collagen production. Previous reports concerning a specific stimulation of collagen synthesis by TGF-β1 have been conflicting. For example, TGF-β1 stimulates protein synthesis in human gingival fibroblasts, but with no increase in the proportion of collagen synthesis;[80,81] other groups, however, have observed a preferential stimulation of collagen synthesis.[45,82] Overall, these results suggest that TGF-β1 does not act merely as an activator of protein synthesis in general, but can have effects on specific proteins (i.e., collagens), a conclusion supported by the finding that the growth factor may stimulate the promoter for the type I collagen gene.[83] In monolayer cultures the proportion of collagen in total proteins synthesized was doubled under the influence of TGF-β1, but in collagen lattices the specific stimulation of collagen synthesis was lower. These results indicate that the regulation of collagen synthesis of cells in a three-dimensional collagen lattice is more complex than in monolayer cultures. It is reported that TGF-β1 has no effect on the steady-state RNA levels of types I and III collagen in cultured rabbit smooth muscle cells.[84] Taking into account that the TGF-β1-induced increase in collagen synthesis correlates with an increase in mRNA levels of types I and III collagen in other mesenchymal cells,[85] these data appear to contrast with our results on collagen synthesis. However, this discrepancy may be due to species differences or to a difference in the regulation pattern of posttranslational modification in the lattice-cultured smooth muscle cells exposed to TGF-β1.

It has been reported that TGF-β1 is effective only in association with EGF.[47] However, we and others[82] have shown that protein and collagen synthesis is stimulated by TGF-β1 alone. Nevertheless, in collagen lattice cultures we found that simultaneous incubation with TGF-β1 and EGF resulted in increased protein synthesis, as compared to TGF-β1 alone. EGF alone influenced smooth muscle cells in a different manner, depending on the culture systems used. EGF decreased protein synthesis of smooth muscle cells cultured as monolayers and increased protein synthesis of smooth muscle cells cultured in collagen lattices. The reason for this phenomenon is unclear at present. One possible explanation is that receptors for growth factors become more accessible as a result of substrate-induced changes of cell shape.[86] Furthermore, extracellular matrix proteins might influence EGF-induced effects on smooth muscle cells; thrombospondin,[87,88] for example, might optimize the mitogenic effects of EGF on smooth muscle cells.[89] Multidomain proteins of the extracellular matrix, like thrombospondin and laminin, have been found to possess domains with sequence homologies to EGF,[90] and such domains with growth factor activity may interact with target cells, thus influencing the response to EGF.[90] Proteins such as thrombospondin and laminin could act and/or be accessible in a different manner in the two culture systems, leading to a different smooth muscle cell response to EGF.

Further experiments with collagen lattice cultures may help to define in more detail the role of growth factors and the role of various extracellular matrix proteins in the cooperative regulation of gene expression in smooth muscle cells during the pathogenesis of atherosclerosis.

ACKNOWLEDGMENTS

I am indebted to Dr. W. Schlumberger and U. Falken for providing data from growth factor experiments and for helpful discussion. I also wish to thank my colleagues Dr. R. Semich, Dr. H. Volmer, and Prof. Dr. H. Kresse for their support during the course of this study. I thank C. Fabritius, R. Fischer, I. Meyer, M. Opalka, S. Otter, A. Pallas, and K. Schlattmann for their skillful technical assistance. Furthermore, Dr. B. Harrach and Dr. N.J. Severs provided helpful criticism of the manuscript and assistance in its preparation. I am grateful to Prof. Dr. H. Robenek and Prof. Dr. J. Rauterberg for their continuous encouragement and for the opportunity to use their laboratories at the Institute for Arteriosclerosis Research, Münster. This work was supported by SFB 223 from the Deutsche Forschungsgemeinschaft.

REFERENCES

1. **Burke, J.M., Balian, G., Ross, R., and Bornstein, P.,** Synthesis of types I and III procollagen and collagen by monkey aortic smooth muscle cells *in vitro*, *Biochemistry*, 16, 3243, 1977.
2. **Layman, D.L., Epstein, E.H., Dodson, R.F., and Titus, J.L.,** Biosynthesis of type I and III collagens by cultured smooth muscle cells from human aorta, *Proc. Natl. Acad. Sci. U.S.A.*, 74, 671, 1977.
3. **Mayne, R., Vail, M.S., and Miller, E.J.,** Characterization of the collagen chains synthesized by cultured smooth muscle cells derived from rhesus monkey thoracic aorta, *Biochemistry*, 17, 466, 1978.

4. **Mayne, R., Vail, M.S., Miller, E.J., Blose, S.H., and Chacko, S.,** Collagen polymorphism in cell cultures derived from guinea pig aortic smooth muscle: comparison with three populations of fibroblasts, *Arch. Biochem. Biophys.*, 181, 462, 1977.
5. **Ross, R. and Klebanoff, S.J.,** The smooth muscle cell. I. In vivo synthesis of connective tissue proteins, *J. Cell Biol.*, 50, 159, 1971.
6. **Abraham, P.A., Smith, D.W., and Carnes, W.H.,** Synthesis of soluble elastin by aortic medial cells in culture, *Biochem. Biophys. Res. Commun.*, 58, 597, 1974.
7. **Asundi, V., Cowan, K., Matzura, D., Wagner, W., and Dreher, K.L.,** Characterization of extracellular matrix proteoglycan transcripts expressed by vascular smooth muscle cells, *Eur. J. Cell Biol.*, 52, 98, 1990.
8. **Dreher, K.L., Asundi, V., Matzura, D., and Cowan, K.,** Vascular smooth muscle biglycan represents a highly conserved proteoglycan within the arterial wall, *Eur. J. Cell Biol.*, 53, 296, 1990.
9. **Wight, T.N. and Ross, R.,** Proteoglycans in primate arteries. I. Ultrastructural localization and distribution in the intima, *J. Cell Biol.*, 67, 660, 1975.
10. **Wight, T.N. and Ross, R.,** Proteoglycans in primate arteries. II. Synthesis and secretion of glycosaminoglycans by arterial smooth muscle cells in culture, *J. Cell Biol.*, 67, 675, 1975.
11. **Schwartz, E., Bienkowski, R.S., Coltoff-Schiller, B., Goldfischer, S., and Blumenfeld, O.,** Changes in the components of extracellular matrix and in growth properties of cultured aortic smooth muscle cells upon ascorbate feeding, *J. Cell Biol.*, 92, 462, 1982.
12. **Cayatte, A.J., Ashraf, M., and Subbiah, M.T.R.,** Morphological and smooth muscle cell phenotypic changes in fetal rabbit aorta during early development, *Basic Res. Cardiol.*, 84, 259, 1989.
13. **Burke, J.M. and Ross, R.,** Synthesis of connective tissue macromolecules by smooth muscle, *Int. Rev. Connect. Tissue Res.*, 8, 119, 1979.
14. **Campbell, G.R. and Campbell, J.H.,** Phenotypic modulation of smooth muscle cells in primary culture, in *Vascular Smooth Muscle in Culture*, Vol. I, Campbell, J.H. and Campbell, G.R., Eds., CRC Press, Boca Raton, FL, 1987, 39.
15. **Campbell, G.R., Campbell, J.H., Ang, A.H., Campbell, I.L., Horrigan, S., Manderson, J.A., Mosse, P.R.L., and Rennick, R.E.,** Phenotypic changes in smooth muscle cells of human atherosclerotic plaques, in *Pathobiology of the Human Atherosclerotic Plaque*, Glagov, S., Newman, W.P., III, and Schaffer, S.A., Eds., Springer-Verlag, New York, Berlin, Heidelberg, 1990, 69.
16. **Ang, A.K., Tachas, G., Campbell, J.H., Bateman, J.F., and Campbell, G.R.,** Collagen synthesis by cultured rabbit aortic smooth muscle cells. Alteration with phenotype, *Biochem. J.*, 265, 461, 1990.
17. **Campbell, G.R. and Campbell, J.H.,** Smooth muscle phenotypic changes in arterial wall homeostasis. Implications for the pathogenesis of atherosclerosis, *Exp. Mol. Pathol.*, 42, 139, 1985.
18. **Chamley-Campbell, J.H., Campbell, G.R., and Ross, R.,** Phenotype-dependent response of cultured aortic smooth muscle to serum mitogens, *J. Cell Biol.*, 89, 379, 1981.
19. **Chamley-Campbell, J., Campbell, G.R., and Ross, R.,** The smooth muscle cell in culture, *Physiol. Rev.*, 59, 1, 1979.
20. **Ross, R.,** The pathogenesis of atherosclerosis — an update, *N. Engl. J. Med.*, 14, 488, 1986.
21. **Thyberg, J., Nilsson, J., Palmberg, L., and Sjölund, M.,** Adult human arterial smooth muscle cells in primary culture. Modulation from contractile to synthetic phenotype, *Cell Tissue Res.*, 239, 69, 1985.
22. **Mayne, R.,** Collagenous proteins of blood vessels, *Arteriosclerosis*, 6, 585, 1986.
23. **McCullagh, K.G. and Balian, G.,** Collagen characterization and cell transformation in human atherosclerosis, *Nature*, 258, 73, 1975.
24. **McCullagh, K.G., Duance, V.C., and Bishop, K.A.,** The distribution of collagen types I, III and V (AB) in normal and atherosclerotic human aorta, *Am. J. Pathol.*, 130, 45, 1980.

25. **Murata, K., Motayama, T., and Kotake, C.,** Collagen types in various layers of the human aorta and their changes with the atherosclerotic process, *Atherosclerosis*, 60, 251, 1986.
26. **Wagner, W.D.,** Modification of collagen and elastin in the human atherosclerotic plaque, in *Pathobiology of the Human Atherosclerotic Plaque*, Glagov, S., Newman, W.P., III, and Schaffer, S.A., Eds., Springer-Verlag, New York, Berlin, Heidelberg, 1990, 167.
27. **Barnes, M.J.,** Collagens in atherosclerosis, *Coll. Relat. Res.*, 5, 65, 1985.
28. **Labat-Robert, J., Bihari-Varga, M., and Robert, L.,** Extracellular matrix, *FEBS Lett.*, 268, 386, 1990.
29. **Bell, E., Ivarsson, B., and Merrill, C.,** Production of a tissue-like structure by contraction of collagen lattices by human fibroblasts of different proliferative potential *in vitro*, *Proc. Natl. Acad. Sci. U.S.A.*, 76, 1274, 1979.
30. **Weinberg, C.B. and Bell, E.,** A blood vessel model constructed from collagen and cultured vascular cells, *Science*, 231, 397, 1986.
31. **Nakagawa, S., Pawelek, S., and Grinnell, F.,** Long-term culture of fibroblasts in contracted collagen gels: effects on cell growth and biosynthetic activity, *J. Invest. Dermatol.*, 93, 792, 1989.
32. **Schlumberger, W., Thie, M., Rauterberg, J., Kresse, H., and Robenek, H.,** Deposition and ultrastructural organization of collagen and proteoglycans in the extracellular matrix of gel cultured fibroblasts, *Eur. J. Cell Biol.*, 50, 100, 1989.
33. **Thie, M., Schlumberger, W., Semich, R., Rauterberg, J., and Robenek, H.,** Aortic smooth muscle cells in collagen lattice culture: effects on ultrastructure, proliferation and collagen synthesis, *Eur. J. Cell Biol.*, 55, 295, 1991.
34. **Schlumberger, W., Thie, M., Rauterberg, J., and Robenek, H.,** Collagen synthesis in cultured aortic smooth muscle cells: modulation by collagen lattice culture, transforming growth factor-β1 and epidermal growth factor, *Arteriosclerosis*, in press, 1991.
35. **Nusgens, B., Merrill, C., Lapiere, C., and Bell, E.,** Collagen biosynthesis by cells in a tissue equivalent matrix *in vitro*, *Coll. Relat. Res.*, 4, 351, 1984.
36. **Vindelov, LL., Christensen, I.J., and Nissen, N.J.,** A detergent-trypsin method for the preparation of nuclei for flow cytometric DNA analysis, *Cytometry*, 3, 323, 1983.
37. **Thie, M., Schlumberger, W., Rauterberg, J., and Robenek, H.,** Mechanical confinement inhibits collagen synthesis in gel-cultured fibroblasts, *Eur. J. Cell Biol.*, 48, 294, 1989.
38. **Krieg, T., Hörlein, D., Wiestner, M., and Müller, P.K.,** Aminoterminal extension peptides from type I procollagen normalize excessive collagen synthesis of scleroderma fibroblasts, *Arch. Dermatol. Res.*, 263, 171, 1978.
39. **Johnson-Wint, B.,** A quantitative collagen film collagenase assay for large numbers of samples, *Anal. Biochem.*, 104, 175, 1980.
40. **Rice, R.H. and Means, G.E.,** Radioactive labeling of proteins *in vitro*, *J. Biol. Chem.*, 246, 831, 1971.
41. **Tsukada, T., Rosenfeld, M.E., Gown, A.M., and Ross, R.,** Immunocytochemical analysis of the cellular composition of atherosclerotic lesions in the human aorta, in *Pathobiology of the Human Atherosclerotic Plaque*, Glagov, S., Newman, W.P., III, and Schaffer, S.A., Eds., Springer-Verlag, New York, Berlin, Heidelberg, 1990, 349.
42. **Sporn, M.B., Roberts, A.B., Wakefield, L.M., and de Crombrugghe, B.,** Some recent advances in the chemistry and biology of transforming growth factor-β, *J. Cell Biol.*, 105, 1039, 1987.
43. **Rizzino, A.,** Transforming growth factor-β: multiple effects on cell differentiation and extracellular matrices, *Dev. Biol.*, 130, 411, 1988.
44. **Ignotz, R. and Massague, J.,** Transforming growth factor-beta stimulates the expression of fibronectin and collagen and their incorporation into the extracellular matrix, *J. Biol. Chem.*, 261, 4337, 1986.
45. **Varga, J., Rosenbloom, J., and Jimenez, S.A.,** Transforming growth factor β (TGF-β) causes a persistent increase in steady-state amounts of type I and type III collagen and fibronectin mRNAs in normal human dermal fibroblasts, *Biochem. J.*, 247, 597, 1987.

46. **Chen, J.K., Hoshi, H., and McKeehan, W.L.,** Transforming growth factor type β specifically stimulates synthesis of proteoglycans in human adult arterial smooth muscle cells, *Proc. Natl. Acad. Sci. U.S.A.*, 84, 5287, 1987.
47. **Roberts, A.B., Anzano, M.A., Lamb, L.C., Smith, J.M., and Sporn, M.B.,** New class of transforming growth factors potentiated by epidermal growth factor: isolation from non-neoplastic tissues, *Proc. Natl. Acad. Sci. U.S.A.*, 78, 5339, 1981.
48. **Chamley-Campbell, J.H. and Campbell, G.R.,** What controls smooth muscle phenotype? *Atherosclerosis*, 40, 347, 1981.
49. **Ross, R.,** The smooth muscle cell. II. Growth of smooth muscle cell in culture and formation of elastic fibers, *J. Cell Biol.*, 50, 172, 1971.
50. **Thyberg, J., Palmberg, L., Nilsson, J., Ksiazek, T., and Sjölund, M.,** Phenotype modulation in primary cultures of arterial smooth muscle cells. On the role of platelet-derived growth factor, *Differentiation*, 25, 156, 1983.
51. **Barnes, M.J., Morton, L.F., and Levene, C.J.,** Synthesis of collagen types I and III by pig medial smooth muscle cells in culture, *Biochem. Biophys. Res. Commun.*, 70, 339, 1976.
52. **Campbell, J.H., Campbell, G.R., Kocher, O., and Gabbiani, G.,** Cell biology of smooth muscle in culture: implications for atherogenesis, *Inter. Angiol.*, 6, 73, 1987.
53. **Leung, D.Y.M., Glagov, S., and Mathews, M.,** Cyclic stretching stimulates synthesis of matrix components by arterial smooth muscle cells *in vitro*, *Science*, 191, 475, 1976.
54. **Rauterberg, J., Allam, S., Brehmer, U., Wirth, W., and Hauss, W.H.,** Characterization of the collagen synthesized by cultured human smooth muscle cells from fetal and adult aorta, *Hoppe Seylers Z. Physiol. Chem.*, 358, 401, 1977.
55. **Babaev, V.R., Bobryshev, Y.V., Stenina, O.V., Tararak, E.M., and Gabbiani, G.,** Heterogeneity of smooth muscle cells in atheromatous plaque of human aorta, *Am. J. Pathol.*, 136, 1031, 1990.
56. **Dartsch, P.C., Bauriedel, G., Schinko, I., Weiss, H.D., Höfling, B., and Betz, E.,** Cell constitution and characteristics of human atherosclerotic plaques selectively removed by percutaneous atherectomy, *Atherosclerosis*, 80, 149, 1989.
57. **Manderson, J.A., Mosse, P.R.L., Safstrom, S.B., Young, G.R., and Campbell, G.R.,** Balloon catheter injury to rabbit carotid artery. I. Changes in smooth muscle phenotype, *Arteriosclerosis*, 9, 289, 1989.
58. **Greve, H., Blumberg, P., Schmidt, G., Schlumberger, W., Rauterberg, J., and Kresse, H.,** Influence of collagen lattice on the metabolism of small proteoglycan II by cultured fibroblasts, *Biochem. J.*, 269, 149, 1990.
59. **Lark, M. and Wight, T.N.,** Modulation of proteoglycan metabolism by aortic smooth muscle cells grown on collagen gels, *Arteriosclerosis*, 6, 638, 1986.
60. **Hedin, U., Bottger, B.A., Forsberg, E., Johansson, S., and Thyberg, J.,** Diverse effects of fibronectin and laminin on phenotypic properties of cultured arterial smooth muscle cells, *J. Cell Biol.*, 107, 307, 1988.
61. **Mauch, C., Adelmann-Grill, B., Hatamochi, A., and Krieg, T.,** Collagenase gene expression in fibroblasts is regulated by a three-dimensional contact with collagen, *FEBS Lett.*, 250, 301, 1989.
62. **Paye, M., Nusgens, B.V., and Lapiere, C.M.,** Modulation of cellular biosynthetic activity in the retracting collagen lattice, *Eur. J. Cell Biol.*, 45, 44, 1987.
63. **Mochitate, K., Pawelek, P., and Grinnell, F.,** Stress relaxation of contracted collagen gels: disruption of actin filament bundles, release of cell surface fibronectin, and down-regulation of DNA and protein synthesis, *Exp. Cell Res.*, 193, 198, 1991.
64. **Paglia, L.M., Wilczek, J., Diaz de Leon, L., Martin, G.R., Hörlein, D., and Müller, P.,** Inhibition of procollagen cell-free synthesis by amino-terminal extension peptides, *Biochemistry*, 18, 5030, 1979.
65. **Schlumberger, W., Thie, M., Volmer, H., Rauterberg, J., and Robenek, H.,** Binding and uptake of Col 1(I), a peptide capable of inhibiting collagen synthesis in fibroblasts, *Eur. J. Cell Biol.*, 46, 244, 1988.

66. **Kindy, M.S., Chang, C.J., and Sonnenshein, G.E.,** Serum deprivation of vascular smooth muscle cells enhances collagen gene expression, *J. Biol. Chem.*, 263, 11,426, 1988.
67. **Gadeau, A.P., Campan, M., and Desgranges, C.,** Induction of cell cycle-dependent genes during cell cycle progression of arterial smooth muscle cells in culture, *J. Cell. Physiol.*, 146, 356, 1991.
68. **Weinberg, C.B. and Bell, E.,** Regulation of proliferation of bovine aortic endothelial cells, smooth muscle cells, and adventitial fibroblasts in collagen lattices, *J. Cell. Physiol.*, 122, 410, 1985.
69. **Nishiyama, T., Tsunenaga, M., Nakayama, Y., Adachi, E., and Hayashi, T.,** Growth rate of human fibroblasts is repressed by the culture within reconstituted collagen matrix but not by the culture on the matrix, *Matrix*, 9, 193, 1989.
70. **Yoshida, Y., Mitsumata, M., Yamane, N., Tomikawa, M., and Nishida, K.,** Morphology and increased growth rate of atherosclerotic intimal smooth muscle cells, *Arch. Pathol. Lab. Med.*, 112, 987, 1988.
71. **Nakagawa, S., Pawelek, P., and Grinnell, F.,** Long-term culture of fibroblasts in contracted collagen gels: effects on cell growth and biosynthetic activity, *J. Invest. Dermatol.*, 93, 792, 1989.
72. **Tiollier, J., Dumas, H., Tardy, M., and Tayot, J.L.,** Fibroblast behavior on gels of type I, III, and IV human placental collagens, *Exp. Cell Res.*, 191, 95, 1990.
73. **Ingber, D.E., Prusty, D., Frangioni, J.V., Cragoe, E.J., Lechene, C., and Schwartz, M.A.,** Control of intracellular pH and growth by fibronectin in capillary endothelial cells, *J. Cell Biol.*, 110, 1803, 1990.
74. **Yaoi, Y., Hashimoto, K., Takahara, K., and Kato, I.,** Insulin binds to type V collagen with retention of mitogenic activity, *Exp. Cell Res.*, 194, 180, 1991.
75. **Paralkar, V.M., Vukicevic, S., and Reddi, A.H.,** Transforming growth factor β type 1 binds to collagen IV of basement membrane matrix: implications for development, *Dev. Biol.*, 143, 303, 1991.
76. **Terrell, G.E. and Swain, J.L.,** Indirect angiogenic agents do not release fibroblast growth factors from extracellular matrix, *Matrix*, 11, 108, 1991.
77. **Assoian, R.K., Komoriya, A., Meyers, C.A., Miller, D.M., and Sporn, M.B.,** Transforming growth factor-β in human platelets, *J. Biol. Chem.*, 258, 7155, 1983.
78. **Assoian, R.K., Fleurdelys, B.E., Stevenson, H.C., Miller, P.J., Madtes, D.K., Raines, E.W., Ross, R., and Sporn, M.B.,** Expression and secretion of type β transforming growth factor by activated human macrophages, *Proc. Natl. Acad. Sci. U.S.A.*, 84, 6020, 1987.
79. **Kehrl, J.H., Wakefield, L.M., Roberts, A.B., Jakowlew, S., Alvarez-Mon, M., Derynck, R., Sporn, M.B., and Fauci, A.S.,** Production of transforming growth factor β by human T lymphocytes and its potential role in the regulation of T cell growth, *J. Exp. Med.*, 163, 1037, 1986.
80. **Narayanan, S., Page, R.C., and Swanson, J.,** Collagen synthesis by human fibroblasts: regulation by transforming growth factor-β in the presence of other inflammatory mediators, *Biochem. J.*, 260, 463, 1989.
81. **Redini, F., Galera, P., Mauviel, A., Loyau, G., and Pujol, J.P.,** Transforming growth factor β stimulates collagen and glycosaminoglycan biosynthesis in cultured rabbit articular chondrocytes, *FEBS Lett.*, 234, 172, 1988.
82. **Fine, A. and Goldstein, R.H.,** The effect of transforming growth factor-β on cell proliferation and collagen formation by lung fibroblasts, *J. Biol. Chem.*, 262, 3897, 1987.
83. **Rossi, P., Karsenty, G., Roberts, A.B., Roche, N.S., Sporn, M.B., and de Crombrugghe, B.,** A nuclear factor 1 binding site mediates the transcriptional activation of a type I collagen promoter by transforming growth factor-β, *Cell*, 52, 405, 1988.
84. **Liau, G. and Chan, L.M.,** Regulation of extracellular matrix RNA levels in cultured smooth muscle cells, *J. Biol. Chem.*, 264, 10, 315, 1989.

85. **Rhagow, R., Postlethwaite, A.E., Keski-Oja, J., Moses, H.L., and Kang, A.H.,** Transforming growth factor-β increases steady state levels of type I procollagen and fibronectin messenger RNAs posttranscriptionally in cultured human dermal fibroblasts, *J. Clin. Invest.*, 79, 1285, 1987.
86. **Kleinman, H.L., Cannon, F.B., Laurie, G.W., Hassell, J.R., Aumailley, M., Terranova, V.P., Martin, G.R., and DuBois-Dalcq, M.,** Biological activities of laminin, *J. Cell Biochem.*, 27, 317, 1985.
87. **Raugi, G.J., Mumby, S.M., Abbott-Brown, D., and Bornstein, P.,** Thrombospondin: synthesis and secretion by cells in culture, *J. Cell Biol.*, 95, 351, 1982.
88. **Majack, R.A., Cook, S.C., and Bornstein, P.,** Platelet-derived growth factor and heparin-like glycosaminoglycans regulate thrombospondin synthesis and deposition in the matrix by smooth muscle cells, *J. Cell Biol.*, 101, 1059, 1985.
89. **Majack, R.A., Cook, S.C., and Bornstein, P.,** Control of smooth muscle cell growth by components of the extracellular matrix: autocrine role for thrombospondin, *Proc. Natl. Acad. Sci. U.S.A.*, 83, 9050, 1986.
90. **Engel, J.,** EGF-like domains in extracellular matrix proteins: localized signals for growth and differentiation? *FEBS Lett.*, 251, 1, 1989.

Chapter 6

VASCULAR CELL RESPONSES TO INJURY: MODULATION BY EXTRACELLULAR MATRIX AND SOLUBLE FACTORS

Joseph A. Madri and Leonard Bell

TABLE OF CONTENTS

ISBN 0-8493-5505-2

I. INTRODUCTION

Atherosclerosis is a major health concern in the Western world. In 1986 there were over 250 million patient discharges for angina and 900,000 patient discharges with the diagnosis of acute myocardial infarction.[1] In addition, 400,000 patients required coronary revascularization in 1986; it is expected that 30 to 50% of the lesions treated with balloon dilatation will restenose in 3 to 6 months,[2] while two thirds of the saphenous vein grafts will either be totally occluded or severely diseased at 7 to 11 years.[3] It is therefore of major importance to understand the responses of vascular cells to injury (endothelial and smooth muscle proliferation, migration, and matrix synthetic profiles) and to be able to modulate these responses beneficially to effect maximal healing with minimal restenosis.

The two major cell types of the vessel wall, the endothelial cell and the smooth muscle cell, respond to injury (angioplasty, endarterectomy, synthetic and saphenous vein grafting) in very different ways and influence each other's response profiles via complex pathways. The vascular system is lined by metabolically active, mitotically quiescent endothelial cells, which provide a nonthrombogenic surface for blood flow, in addition to having a broad range of metabolic activities. Smooth muscle cells are found beneath the endothelium, in the media of large vessels. Vascular smooth muscle cells play significant roles in maintaining vessel wall integrity, including maintaining the connective tissues of the vessel wall, controlling vascular tone, and influencing endothelial cell behavior.[4,5] Vascular cells (large vessel-derived endothelial and smooth muscle cells) have been found to respond to injury in specific ways, depending upon the vascular bed and the cell type(s) injured.

In this review we will consider responses to iatrogenic injuries such as those occurring to the vessel wall during angioplasty, autologous and synthetic grafting, and endarterectomy. Following the denudation injury evoked by these therapies, large vessel endothelial cells adjacent to affected areas exhibit rapid sheet migration over the exposed extracelluar matrix (or synthetic graft material) and proliferate in order to reconstitute a continuous endothelial cell lining.[6,7] Medial smooth muscle cells of large and medium-sized vessels respond to these treatments by proliferation and migration into the intima, where they synthesize matrix components, which results in the formation of an expanded neointimal compartment that narrows the vessel lumen (Figure 1).[8]

In this paper we will examine the responses to injury of large vessel endothelial and smooth muscle cells, and their modulation by the composition and organization of the extracellular matrix and the presence of selected soluble autocrine and paracrine factors.

II. EXTRACELLULAR MATRIX COMPONENTS AND SOLUBLE FACTORS COORDINATELY MODULATE AORTIC ENDOTHELIAL AND SMOOTH MUSCLE CELL MIGRATION *IN VITRO* AND *IN VIVO*

Several of the currently available invasive treatments of large and medium vessel occlusive disease (angioplasty, autologous and synthetic grafting, endarterectomy, atherectomy, and laser ablation) result in significant deendothelialization and medial injury of vascular segments, and subsequent platelet adhesion, aggregation, and release of soluble factors. This platelet releasate has been proposed to play an important role in the development of arteriosclerosis,[4,5,9] since it contains growth factors and vasoactive agents, such as platelet-derived growth factor (PDGF), transforming growth factor beta 1 (TGF-β1), serotonin, norepinephrine, and histamine, which modulate vascular endothelial and smooth muscle cell chemotaxis, migration, and proliferation. In addition to eliciting a dramatic platelet response, the extracellular matrix components of the vessel wall also have profound effects on the behavior of vascular endothelial and smooth muscle cells, eliciting changes in extracellular matrix metabolism and the synthesis, secretion, and activation of a variety of vascular cell-derived soluble factors.

III. ENDOTHELIAL CELLS — *IN VITRO* STUDIES

A. Extracellular Matrix Effects on Endothelial Cell Behavior

Matrix components have been found to have a significant effect on the migration of bovine aortic endothelial cells *in vitro*. Using an *in vitro* migration assay developed in our laboratory, we have shown that interstitial collagens (types I and III) permit/elicit a rapid migration rate, while basement membrane components (types IV and V collagen and laminin) elicit intermediate migratory rates, and fibronectin was found to elicit a low migration rate.[7,10-13] Furthermore, we noted that migratory behavior on these substrates could be modulated by altering the amount of matrix component coated on the culture dishes.[7] Bovine aortic endothelial cells migrating on substrates that permit/elicit intermediate or high migratory rates, such as collagen types I, III, and IV, and laminin, were also associated with reorganization of several cytoskeletal components, including the cortical cytoskeletal components fodrin,[10] protein 4.1,[11] as well as vinculin and α-actinin.[14] However, substrates that elicited a low migratory rate, such as fibronectin, were not associated with reorganization of these cytoskeletal components following stimuli to migrate.[10,11] These data support the concept that existing and newly synthesized extracellular matrix modulates cellular behavior via cell surface matrix-binding proteins (integrins and nonintegrin moieties), in

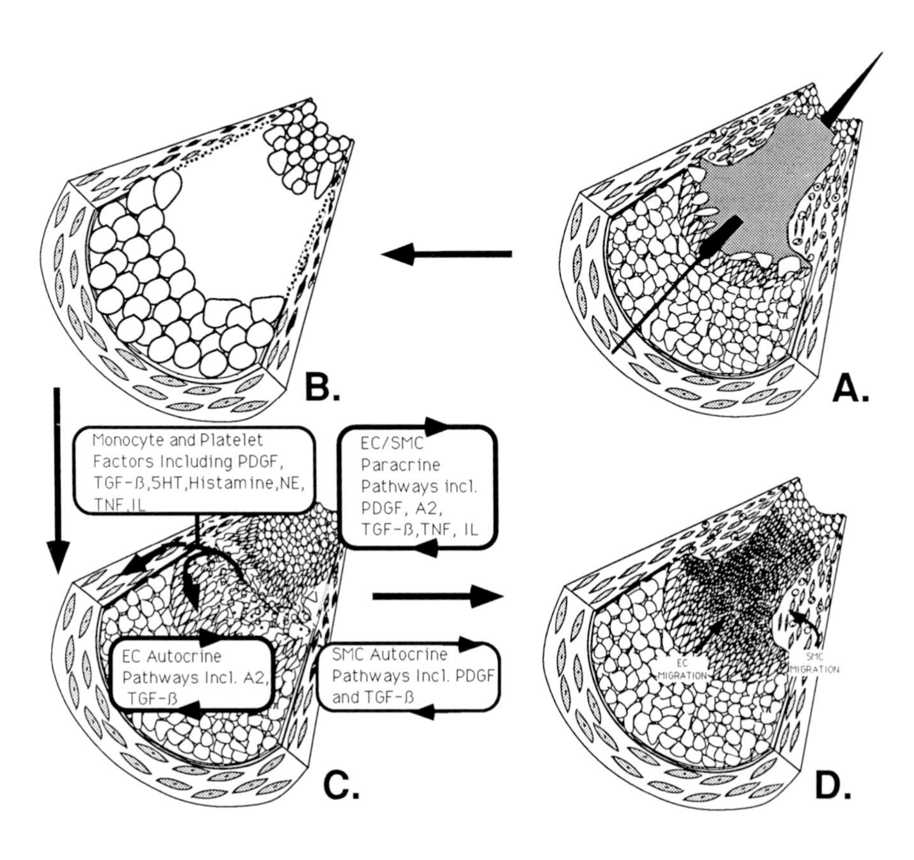

FIGURE 1. Schematic representation of the autocrine and paracrine regulation of vascular cell responses to iatrogenic injuries such as angioplasty, endarterectomy, or autologous or vein grafting. (A) A typical arterial segment containing an atherosclerotic lesion which exhibits luminal narrowing due to neointimal thickening is subjected to angioplasty. (B) This creates a denuded segment of artery at the angioplasty site, which also exhibits marked medial injury and subsequent necrosis. (C) This iatrogenic injury (or the placement of a synthetic graft or a denuded saphenous vein segment) elicits platelet adhesion, aggregation, and release, as well as monocytic cell adhesion, at the site of the injury. The released soluble factors elicit local smooth muscle cell migration into the injured area, proliferation, and changes in extracellular matrix metabolism. These factors also inhibit endothelial cell migration from the margins of the injury and alter endothelial cell matrix metabolism, prolonging the platelet/monocytic response. In addition, smooth muscle and endothelial cell autocrine and paracrine pathways become activated, further enhancing smooth muscle migration, proliferation, and matrix production, while inhibiting endothelial cell migration and altering endothelial cell matrix production, favoring continued monocytic adhesion. (D) The repair of the iatrogenic injuries caused by angioplasty and grafting procedures can be thought of as a process involving the competitive responses of medial smooth muscle cells and the overlying endothelial cells. Stimuli that inhibit or retard endothelial migration and enhance smooth muscle cell migration at the injury site will lead to the development of a stenotic lesion at the injury site. Thus, iatrogenic injury produced during invasive therapy of atherosclerotic vascular disease can itself initiate and maintain smooth

part, by affecting organization and dynamics of the cytoskeleton.[7,10-14] Since endothelial cells have been found to be capable of interacting with a wide variety of matrix components, the findings that they express several integrins ($\alpha1\beta1$, $\alpha2\beta1$, $\alpha3\beta1$, $\alpha5\beta1$, and $\alpha v\beta3$) on their surfaces is expected.[15-17] We have demonstrated that integrin matrix binding proteins participate in adhesion and migration of bovine aortic endothelial cells *in vitro*.[17,18] Attachment assays on fibrinogen, laminin, and fibronectin substrates using arginine-glycine-aspartate (RGD)-containing peptides as inhibitors illustrated dose-dependent RGD sensitivity and specificity, as arginine-glycine-glutamate (RGE)-containing peptides were noted to have no effect on endothelial cell attachment. Immunofluorescence studies during the attachment and spreading process revealed rapid matrix protein-specific organization of beta 1 and beta 3 integrins within 1 h of plating. Beta 1 integrin dimers were observed as linear stress fiber-type arrays on fibronectin and laminin substrates, while $\beta3$ integrins were noted to organize into punctate patterns on fibrinogen substrates. A nonintegrin laminin binding protein (LB69) was observed not to organize on any of these matrix components at early time points.[17,18] However, at later times LB69 was noted to organize following plating, suggesting a possible spatiotemporal segregation of integrin and nonintegrin binding proteins during cell attachment and spreading. Recent studies have demonstrated that while the underlying matrix elicits specific patterns of integrin organization in bovine aortic endothelial cells, matrix components do not elicit changes in the sizes of cell surface pools of $\beta1$ or $\beta3$ integrins.[19] In contrast, selected soluble factors found in platelet releasates and produced by endothelial cells themselves, such as TGF-$\beta1$, which causes a dramatic decrease in migration of bovine aortic endothelial cells,[9,20] were noted to increase the sizes of the cell-surface pools of $\beta1$ (specifically a5$\beta1$) and $\beta3$ integrins, but appeared not to alter integrin organization *in vitro*.[19]

B. Effects of Soluble Factors on Endothelial Cell Behavior

In more recent studies, we have further explored the mechanisms of action of TGF-$\beta1$, which, in addition to being a major component of platelet releasates, is also synthesized and secreted by endothelial and smooth muscle cells.[21,22] TGF-$\beta1$-treated large vessel endothelial cells have been found to have increased fibronectin mRNA levels and exhibit increased synthesis and deposition of fibronectin.[20] In addition, using differential library screening techniques, Southern and Northern blots and polymerase chain reaction analyses comparing cultured aortic endothelial cells grown in the absence and presence of TGF-$\beta1$,

---

muscle and endothelial cell behavior that will elicit the reformation of another stenotic lesion at the injury site. (EC = endothelial cells; SMC = smooth muscle cells; PDGF = platelet-derived growth factor; TGF-$\beta1$ = transforming growth factor beta 1; 5HT = 5 hydroxytryptamine; NE = norepinephrine; TNF = tumor necrosis factor; IL = interleukins; A2 = angiotension II).

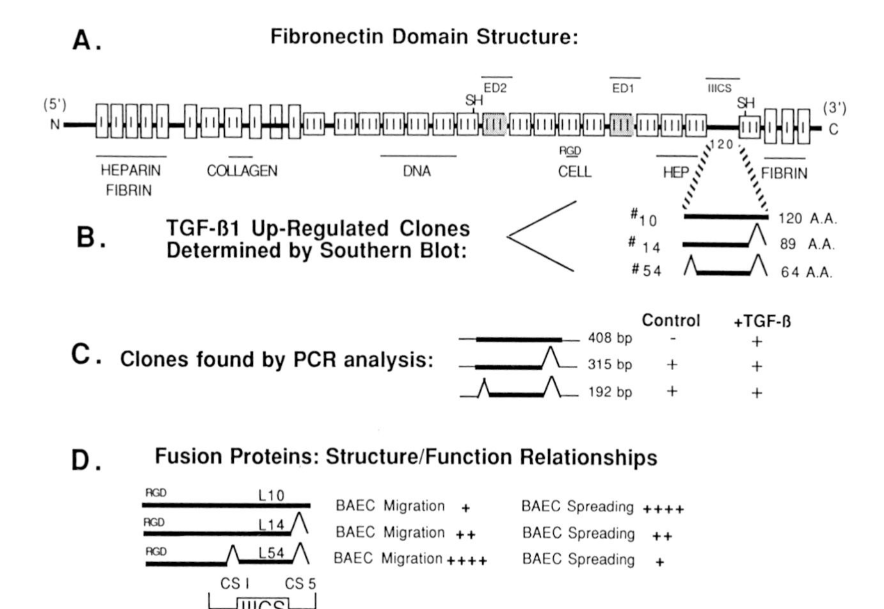

FIGURE 2. Fibronectin domain structure/function relationships. (A) Schematic representation of the fibronectin molecule modeled after Kornblitt et al., illustrating the type I (I), type II (II), and type III (III) repeats. Heparin, fibrin, collagen, DNA, and cell binding domains are noted by underlining, as are the alternatively spliced type III domains, IIICS, and ED1 and ED2, which are also represented as shadowed boxes. [N = amino terminal; (5′) = 5′region; C = carboxy terminal; (3′) = 3′ region.] (B) Schematic representation of the alternatively spliced IIICS domains that are modulated by TGF-β1 treatment of cultured bovine aortic endothelial cells as determined by Southern blot analyses of clones isolated from differential library screening of control and TGF-β1-treated cultures. #10 — = complete, unspliced IIICS domain; #14 —∧ = IIICS domain with the CS V region spliced out; #54 ∧—∧ = IIICS domain with both the CS I and the CS V regions spliced out. Clone #10 was maximally upregulated with TGF-β1 treatment, while clones #14 and #54 were only modestly upregulated by TGF-β1 treatment. (C) Schematic representation of the TGF-β1 modulation of alternatively spliced IIICS isoforms of fibronectin, as determined by polymerase chain-reaction analyses using oligonucleotides flanking the IIICS domain. The 408-bp fragment coding for the complete IIICS domain was found only in the TGF-β1-treated samples, while the 315-bp CS-V-spliced and the 192 bp CS-V- and CS-I-spliced fragments were noted in control and TGF-β1-treated samples. (D) Schematic representation of the structure/function relationships among fibronectin fusion proteins containing the RGD domain and the complete IIICS domain (L10); the RGD domain and the CS-V-spliced isoform of the IIICS domain (L14); and the RGD domain and the CS-V- and CS-I-spliced isoform of the IIICS domain (L54) and aortic endothelial cell migration and cell spreading. (+ = lowest migration or cell spreading; ++++ = highest migration or cell spreading).

we noted differences in the mRNA levels for this alternatively spliced molecule.[23] Namely, changes in the relative amounts of selected mRNA species coding for isoforms spliced in the IIICS region of fibronectin were documented (Figure 2). In control cultures the isoforms of fibronectin produced contained both ED domains and either CS1, or CS1 and CS5, spliced domains of the IIICS

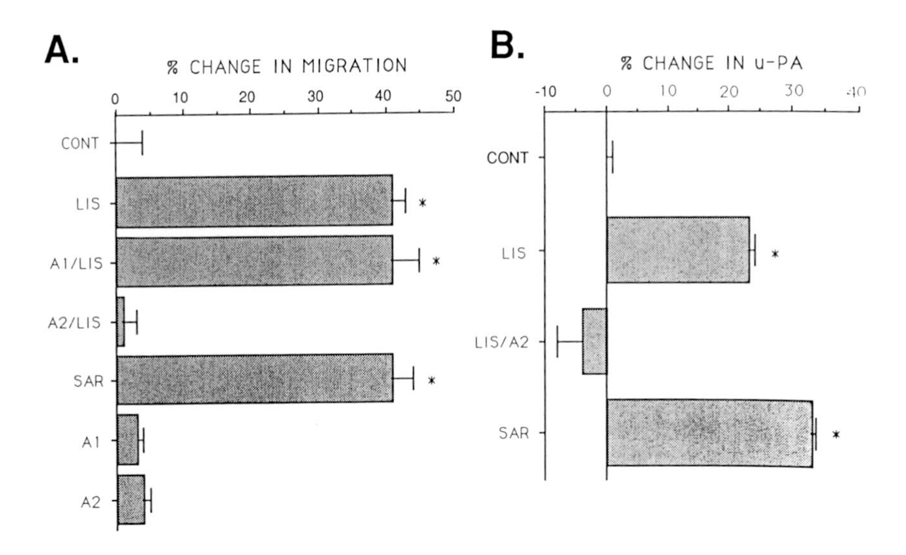

FIGURE 3. Endothelial cell migratory and protease responses to angiotensin II, angiotensin II antagonists, and angiotensin-converting enzyme inhibitors. (A) Changes in bovine aortic endothelial cell migration in the presence of lisinopril (LIS); angiotensin I and lisinopril (AI/LIS); angiotensin II and lisinopril (AII/LIS); sar^1, ile^8-angiotensin II (SAR); angiotensin I (AI); angiotensin II (AII) compared to control (CONT) conditions. (* = $p < 0.001$ vs. control.) (B) Changes in bovine aortic endothelial cell plasminogen activator (u-PA) activity in the presence of lisinopril (LIS); angiotensin II and lisinopril (AII/LIS); and sar^1, ile^8-angiotensin II (SAR) compared to control (CONT) conditions (19 ± 3 mPU/μgm). (* = $p < 0.001$ vs. control.) (Graphs based on data taken from Reference 24).

domain. In contrast, in the presence of TGF-β1, cultures produced increased amounts of fibronectin mRNAs containing both ED domains and the complete unspliced IIICS domain, with lesser amounts of the CS1, or CS1 and CS5, spliced regions of the IIICS domain (Figure 2).[23] Furthermore, we have demonstrated that the proteins coded for by these mRNA species modulate aortic endothelial cell behavior. Specifically, endothelial migration is decreased in the presence of fibronectin molecules containing the CS5 and CS1 regions of the IIICS domain, which is consistent with the observed biological effects of TGF-β1 on aortic endothelial cells, which are decreased migration rates, increased cell spreading, and increased fibronectin mRNA and protein deposition (Figure 2).[20,23]

Another autocrine/paracrine effector system involved in modulation of endothelial cell behavior following injury that we are actively investigating is the angiotensin system. Recently, we have observed that endogenous angiotensin II (AII) functions to downregulate aortic endothelial cell migration *in vitro* (Figure 3).[24] Relief of endogenous AII stimulation, either by inhibition of conversion of endogenous AI to AII by endothelial cell surface angiotensin-converting enzyme (ACE) or direct antagonism at the endothelial cell AII receptor, elicits dramatic increases in migration rates. This is associated with increases in mRNA

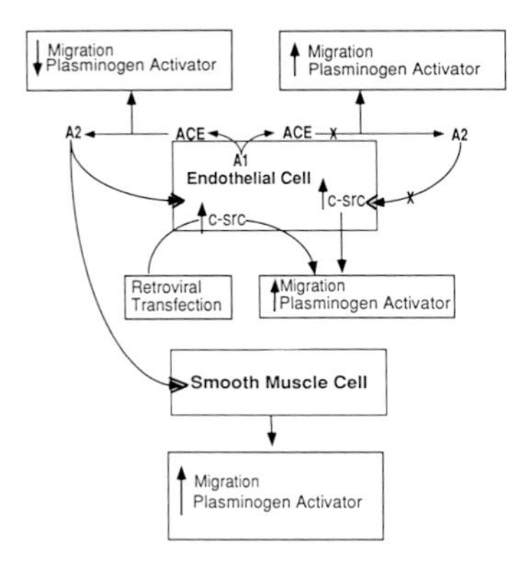

FIGURE 4. Schematic diagram depicting the relationship between the angiotensin system, c-*src* levels, protease levels, and vascular cell migration. Endogenous AII produced by endothelial cells reduces the endothelial cell migration rate and enhances the smooth muscle migration rate, in part, via modulation of plasminogen activator levels. Reduction of endogenous AII levels by inhibition of endothelial cell surface angiotensin-converting enzyme or the use of AII antagonists elicits enhancement of endothelial cell migration while reducing smooth muscle cell migration, again, in part, via modulation of plasminogen activator levels. Relief of endogenous AII stimulation is also associated with increases in c-*src* mRNA, protein, and kinase activity levels in endothelial cells. Retroviral transfection of endothelial cells with vectors containing the c-*src* gene results in stable and increased expression of c-*src* mRNA, protein, and kinase activity and elicits dramatic increases in the endothelial cell migration rate and plasminogen activator activity. Thus, modulation of the angiotensin system in the vascular wall, using conventional drug therapy and/or potentially retroviral gene delivery systems,[25,35] may have significant effects upon the outcome of postangioplasty and saphenous vein graft procedures.

to the protooncogene c-*src*, and in protein and c-*src* kinase activity, decreases in fibronectin deposition, and increases in urokinase plasminogen activator (u-PA) activity, suggesting a possible causal relationship between c-*src* levels and migratory rates.[24,25] Additionally, we have found that stable and elevated expression of c-*src* in bovine aortic endothelial cells accomplished by transfection with retroviral vectors results in increased c-*src* mRNA, protein, and kinase levels, as well as increased u-PA levels and activity, decreased fibronectin deposition, and dramatically increased migration rates (Figure 4).[25] However, bovine aortic endothelial cell transfectants containing a c-*src* kinase-negative mutant mRNA do not exhibit increased migration rates or u-PA activity, and inhibition of u-PA with an anticatalytic antibody directed against bovine u-PA abolishes the c-*src*-induced increases in migration.[25] These data support the concept that there is a causal relationship between c-*src* levels, u-PA activity, and migration rates in aortic endothelial cells.

C. Protease-Protease Inhibitor Systems Affecting Endothelial Cell Migration

Lastly, studies investigating possible mechanisms by which extracellular matrix components and TGF-β1 mediate changes in migratory behavior of bovine aortic endothelial cells have shown that both soluble factors and matrix composition modulate levels of plasminogen activator and plasminogen activator inhibitor-1 in these cells, and inhibitors of serine proteases decrease their migration rates. These data are consistent with the hypothesis that changes in protease-protease inhibitor systems may also play a role in modulating migratory behavior. Specifically, plasminogen activator activity levels were found to be decreased, while plasminogen activator inhibitor-1 levels were increased, by treatment with TGF-β1[26] (L. Bell and J.A. Madri, in preparation).

IV. SMOOTH MUSCLE CELLS — *IN VITRO* STUDIES

A. Effects of Soluble Factors on Smooth Muscle Cell Behavior

While endothelial loss following denudation, and its subsequent reconstitution following injury, is an important factor in the maintenance of luminal patency, the responses of local vascular smooth muscle cells is equally important.[27-29] Stimulated by platelet releasate (TGF-β1, PDGF, serotonin, norepinephrine, heparin), endothelial-derived factors (angiotensin II, TGF-β1, PDGF), as well as autocrine and paracrine factors derived from the smooth muscle cells themselves, smooth muscle cells in proximity to the lesion respond by migrating into the injured area (forming the neointima), synthesizing and depositing extracellular matrix components, and proliferating. This response is thought to lead to intimal thickening and ultimately restenosis.[27-31] In previous studies we have found that many of the same soluble factors that inhibit endothelial cell migration enhance vascular smooth muscle migration in a process that is independent of proliferation (Figure 5).[4,14,19,24] Changes in extracellular matrix component synthesis and deposition, and integrin expression, have been documented following stimulation with soluble factors such as TGF-β1, and may contribute to the changes in migratory rate.[19] Specifically, our studies have revealed that in addition to increasing migratory rate of bovine aortic smooth muscle cells, TGF-β1 treatment increases fibronectin mRNA and protein deposition in both confluent and migrating cultures of the cells. In addition we have noted that TGF-β1 and PDGF treatment cause an increase in the surface expression of β3 integrins, supporting the concept that selected soluble autocrine and paracrine factors may be eliciting their effects on smooth muscle cell migration, in part, via the selective modulation of matrix components and cell-surface matrix receptors, including the integrins.[19,31]

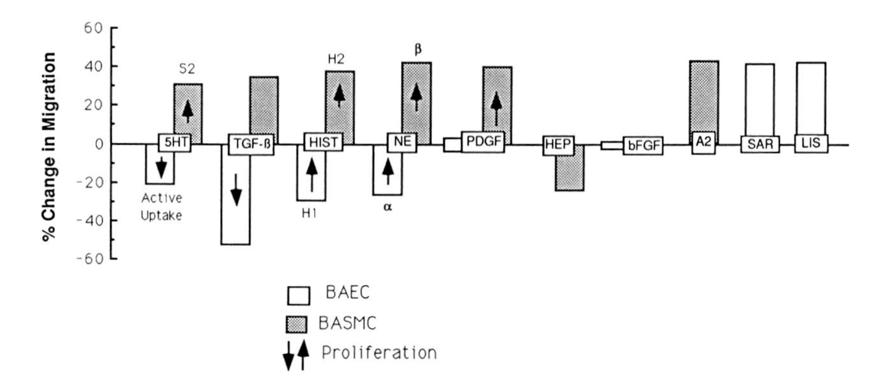

FIGURE 5. Smooth muscle and endothelial cell responses to soluble autocrine and paracrine factors produced following denudation injury. Relative changes in migration rates of bovine-specific endothelial cells (BAEC) and bovine aortic smooth muscle cells (BASMC) cultured on type I collagen and treated with serotonin (5HT), transforming growth factor β1 (TGF-β), histamine (HIST), norepinephrine (NE), platelet-derived growth factor (PDGF), heparin (HEP), basic fibroblast growth factor (bFGF), or endogenous BAEC or exogenous BASMC angiotensin II (A2). All the tested factors, except heparin, PDGF, and basic fibroblast growth factor, decreased BAEC migration rates and increased BASMC migration rates. Basic fibroblast growth factor had no observable effect on BAEC migration, and heparin was observed to decrease BASMC migration. The activities of these factors was dose dependent and specifically blocked by antagonists or blockers. In BAEC cultures diphenhydramine (H1 receptor specific) blocked the HIST effect, imipramine (an uptake inhibitor) blocked the 5HT effect, phenoxybenzamine (an α blocker) blocked the NE effect; saralasin (an A2 antagonist) and lisinopril (an inhibitor of angiotensin-converting enzyme) increased BAEC migration. In BASMC cultures cimetidine (H2 receptor specific) blocked the HIST effect, ketanserin (S2 receptor specific) blocked the 5HT effect, propanolol (a β blocker) blocked the NE effect, saralasin (an A2 antagonist) blocked the A2 effect. All factors manifested their effects on cell migration independently of their effects on cell proliferation. (Reproduced with permission from Reference 4).

We are also actively investigating the role(s) of the angiotensin system in the modulation of smooth muscle cell behavior following injury. We have observed that angiotensin II functions to enhance aortic smooth muscle cell migration and u-PA activity *in vitro* (Figure 6).[24] The use of AII antagonists inhibits the AII effect, while inhibitors of angiotensin-converting enzyme do not affect smooth muscle cell migration rates. This suggests that smooth muscle cells have AII surface receptors, but, unlike endothelial cells, are not capable of producing endogenous AII.[24]

B. Protease-Protease Inhibitor Systems Affecting Smooth Muscle Cell Migration

Lastly, as noted for aortic endothelial cells, both matrix composition and soluble factors modulate plasminogen activator levels in aortic smooth muscle cell cultures. However, in contrast to the endothelial cells, plasminogen activator levels in the smooth muscle cells are increased in response to TGF-β1 (L. Bell and J.A. Madri, in preparation).

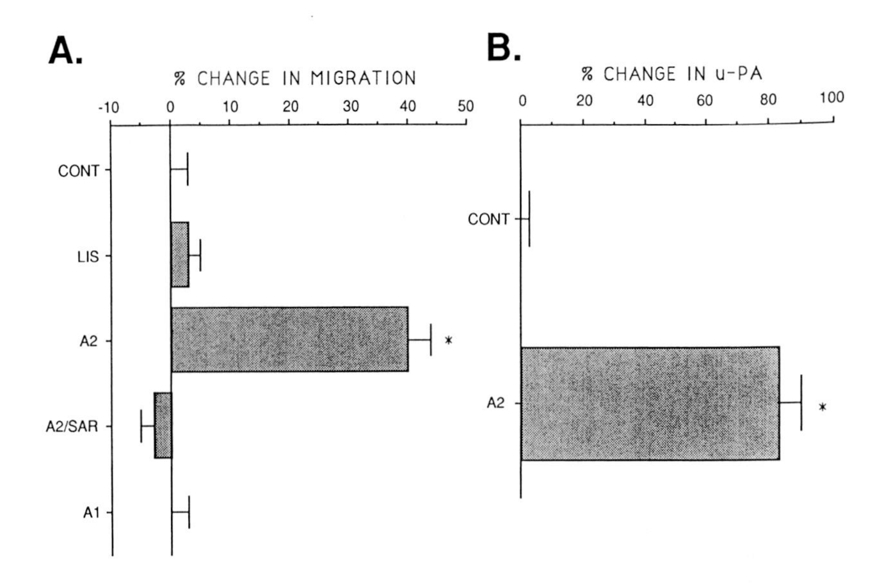

FIGURE 6. Smooth muscle cell migratory responses to angiotensin II, angiotensin II antagonists, and angiotensin-converting enzyme inhibitors. (A) Changes in bovine aortic medial smooth cell migration in the presence of lisinopril (LIS); angiotensin II (AII); angiotensin I and lisinopril (AI/LIS); angiotensin II and sar^1, ile^8-angiotensin II (AII/SAR); angiotensin I (AI); compared to control (CONT) conditions. (* = $p < 0.001$ vs. control.) (B) Changes in bovine aortic medial smooth cell plasminogen activator (u-PA) activity in the presence of angiotensin II (AII) compared to control (CONT) conditions (9 ± 1 mPU/μgm). (* = $p < 0.001$ vs. control.) (Graphs based on data taken from Reference 24).

These *in vitro* studies suggest that selected matrix components and soluble factors elicit different effects in aortic endothelial and smooth muscle cells, inhibiting migration in one case (endothelial cells), while enhancing migration in the other (smooth muscle cells), which may contribute to the development of restenosis following angioplasty or saphenous vein bypass.

V. *IN VIVO* STUDIES

In addition to the above-mentioned *in vitro* studies, we have also investigated vascular cell-extracellular-matrix and cell-cell interactions in a rat carotid balloon deendothelialization model.[19,20] In this model there is incomplete reendothelialization of the initially deendothelialized area, and increased fibronectin and TGF-β1 staining throughout the media and luminal surface of the chronically deendothelialized region of the injured vessel. Our findings using this *in vivo* model correlate well with our *in vitro* studies in which endothelial cell migration on a type I collagen coating was inhibited, and smooth muscle cell migration was enhanced by (1) addition of soluble fibronectin to the culture media, (2) migration on insoluble fibronectin, and (3) TGF-β1 treatment of migrating cells in which there were significant increases in fibronectin mRNA and protein levels.[31]

Recent studies using this *in vivo* model also revealed distinct differences in the localization of β1 and β3 integrins in the neointimal regions and luminal lining cells of both the chronically deendothelialized and reendothelialized areas.[19] Specifically, in the chronically deendothelialized areas the neointimal smooth muscle cells exhibited increased staining intensity for β3 integrins, with no changes in β1 integrins compared with uninjured and postinjury medial smooth muscle cells. These findings are consistent with the effects of TGF-β1 and PDGF on aortic smooth muscle cells noted *in vitro*. In contrast, in the reendothelialized areas, the neointimal smooth muscle cells nearer the lumen displayed much less intense staining for β3 integrins (and for fibrinogen) than those deeper in the neointima, supporting the well-accepted concept of endothelial cell modulation of smooth muscle phenotype.[19]

The role of the angiotensin system in modulating vascular cell behavior *in vivo* has also been investigated. Consistent with predictions based upon our *in vitro* observations in which interruption of the angiotensin system would be expected to improve the vessel wall response to injury via increased endothelial cell migration and increased smooth muscle cell migration, these studies have suggested that interruption of the angiotensin system improves the vessel wall response to injury *in vivo*. Treatment with inhibitors of angiotensin-converting enzyme dramatically reduced the neointima formation following balloon denudation injury in a rat angioplasty model,[32] specifically reduced medial thickening in a spontaneously hypertensive rat model,[33] and reduced neointima formation in a hyperlipidemic rabbit model.[34]

Thus, both *in vivo* and *in vitro*, the presence of soluble factors as well as the composition of existing and newly synthesized extracellular matrix appear to have profound effects upon the migratory behavior of large vessel endothelial and medial smooth muscle cells.

VI. CONCLUSIONS

These data, taken together, support the concepts that modulation of vascular cell migratory behavior is complex and involves several mechanisms, including complex autocrine and paracrine pathways that modulate protease-protease inhibitor systems, changes in selected matrix component synthesis and organization, changes in expression and organization of cell surface matrix binding proteins (integrins and nonintegrins), and cytoskeletal reorganization and altered expression of specific protooncogenes. Further, dissection of these complex interactions may allow the employment of specific therapies designed to regulate vascular cell behavior and thereby improve the vessel wall response to the injuries associated with coronary revascularization.

ACKNOWLEDGMENTS

We would like to thank all the students and fellows who have been associated with our laboratories over the past several years, for their interest, enthusiasm, and contributions. This research was supported in part by USPHS grants RO1-HL-28373 (JAM) and Physician Scientist Award K11-HL-02351.

REFERENCES

1. **Feinleib, M., Havlik, R., Gillum, R., Pokras, R., McCarthy, E., and Moien, M.,** Coronary heart disease and related procedures. National hospital discharge survey data, *Circulation*, 79, I13, 1989.
2. **Beatt, K.J., Erruys, P., and Hugenholtz, P.J.,** Restenosis after coronary angioplasty: new standards for clinical studies, *Am. Coll. Cardiol.*, 15, 491, 1990.
3. **Bourassa, M., Fisher, L., Campeau, L., Gillespie, M., McConney, M., and Lesperance, N.,** Long-term fate of bypass grafts: the coronary artery surgery study (CASS) and Montreal Heart Institute experiences, *Circulation*, 72, V71, 1985.
4. **Madri, J. A., Bell, L., Marx, M., Merwin, J.R., Basson, C.T., and Prinz, C.,** The effects of soluble factors and extracellular matrix components on vascular cell behavior *in vitro* and *in vivo*: models of de-endothelialization and repair, *J. Cell. Biochem.*, 45, 1, 1991.
5. **Madri, J.A., Merwin, J.R., Bell, L., Basson, C.T., Kocher, O., Perlmutter, R., and Prinz, C.,** Interactions of matrix components and soluble factors in vascular cell responses to injury: modulation of cell phenotype, in *Endothelial Cell Dysfunction*, Simionescu, N. and Simionescu, M., Eds., Plenum Press, New York, in press, 1991.
6. **Madri J.A.,** The extracellular matrix as a modulator of neovascularization, in *Cardiovascular Disease: Molecular and Cellular Mechanisms, Prevention, Treatment*, Gallo, M., Ed., Plenum Press, New York, 1987, 177.
7. **Madri, J.A., Pratt, B.M., and Yanniarello-Brown, J.,** Matrix-driven cell size changes modulate aortic endothelial cell proliferation and sheet migration, *Am. J. Pathol.*, 132, 18, 1988.
8. **Ross, R.,** Endothelial injury and atherosclerosis, in *Endothelial Cell Biology in Health and Disease*, Simionescu, N. and Simionescu, M., Eds., Plenum Press, New York, 1988, 371–384.
9. **Bell, L. and Madri, J.A.,** Effect of platelet factors on migration of cultured bovine aortic endothelial and smooth muscle cells, *Circ. Res.*, 65, 1057, 1989.
10. **Pratt, B.M., Harris, A.S., Morrow, J.S., and Madri, J.A.,** Mechanisms of cytoskeletal regulation: modulation of aortic endothelial cell spectrin by the extracellular matrix, *Am. J. Pathol.*, 117, 337, 1984.
11. **Leto, T.L., Pratt, B.M., and Madri, J.A.,** Mechanisms of cytoskeletal regulation: modulation of aortic endothelial cell protein band 4.1 by the extracellular matrix, *J. Cell Physiol.*, 127, 423, 1986.
12. **Madri, J.A., Pratt, B.M., and Yannariello-Brown, J.,** Endothelial cell-extracellular matrix interactions: matrix as a modulator of cell function, in *Endothelial Cell Biology in Health and Disease*, Simionescu, N. and Simionescu, M., Eds., Plenum Press, New York, 1988, 167.

13. **Pratt, B.M., Form, D., and Madri, J.A.,** Endothelial cell-extracellular matrix interactions, in *Biology, Chemistry and Pathology of Collagen*, Fleishmajer, R., Olsen, B. and Kuhn, K., Eds., *Ann. N.Y. Acad. Sci.*, 460, 274, 1985.
14. **Madri, J.A., Kocher, O., Merwin, J.R., Bell, L., and Yannariello-Brown, J.,** The interactions of vascular cells with solid phase (matrix) and soluble factors, *J. Cardiovasc. Pharmacol.*, 14, S70, 1989.
15. **Albelda, S.M. and Buck, C.A.,** Integrins and other cell adhesion molecules, *FASEB J.*, 4, 2868, 1991.
16. **Fitzgerald, L.A., Charo, I.F., and Phillips, D.R.,** Human and bovine endothelial cells synthesize membrane proteins similar to human platelet glycoproteins IIb and IIIa, *J. Biol. Chem.*, 260, 10,893, 1985.
17. **Basson, C.T., Knowles, W.J., Bell, L., Albelda, S.M., Castronovo, V., Liotta, L., and Madri, J.A.,** Spatiotemporal segregation of endothelial cell integrin and non-integrin extracellular matrix binding proteins during adhesion events, *J. Cell Biol.*, 110, 789, 1990.
18. **Yannariello-Brown, J., Wewer, U., Liotta, L., and Madri, J.A.,** Distribution of a 69kD laminin binding protein in aortic and microvascular endothelial cells: modulation during cell attachment, spreading and migration, *J. Cell Biol.*, 106, 1773, 1988.
19. **Basson, C.T., Kocher, O., Basson, M.D., Asis, A., and Madri, J.A.,** Differential modulation of vascular cell integrin and extracellular matrix expression *in vitro* by TGF-β1 correlates with reciprocal effects on cell migration, *J. Cell. Physiol.*, in press, 1992.
20. **Madri, J.A., Reidy, M.A., Kocher, O., and Bell, L.,** Endothelial cell behavior following denudation injury is modulated by TGF-β1 and fibronectin, *Lab. Invest.*, 60, 755, 1989.
21. **Antonelli-Orlidge, A., Saunders, K.B., Smith, S.R., and D'Amore, P.A.,** An activated form of transforming growth factor beta is produced by cocultures of endothelial cells and pericytes, *Proc. Natl. Acad. Sci. U.S.A.*, 86, 4544, 1989.
22. **Roberts, A.B. and Sporn, M.B.,** The transforming growth factor-betas. Peptide growth factors and their receptors, in *Handbook of Experimental Pharmacology*, Vol. 95, Sporn, M.B. and Robers, A.B., Eds., Springer-Verlag, Heidelberg, 1990, 419.
23. **Kocher, O., Kennedy, S., and Madri, J.A.,** Alternative splicing of fibronectin mRNA in the IIICS region in endothelial cells: functional significance, *Am. J. Pathol.*, 137, 1509, 1990.
24. **Bell, L. and Madri, J.A.,** Influence of the angiotension system on endothelial and smooth muscle cell migration *in vitro*, *Am. J. Pathol.*, 137, 7, 1990.
25. **Bell, L., Luthringer, D.J., Madri, J.A., and Warren, S.L.,** Autocrine angiotensin system regulation of endothelial cell behavior involves modulation of $pp60^{c\text{-}src}$ expression, *J. Clin. Invest.*, 89: 315–320, 1991.
26. **Saksela, O., Moscatelli, D., and Rifkin, D.B.,** The opposing effects of basic fibroblast growth factor and transforming growth factor beta on the regulation of plasminogen activator activity in capillary endothelial cells, *J. Cell Biol.*, 105, 957, 1987.
27. **Fishman, A.P., Ed.,** *Endothelium*, The New York Academy of Sciences, New York, 1982.
28. **Munro, J.M. and Cotran, R.S.,** The pathogenesis of atherosclerosis: atherogenesis and inflammation, *Lab. Invest.*, 58, 249, 1988.
29. **Ross, R.,** Medical progress: the pathogenesis of atherosclerosis — an update, *N. Engl. J. Med.*, 314, 488, 1986.
30. **Hedin, U., Bottger, B.A., Forsberg, E., Johansson, S., and Tyberg, J.,** Diverse effects of fibronectin and laminin on phenotypic properties of cultured arterial smooth muscle cells, *J. Cell Biol.*, 107, 307, 1988.
31. **Madri, J.A., Kocher, O., Merwin, J.R., Basson, C.T., and Bell, L.,** The interactions of vascular cells with transforming growth factor β, in *Transforming Growth Factor-βs: Chemistry, Biology and Therapeutics*, Piez, K.A. and Sporn, M.B., Eds., The New York Academy of Sciences, 593, 243, 1990.

32. **Powell, J.S., Clozel, J.-P., Muller, R.K.M., Kuhn, H., Hefti, F., Hosang, M., and Baumgartner, H.R.,** Inhibitors of angiotensin-converting enzyme prevent myointimal proliferation after vascular injury, *Science*, 245, 186, 1989.
33. **Owens, G.K.,** Influence of blood pressure on development of aortic medial smooth muscle hypertrophy in spontaneously hypertensive rats, *Hypertension*, 9, 178, 1987.
34. **Chobanian, A.V., Haudenschild, C.C., Nickerson, C., and Drago, R.,** Antiatherogenic effect of captopril in the Watanabe heritable hyperlipidemic rabbit, *Hypertension*, 15, 327, 1990.
35. **Verma, I.M.,** Gene therapy, in *Sci. Am.*, 263, 68, 1990.

Chapter 7

CYTOCHEMICAL CHARACTERIZATION AND MAPPING OF GLYCOCONJUGATES IN ARTERIAL TISSUES AND VASCULAR CELL CULTURES BY LIGHT AND ELECTRON MICROSCOPY

Wolfgang Völker

TABLE OF CONTENTS

ISBN 0-8493-5505-2

I. INTRODUCTION

The pathogenic mechanisms that foster the development of an atherosclerotic lesion involve complex interactions between vascular wall cells, extracellular matrix, and various components of the blood.[1] In order to obtain more insight into the process of atherosclerosis, it is instructive to explore cell-matrix interactions and the organization and changes of the extracellular matrix of the vascular wall, by applying appropriate cytochemical methods for light and electron microscopy. This review focuses on recent data obtained by cuprolinic blue staining, colloidal gold labeling, and silver enhancement techniques, applied to the study of proteoglycans and thrombospondin. Apart from collagen, elastin, and fibronectin, these molecules are major components of normal and atherosclerotic vascular tissues, but their precise role in connective tissue organization, regulation of cell activities, and atherosclerosis is less well understood.

Proteoglycans consist of a filamentous core protein to which sulfated glycosaminoglycan side chains are covalently attached. The different proteoglycans are distinguished biochemically by their contents of the glycosaminoglycans, such as chondroitin sulfate (CS), dermatan sulfate (DS), and heparan sulfate (HS).[2] Great interest has been focused on proteoglycans in the vascular wall since Hollander[3] presented a concept on the interaction of lipoproteins with acid mucopolysaccharides (i.e., proteoglycans). Proteoglycans are thought to play an important, yet still poorly understood, role in the formation of fatty streaks and fibrous plaques during early stages of atherosclerosis. The idea that proteoglycans have such a role has been developed and extended by recent investigations from Berenson et al.[4] Proteoglycans may form complexes with altered low-density lipoprotein, which are taken up by macrophages by a high-affinity process devoid of feedback control,[5] thereby converting macrophages to foam cells that remain in the artery wall. Furthermore, it is known that CS-proteoglycans are secreted to a far greater extent by activated smooth muscle cells than by quiescent cells.[6] In human atherosclerotic arteries, CS-proteoglycans are accumulated as a pericellular layer around activated smooth muscle cells, as demonstrated by electron microscopy after positive staining of proteoglycans with cuprolinic blue *in situ.*[7] HS-proteoglycan, by contrast, acts as an endogenous inhibitor of cell proliferation and is markedly reduced in atherosclerotic arteries.[8] As HS is essential for maintaining smooth muscle cells quiescent in the G_0 phase, its loss might contribute to the stimulation of smooth muscle cell proliferation.[9]

Thrombospondin, a glycoprotein originally described as a component of α-granules in platelets, is also secreted by vascular endothelial cells.[10-12] In atherosclerotic tissues, in experimentally injured arteries, and in cell cultures, thrombospondin is enriched and deposited around smooth muscle cells.[13-15] Thrombospondin strongly and specifically binds HS-proteoglycan in the pericellular matrix and on the plasma membrane and is functionally essential for smooth muscle cell proliferation.[16] These few data alone indicate that both proteoglycans and thrombospondin are involved in cell regulation activities such as migration and proliferation, both of which represent central events in the early stages of atherosclerosis.

Further insight into the roles of proteoglycans and thrombospondin in normal and atherosclerotic processes is gained by cytochemical analysis of mammalian arterial tissues and cultured vascular cells. Comparison of normal and atherosclerotic material obtained from human autopsies and from experimentally injured arteries of animals allows study of pathological alterations in the organization of extracellular matrix. Cell cultures provide an excellent system for investigating binding and degradation of matrix components. Both these approaches are used in this review.

II. DETECTION OF PROTEOGLYCANS IN ARTERIAL TISSUES WITH CUPROLINIC BLUE

A. Selective Staining of Proteoglycans with Cuprolinic Blue

Much of the recent progress in cytochemistry is attributable to advances in the use of selective dyes in combination with other new experimental methods. Cuprolinic blue staining has proven to be the method of choice for *in situ* detection of proteoglycans by electron microscopy of ultrathin-sectioned connective tissues.[17,18] This dye has major advantages over ruthenium red staining, permitting discrimination of the major types of proteoglycans *in situ* as well as allowing the detection of structural and topographical changes of these molecules in atherosclerotic tissues.[19-21]

Cuprolinic blue is a tetracationic phtalocyanin-like dye and so has a high affinity for anionic substances.[22] The most anionic molecules in the extracellular matrix are represented by the proteoglycan molecules.[23,24] As mentioned earlier, these molecules are characterized by various numbers of polysulfated glycosaminoglycan side chains along their filamentous protein axis.[2,25-27] Binding of cuprolinic blue mainly occurs at sulfate-containing polyanions (the glycosaminoglycan side chains of the proteoglycans) in the presence of a critical electrolyte concentration of 0.3 *M* $MgCl_2$ at pH 5.6.[28,29] Under these conditions polycarboxylates and polyester phosphates remain unstained. A secondary staining with Na_2WO_4 is required for electron microscopic visualization of the proteoglycan/cuprolinic blue precipitates. The electron-dense WO_4^{--} anions strongly interact with preadsorbed cuprolinic blue. Proteoglycan/cuprolinic blue complexes then appear as long-shaped precipitates in the extracellular matrix. Other structures of the extracellular matrix remain virtually unstained when osmication is omitted. Osmium tetroxide stains most tissue structures, with the exception of proteoglycans, which remain poorly stained. As a secondary stain following cuprolinic blue staining, however, osmium tetroxide is useful for bringing into view structures that are associated with proteoglycans. Another useful staining combination is to use cuprolinic blue followed by the sequence osmium tetroxide-thiocarbohydrazide-osmium tetroxide to preserve both proteoglycan/cuprolinic blue precipitates and lipid droplets in atherosclerotic plaques.[7,30-32] However, under these rigid conditions of fixation, the proteoglycan/cuprolinic blue precipitates are circularly shaped and appear similar to proteoglycans stained with ruthenium red.[19,20,32] In cell cultures, cuprolinic blue staining is easily combined with gold-labeling experiments and is extremely useful for demonstrating interactions between proteoglycans and other glycoconjugates, matrix components, and cell structures.[33,34]

B. Morphological Discrimination of Proteoglycan/Cuprolinic Blue Precipitates

Morphometric analysis in various mammalian tissues, such as rat, bovine, and human arteries, consistently reveals distinct size classes of proteoglycan/cuprolinic blue precipitates.[21,35,36] These different classes of proteoglycan/cuprolinic blue precipitate are preferentially localized in different areas of the extracellular matrix, specifically in the amorphous so-called soluble matrix, at the basement membrane-like structures, and at elastic and collagenous fibers (Figure 1a).

Even without the benefit of staining with osmium tetroxide, collagen fibrils are easily detected by their regular decoration with proteoglycan/cuprolinic blue precipitates. These precipitates are typically 40 nm in length (Figure 1b) and, in cuprolinic blue/osmium tetroxide preparations, are seen to connect adjacent collagen d-band regions (Figure 1c). A larger-sized type of proteoglycan/cuprolinic blue precipitate, about 90 to 120 nm in length, is concentrated in areas of soluble matrix (Figure 1a). An additional small type of proteoglycan/cuprolinic blue precipitate, detected after treatment of the tissue with specific glycosaminoglycan-degrading enzymes, is present along the basement membrane-like layers associated with smooth muscle cells and elastic fibers in the extracellular matrix (Figure 2a).

By electron microscopy of purified proteoglycan subfractions adsorbed onto nitrocellulose for cytochemical processing, cuprolinic blue staining shows that the various types of proteoglycan can be discriminated by precipitate size. In arterial tissue this discrimination is further facilitated by the preferential concentration of the rather homogeneous proteoglycan/cuprolinic blue subpopulations in specific compartments of the extracellular matrix (Figure 1a). The observations indicate that proteoglycan/cuprolinic blue precipitates, as seen in electron micrographs, represent individual molecules. The long- or egg-shaped configuration of the precipitates results from collapse of the glycosaminoglycan side chains along the filamentous core protein.

C. Discrimination and Characterization of Proteoglycans *in situ* by Use of Glycosaminoglycan-Degrading Enzymes

Enzymatic degradation of glycosaminoglycan side chains of proteoglycan molecules prevents interaction of these molecules with cuprolinic blue and thereby blocks the formation of proteoglycan/cuprolinic blue precipitates. The microbial enzymes chondroitin sulfate AC lyase, chondroitin sulfate ABC lyase, and heparitinase specifically degrade CS, CS together with DS, and HS, respectively. In combination with cuprolinic blue staining, these enzymes allow precise location and mapping of specific types of proteoglycans *in situ*.

For sufficient infiltration of enzymes and for preservation of the structural integrity of the tissue, glutaraldehyde-prefixed tissues are sliced with a vibracut microtome to produce nondisturbed slices of about 100 μm in thickness. Aldehyde fixatives must be extracted from the tissue slices by thorough washing before giving the enzyme. Tissue samples may be stored in aldehyde fixative for several months before use, as polysaccharides are resistant to aldehyde fixation. The effect of the enzymatic degradation of glycosaminoglycans is seen in electron micrographs as the loss of specific types of proteoglycan/cuprolinic blue precipitate. Chondroitin sulfate AC lyase treatment, for example, reduces the number of large-sized precipitates in areas of soluble matrix, while the small-sized precipitates along collagenous and elastic fibers remain barely altered in number. This indicates that the large-sized precipitates represent CS-rich proteoglycans. After treatment with chondroitin sulfate ABC lyase, which attacks both chondroitin sulfate and dermatan sulfate, the usual precipitates found in the soluble matrix and at the surface of collagenous fibers are no longer observed (Figure 2a). Overall, more than 90% of the precipitates are lost; only on the surface of elastic fibers and along basement-membrane-like layers are precipitates still detectable. This remaining population of proteoglycan is sensitive to heparitinase degradation, indicating that it is characterized by HS glycosaminoglycan side chains.

D. Morphological Changes of Proteoglycan/Cuprolinic Blue Precipitates in Atherosclerotic Tissues

Comparison of human atherosclerotic tissue with “normal” areas in the same individual reveals topographic and morphological changes in proteoglycan/cuprolinic blue precipitates.[7,21,36] In the periphery of medial and neointimal smooth muscle cells from atherosclerotic plaques, an accumulation of large-sized chondroitinase AC-sensitive precipitates is frequently observed (Figure

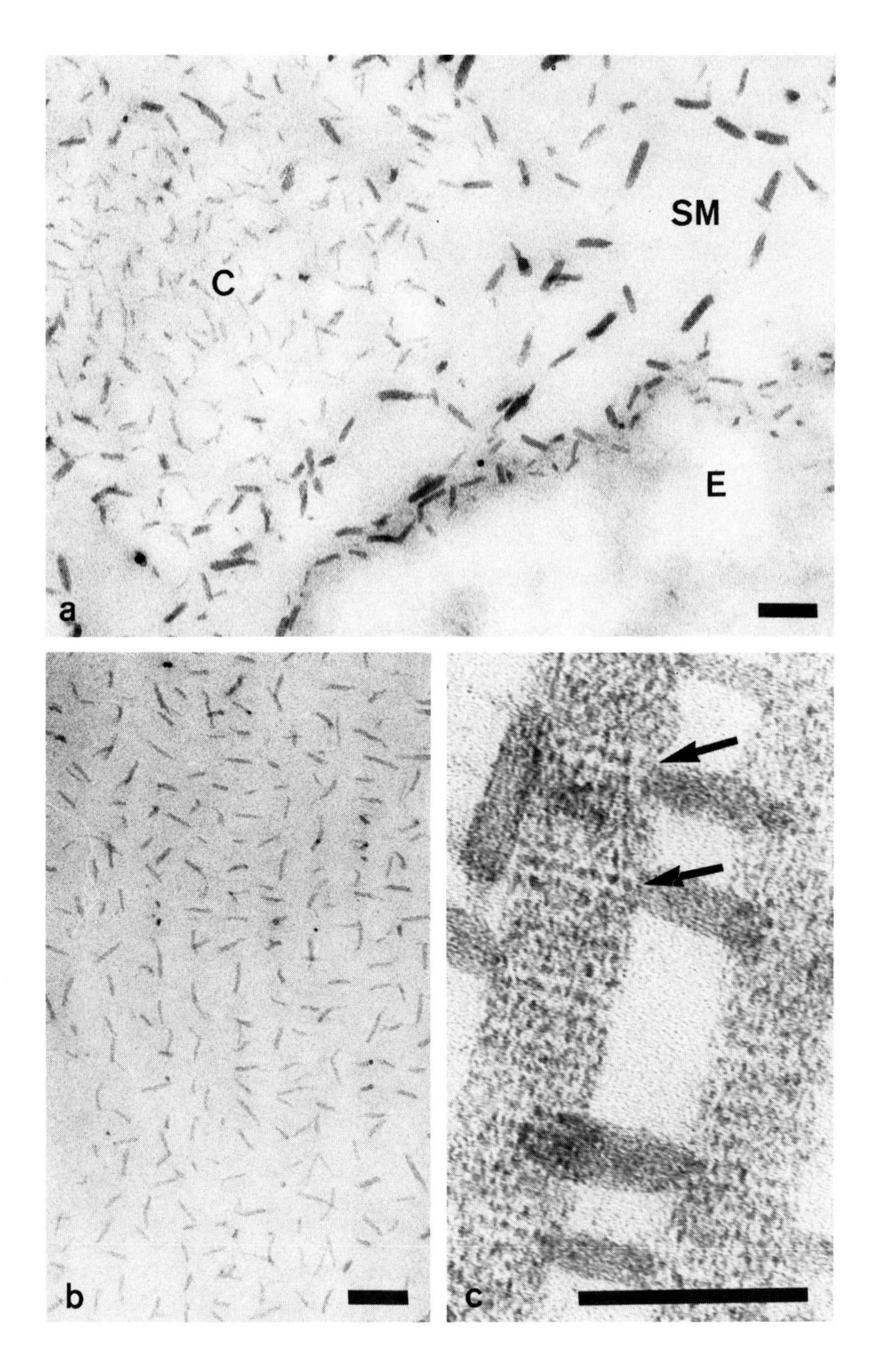

FIGURE 1. Topohistochemical distribution of proteoglycan/cuprolinic blue precipitates in normal vascular tissue. (a) Bovine arterial tissue showing the three types of precipitate of different size. A different type of precipitate is preferentially associated with areas of soluble matrix (SM), collagen

2b). Heparitinase-sensitive precipitates along the basement-membrane-like layer on the smooth muscle cell surface are reduced in number, and their regular arrangement along the interstitial aspect of the layer is lost (Figure 2c). Instead, the large-sized precipitates representing CS-proteoglycans are in close contact with the plasma membrane (Figure 2d). Depositions of mineralized calcium (Figure 3a), extracellular lipid droplets (Figure 3b), and fibrin (Figure 3c) are frequently observed in the neighborhood of the large CS-proteoglycan/cuprolinic blue precipitates.

Besides topohistochemical changes, morphological alterations in the large-sized proteoglycan/cuprolinic blue precipitates are apparent in atherosclerosis. By measurement and comparison of the size of the large precipitates in nine individuals, it turned out that the chondroitinase-AC-sensitive precipitates in plaque regions are significantly longer than in adjacent macroscopically normal-appearing arterial tissue[7] (Figure 4). The increase in length was about 16%. In rat carotids injured by balloon catheterization to stimulate neo/myointimal thickening, a shift towards longer proteoglycan/cuprolinic blue precipitates was also observed.[21] In this experiment in which injured left carotids were compared with normal right carotids in the same animal, the chondroitinase-AC-sensitive precipitates in the medial layer of the injured vessels were as much as 40% longer than those in the noninjured arteries.

E. Summary and Discussion of Results Obtained by Cuprolinic Blue Staining

The present results show that cuprolinic blue staining, in combination with topohistochemical and morphometric analyses, reveals at least three different populations of proteoglycan/cuprolinic blue precipitates in human arteries (Figure 5). From the enzyme-degradation experiments it is concluded that a CS-proteoglycan predominates in areas of soluble matrix situated between the medial smooth muscle cells and collagenous and elastic fibers. DS-proteoglycan is regularly arranged along collagen fibers and is apparently involved in crosslinking adjacent fibrils by bipolar attachment at the collagen d-band regions. HS-proteoglycan is associated with basement-membrane-like layers around smooth muscle cells and with elastic fibers.

"Normal" and plaque tissues differ in the distribution of CS-proteoglycans and in the length of the proteoglycan/cuprolinic blue precipitates. These quantitative, spatial, and structural changes of proteoglycans in atherosclerotic lesions indicate disturbances in the extracellular matrix of the vascular wall. There is strong evidence that molecular changes in both lipoproteins and in the

---

bundles (C), and elastic fibers (E).(b) Unstained longitudinally sectioned bundles of collagen reveal a regular decoration with the characteristic 40-nm form of precipitate. (c) High-magnification micrograph of an osmium-stained specimen, showing attachment of these precipitates at the collagen d-band regions (arrows).(Bars: 0.1 μm).

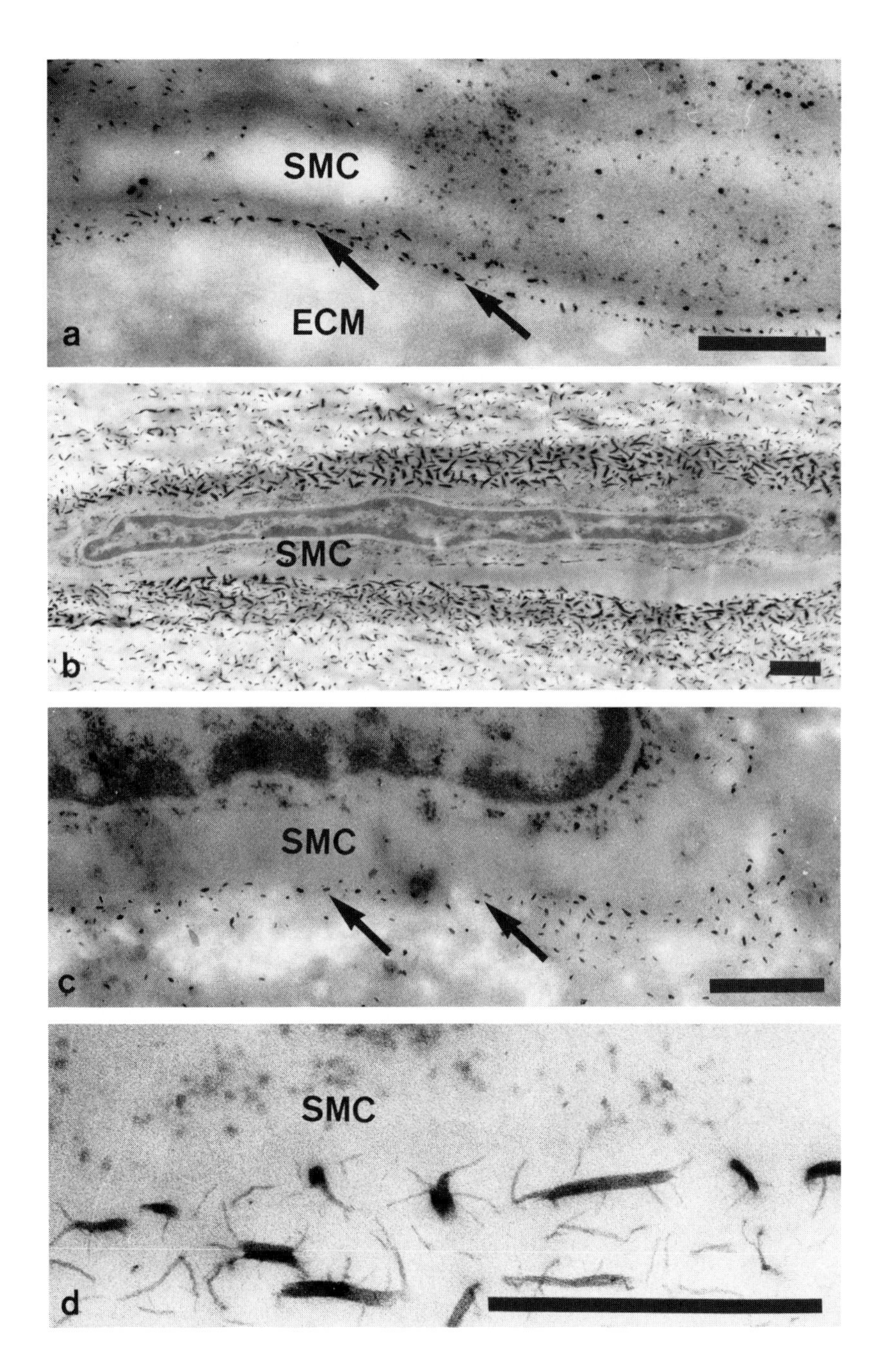

FIGURE 2. Cell surface-associated proteoglycan/cuprolinic blue precipitates in normal and atherosclerotic human arteries. (a) The basement-membrane-like layer on the surface of medial smooth muscle cells (SMC) reveals a layer of chondroitinase-ABC-resistant precipitates (arrows).

glycosaminoglycan side chains of CS-proteoglycans might increase the affinity of these components for one another, leading to immobilization of lipoprotein in the extracellular matrix and deposition of lipid droplets.[4,5,37-43]

Biochemical analysis of atherosclerotic and nonatherosclerotic areas of human aortic samples has shown an age-dependent decrease of HS-proteoglycans from 41 to 20% and an increase in cholesterol content, the loss of HS being higher in plaque tissue than in normal tissue.[8] CS- and DS-glycosaminoglycans, on the other hand, show a marked increase, in line with cholesterol content of the tissue and the age of the subject.[44] The apolipoprotein B content of the tissue is also proportional to that of CS.[45] These observations provide evidence that CS-proteoglycans do indeed play a major role in the retention and accumulation of low-density lipoprotein. The detection of complexes between CS-proteoglycans and lipid droplets in human atherosclerotic tissues in the present cytochemical studies suggests increased mutual affinities of proteoglycans and low-density lipoprotein, as has been shown in binding studies *in vivo* and *in vitro*.[46-50] An increase in the content or length of glycosaminoglycan side chains might explain the formation of elongated proteoglycan-cuprolinic blue precipitates observed in the present study. Strong accumulation of CS-proteoglycans around smooth muscle cells, and loss of HS-proteoglycans, might further promote neointimal outgrowth of smooth muscle cells and, hence, formation of atherosclerotic thickenings.

III. DETECTION OF GLYCOCONJUGATES IN VASCULAR CELL CULTURES WITH COLLOIDAL GOLD

A. Preparation and Application of Colloidal Gold Probes for the Study of Binding and Internalization of Glycoconjugates in Cell Cultures

In the last decade colloidal gold-marker techniques have become increasingly applied in cytochemical studies on an ever-expanding number of biological systems.[51-53] The popularity of the technique is due to the facility with which colloidal gold particles can be coupled to macromolecules by electrostatic interaction, and to the visibility of the markers as clearly defined dense particles when viewed by electron microscopy. The arsenal of gold-labeling techniques for the localization of endogenous macromolecules and related structures

---

These are suggested to represent HS-proteoglycan. In the extracellular matrix (ECM), most precipitates are removed by chondroitinase ABC treatment. (b) Smooth muscle cell (SMC) in atherosclerotic medial tissue surrounded by a dense layer of CS-proteoglycan/cuprolinic blue precipitates (no enzyme treatment). (c) Disorganization of cell-surface-associated HS-proteoglycan/cuprolinic blue precipitates, as visualized after chondroitinase ABC treatment (arrows). (d) High-magnification micrograph of a smooth muscle cell (SMC) in plaque tissue decorated with very large-sized CS-proteoglycan/cuprolinic blue precipitates. The extending filaments evidently represent noncollapsed glycosaminoglycan side chains (no enzyme treatment). (Bars: 0.5 μm).

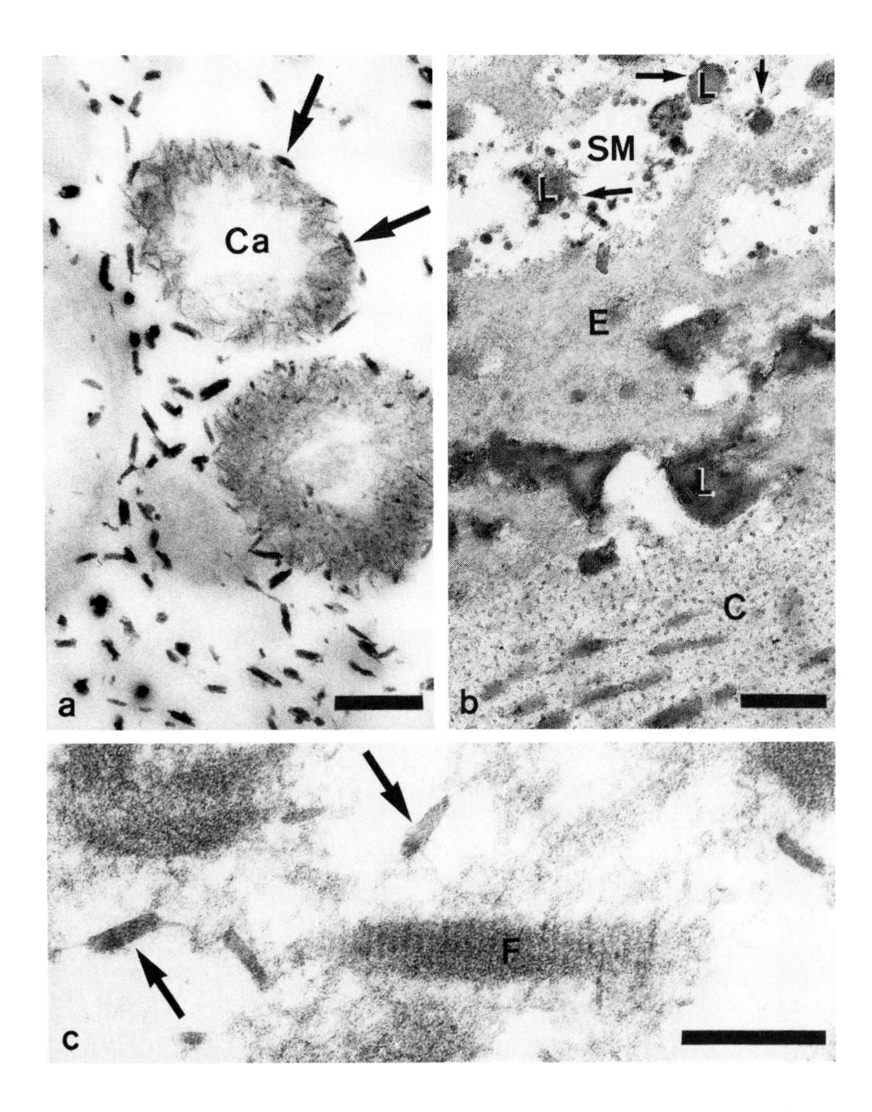

FIGURE 3. Depositions of blood-derived material in extracellular matrix compartments of human atherosclerotic plaque tissue containing CS-proteoglycan/cuprolinic blue precipitates (arrows). (a) Mineralized calcium apatite (Ca) (cuprolinic blue staining). (b) Lipid droplets (L) between collagen (C) and elastin (E) in the soluble matrix (SM) (cuprolinic blue and osmium/thiocarbohydrazide/osmium staining). (c) Fibrin fibrils (F) (cuprolinic blue and osmium staining). (Bars: 0.5 μm).

currently includes methods involving antibodies,[54,55] protein A,[56-58] biotin and avidin,[59,60] lectins,[61,62] enzymes,[63] and nucleotides.[64,65] Gold-conjugated macromolecules can also be applied in living cell cultures to study ligand binding, receptor dynamics, endocytotic uptake, and intracellular degradation.[66-69]

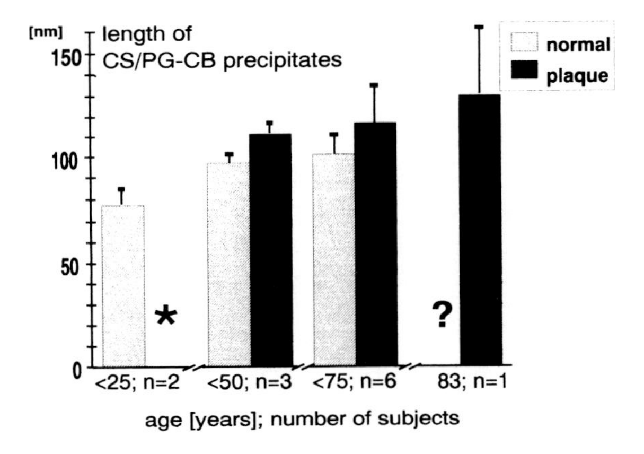

FIGURE 4. Histogram illustrating significant increase in the length of CS/proteoglycan-cuprolinic blue precipitates observed in human arteries. This increase depends on both the severity of arterial damage and the age of the subject. Arterial tissues were obtained by autopsy. In nine of the ten subjects, precipitates could be compared between the same individual by use of normal and plaque tissue from the same vessel. Two subjects were nonatherosclerotic (*), and from one subject (?) only plaque material was available.

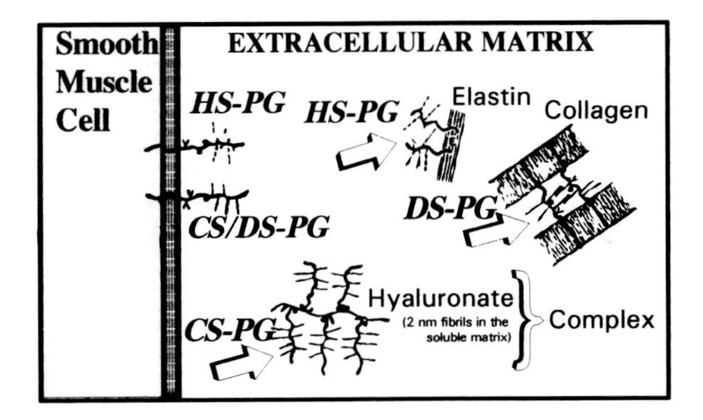

FIGURE 5. Schematic drawing showing sites of proteoglycans in the extracellular matrix of arterial tissue, as deduced from cuprolinic blue staining and from data in the literature. (HS-PG, heparan sulfate proteoglycans; DS-PG, dermatan sulfate proteoglycans; CS-PG, chondroitin sulfate proteoglycans; CS/DS-PG, cell surface-associated proteoglycans containing both chondroitin sulfate and dermatan sulfate.)

A standard procedure for the preparation of homogeneous suspensions of colloidal gold particles is reduction of chloroauric acid, using sodium citrate,[70] tannic acid,[71] or thiocyanate.[72] The size of the resulting particles mainly depends on the concentration of the reducing agent; sizes between 1 and 40 nm are suitable for most purposes.[70,73] General aspects of the preparation of gold suspensions, protein-gold conjugation procedures, and storage of the resulting probes are beyond the scope of the present review and are fully described elsewhere.[51,52,74] Here, a simple method of preparation of protein-gold probes and immunogold markers will be reviewed from our laboratory experience of the preparation of glycoconjugate-gold probes and gold-conjugated secondary antibodies, for use in cell culture experiments. The results obtained by light and electron microscopy demonstrate the distribution and the sites of uptake of proteoglycans and thrombospondin in vascular cells.

Isolated arterial proteoglycans and thrombospondin from platelets were complexed with colloidal gold for subsequent incubation with vascular cell cultures.[34,75,76] The following protocol stresses some of the steps of the preparation procedure that are not in common use in protein-gold couplings.

Briefly, 17-nm colloidal gold particles are prepared according to the method of Frens,[70] concentrated 60-fold by centrifugation (13.000 g), and stored in closed vials.

For protein-gold conjugation, an appropriate aliquot of a few hundred microliters of the very dense gold suspension is rediluted in the ratio of 1:10 in deionized water ($OD_{520} \approx 2.5$). This lowers the concentration of any undesired salts derived from reducing agents and gold chloride. For optimal conjugation the pH of the gold suspension must be adjusted near the isoelectric point of the protein to be coupled. A few microliters of a 200-m*M* citric acid solution for low pH values, and 200 m*M* K_2CO_3 for neutral and basic values, are sufficient for adjustment. The pH is determined with pH-sensitive color-indicator sticks instead of glass electrodes. For proteoglycans and thrombospondin a pH near 5.6 is required; for immunoglobulins the pH should be about 9.0. Both chloride ions and high ionic-strength conditions in the protein sample initially disturb the stability of protein-gold complexes and should be avoided at the beginning of the coupling procedure.

For conjugation with colloidal gold, the protein is added stepwise to the gold suspension, and the stability of the complex checked in small aliquots by testing for flocculation in the presence of 5% NaCl. If the color turns from red to violet or blue, the gold-protein complex is unstable, and more protein is required for saturation. Saturated gold suspensions do not flocculate in the presence of 5% NaCl, and the color remains red. Judgement of the stability of protein-gold suspensions held in small transparent plastic tubes by visual estimation of color is easily carried out on an illumination box, using an oversaturated gold suspension, together with NaCl, as reference. In the case of overtitration with protein, retitration with separately prepared gold colloid is possible.

Having reached optimal saturation of colloidal gold with the protein, a further stabilization of the protein-gold complex is reached by shifting the pH to 7.4 with about 40 m*M* HEPES buffer. The practice of centrifuging the complexes for separating bound and possibly unbound protein should be avoided, especially if self-aggregating or sticky molecules, such as proteoglycans and thrombospondin are used. In our hands, 20 to 40 μg of protein per milliliter of gold suspension were usually required. Proteoglycan-gold and thrombospondin-gold probes were used immediately after preparation in our experiments on cell cultures. The complexes were usually added to the cultures at 4°C, with postincubation at 37°C for various periods of time. Pretreatment with 2% bovine serum albumin was used to reduce nonspecific binding.

One to two hours at 4°C is sufficient to allow binding of gold probes to extracellular matrix components and the cell surface. Although nonspecific binding in the presence of bovine serum albumin is low, a control experiment using bovine serum albumin gold conjugates is routinely done to calculate the extent of any bovine serum albumin-mediated nonspecific binding. In the event of significant nonspecific binding, fish gelatine may be used instead of bovine serum albumin. Competition experiments or preincubation with an excess of nonconjugated protein also allows calculation of the specificity of binding. However, it should be noted that matrix components such as proteoglycans, thrombospondin, or fibronectin are self-aggregating, and so their gold probes may also associate with the preadsorbed nonconjugated material.

After washing we incubate at 37°C for various periods of time to follow the dynamics of cell surface-associated movement of receptor-bound ligands and their intracellular translocation. For electron microscopic investigation, platinum/carbon cell-surface replicas offer the best approach for ligand mapping on top of the cell (Figure 6a). Ultrathin sections are useful for colocalization of receptor-bound ligands with proteoglycan/cuprolinic blue precipitates (Figure 6b) and allow the internalization of ligands in coated pits (Figure 6c) and their subsequent fates in intracellular compartments (Figure 6d) to be detected.

B. Preparation and Application of Immunogold Markers for the Detection of Proteoglycan-Related Structures in Cell Cultures

Primary and secondary antibodies and protein A are increasingly used in immunogold-marking procedures for the electron microscopic detection of antigens.[52,53,74] Gold particles of different sizes are useful for double-labeling experiments in studies on interactions between cell structures, matrix components, and other substances of interest.[77-79] Combinations of colloidal gold and horseradish peroxidase marker systems, with or without silver enhancement, offer alternative approaches for dual labelings.[80]

Another approach to ultrastructural immunocytochemistry that we have developed combines immunogold marking with positive staining using selective dyes. By immunogold labeling of thrombospondin and cuprolinic blue staining of proteoglycans, for example, we have been able to investigate the structural interactions between proteoglycans and thrombospondin.[33] In these experiments endothelial cell cultures were fixed with formaldehyde and labeled with primary antiserum against thrombospondin. Gold-conjugated goat anti-rabbit IgG was used as the immunomarker. The secondary antibody was conjugated to 12-nm colloidal gold particles, using the procedure described in the previous section. After the coupling procedure, 2% bovine serum albumin was added to

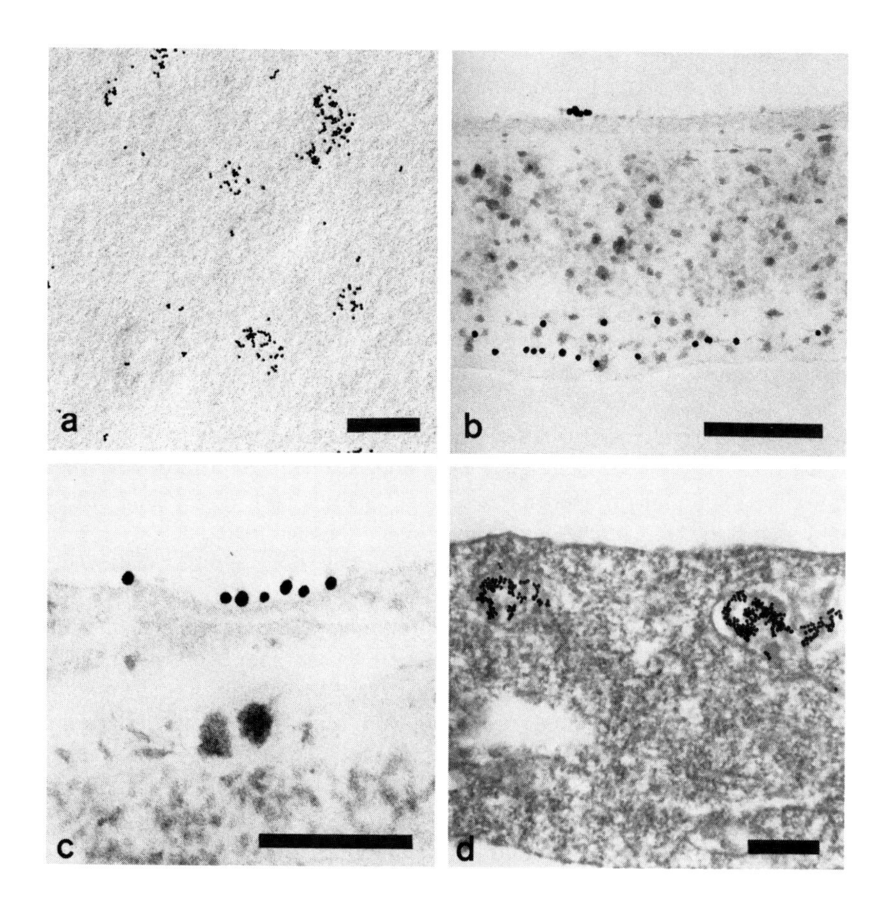

FIGURE 6. Binding and internalization of thrombospondin-gold probes in cultured porcine endothelial cells. (a) Platinum/carbon replica showing thrombospondin-gold clusters on the plasma membrane. (b) Cuprolinic blue staining reveals thrombospondin-gold binding sites not only on the cell surface, but also at proteoglycan/cuprolinic blue precipitates in the subcellular matrix. (c) Binding of thrombospondin-gold in a coated pit. (d) Internalized thrombospondin-gold markers are found in lysosomes after 1 h of incubation at 37°C (osmium staining). (Bars: 0.2 μm).

the complexes, and the pH was adjusted to 7.4 by addition of 40 m*M* HEPES buffer. This highly concentrated gold marker suspension containing about 20 to 40 μg of protein per milliliter can be stored in closed vials in the presence of 0.01% sodium azide for several months. This means that gold markers of different size, coupled to a variety of antibodies, may be prepared and are ready for use at any time.

The gold-marker suspension is routinely diluted 1:20 in appropriate cell culture buffers and added to fixed cell cultures for binding to preadsorbed primary antibodies. After washing, postfixation, cuprolinic blue staining, and secondary staining with Na_2WO_4 are carried out. The samples are then detached from the culture dish with propylene oxide and embedded in epoxy resin. Ultrathin sections are prepared and examined in the transmission electron microscope. Using this combination of methods, cuprolinic blue-positive proteoglycans and immunogold-labeled thrombospondin were colocalized on the surface of endothelial cells.

For intracellular detection of thrombospondin, ultrasmall colloidal gold particles of about 3 nm in size, conjugated with secondary antibodies, were used. These colloidal gold particles were prepared and coupled with protein by the method of Baschong et al.[72] Stability of the yellow and rather transparent gold-protein suspensions was tested with 20% NaCl, but in this case, the barely visible turbidity indicating flocculation was measured in a photometric device at a 560-nm wavelength.

For infiltration of ultrasmall immunogold markers, cytoplasmic membranes were permeabilized with 0.5% saponin in the presence of formaldehyde (Figures 7a, b). The silver enhancement technique may be applied for subsequent enlargement of the marker size for comfortable electron microscopic visualization in intermediate- and low-magnification micrographs of ultrathin sections (Figure 7c) and, when done for longer periods, is useful at the light microscopic level[81-84] (Figure 7d). Principles of the silver enhancement technique, and a schedule of preparation of a stable and easy-to-handle development kit, are described later.

C. Summary and Discussion of Results Obtained by Colloidal Gold Labeling

Electron micrographs of smooth muscle cell cultures treated with proteoglycan-gold conjugates reveal the gold marker particles in strings (Figure 8). What these structures represent are the large, individual filamentous proteoglycan molecules decorated with multiple gold particles along the protein axis. These structures are therefore fundamentally different from globular protein-gold conjugates, such as thrombospondin-gold, in which the gold particles are (ideally) completely enveloped with protein. The DS-proteoglycan-gold complex was found to have an average length of 170 nm. This value is consistent with biochemical data on the relative molecular mass of the core protein, calculated to be about 40 to 50 kDa.[85,86] In bovine arterial smooth muscle cells, we found that the DS-proteoglycan-gold complexes bind to coated, as well as to noncoated, cell-surface membrane areas.[75,76] Binding to the plasma membrane occurs only at one terminal of the proteoglycan-gold strings, consistent with the presence of a linkage region at one end of the protein, for binding to hyaluronic acid, as identified in cartilage proteoglycans.[25-27] After binding to the plasma membrane, proteoglycan-gold conjugates are internalized and translocated to multivesicular bodies and lysosomes. Ultimately, the gold particles are extruded from residual vacuoles.

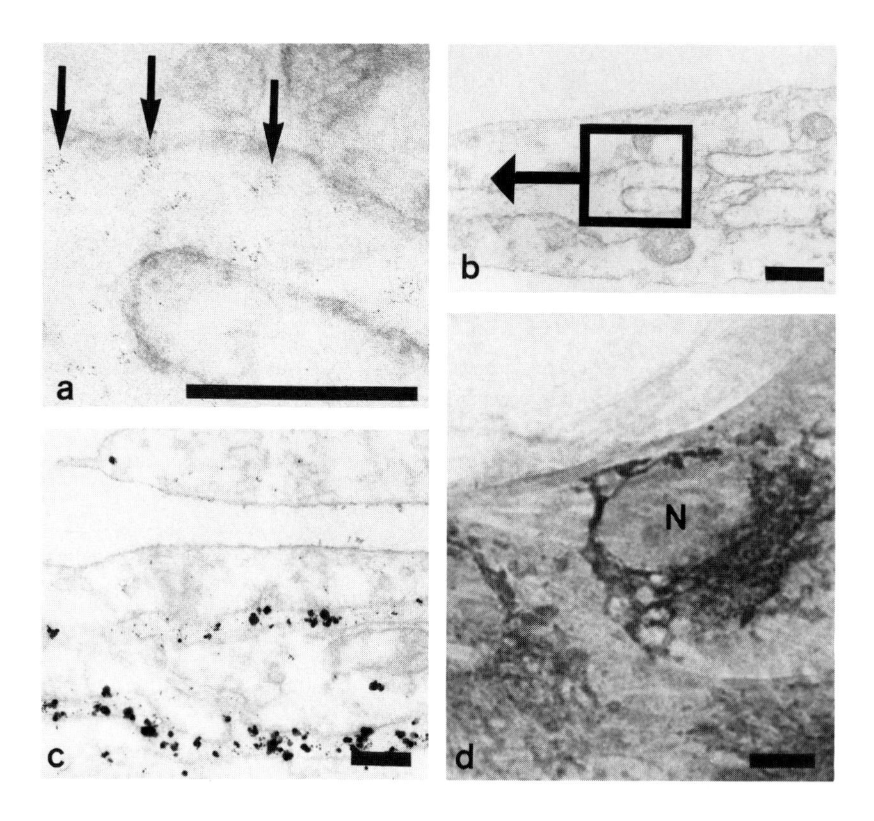

FIGURE 7. Ultrasmall antithrombospondin immunogold marker applied to the study of thrombospondin localization in porcine endothelial cells, as viewed in light and electron micrographs. (a) High-magnification electron micrograph showing goat anti-rabbit 3-nm gold marker (arrows) bound to antibodies to thrombospondin, at sites of thrombospondin synthesis in endoplasmic reticulum. (b) At lower magnification, the ultrasmall gold marker is barely discernible. (c) Silver intensification for about 5 min allows detection of the ultrasmall immunogold marker in low- and intermediate-magnification electron micrographs. (d) Silver intensification for about 20 min even allows the detection of the immunogold marker in light micrographs. (N, nucleus.) (Bars: (a,b,c) 0.1 μm; (d) 10 μm).

In contrast to the proteoglycan-gold probes, thrombospondin-gold conjugates represent globular complexes.[34] A gold particle of about 17 nm in diameter, as used in these studies, is calculated to be covered with about six to seven thrombospondin molecules. This estimate was derived from a calculation based on 5.8×10^{-12} spherical 17-nm gold particles per milliliter, as obtained by the procedure described above, and 4×10^{-13} trimeric thrombospondin molecules (of 450 kDa) per milliliter, from the 30 μg of protein required for saturation.

The most fascinating progress in gold-labeling experiments was made by conjugation of proteolytic fragments of thrombospondin molecules to colloidal gold.[34] This approach allows cytochemical labeling studies of molecular do-

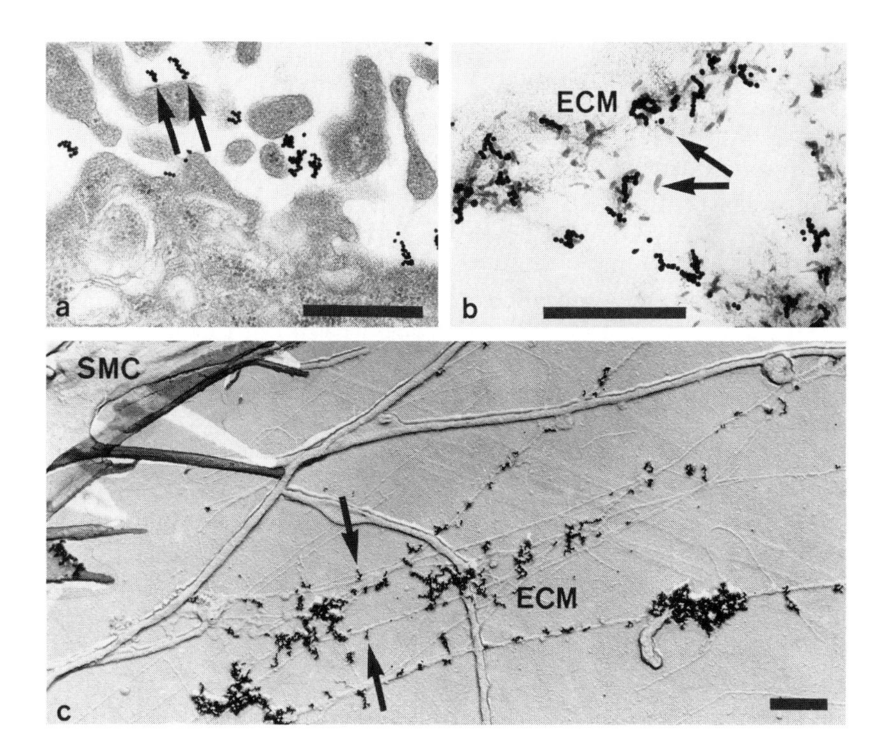

FIGURE 8. Pearl string-like proteoglycan-gold conjugates in cultures of bovine arterial smooth muscle cells (SMC). (a) Binding to the plasma membrane of a smooth muscle cell occurs by only one terminal of the proteoglycan-gold complex (arrows) (osmium staining). (b) Proteoglycan-gold conjugates bind to the extracellular matrix (ECM) close to endogenous proteoglycans that are visualized by cuprolinic blue staining (arrows). (c) In platinum/carbon replicas, proteoglycan-gold conjugates are seen in close contact with fibrillar components of the extracellular matrix on the culture dish bottom (arrows). (Bars: 0.5 μm).

mains, such as the HS-binding region of the thrombospondin molecule, and thereby discrimination of cell surface-associated receptors. While intact thrombospondin requires coupling to colloidal gold at pH 5.6, the N-terminal fragment of the molecule containing the HS-binding domain has a rather high isoelectric point and, consequently, requires coupling at about pH 9.0. The remaining large thrombospondin fragment devoid of the HS-binding region has to be conjugated as with intact thrombospondin at pH 5.6.

In porcine endothelial cells, both intact thrombospondin-gold conjugates and gold-conjugated proteolytic fragments of thrombospondin containing the HS-binding region bind to coated pits and are internalized and accumulated in lysosomes. This binding is inhibited in the presence of heparin, an exogenous sulfated glycosaminoglycan. However, thrombospondin fragments devoid of the HS-binding domain do not bind to coated pits and are not internalized, at least in porcine endothelial cells. This indicates that the HS-binding region, but not other epitopes such as the RGD sequence containing the aminoacids arginine, glycine, and aspartate, or the collagen-binding region, are required for receptor-mediated uptake.

IV. DETECTION OF GLYCOCONJUGATES IN VASCULAR CELL CULTURES WITH THE SILVER ENHANCEMENT METHOD

A. Preparation and Application of a Simple Silver Enhancement Kit

For silver intensification of protein-gold-labeled cell cultures, chemical postfixation and thorough washing with deionized water are necessary to remove any chloride ions and other contamination. The silver developer solutions may be prepared according to published protocols [80,81,83,87,88] or obtained as ready-to-use solutions from commercial distributors.

In cell culture dishes as well as in blotting experiments, large quantities of silver enhancer are required. Commercial enhancement kits (from Janssen/Belgium, Amershan/F.R.G., BioCell/U.K., or AurIon/The Netherlands) offer high reliability and easy handling, but they are rather costly. Laboratory preparation of developer reagents, following published protocols, on the other hand, frequently results in rather unstable and light-sensitive solutions, that can be stored only for a short time. An ideal enhancement kit should combine the advantages of a low-cost laboratory-made enhancer with the simple handling and the low rates of self-nucleation and the stability and insensitivity to light of commercial kits. A protocol detailing how to prepare a kit for use in light and electron microscopic applications, based on our experience, has been briefly described elsewhere.[89] The kit consists of two standard solutions and may be stored for several months; it is rapidly prepared by mixing equal volumes, and the development is stopped simply by washing the specimen with deionized water.

The kit consists of a silver solution and a developer solution. In the silver solution, about 1.5 g of silver nitrate is dissolved in a mixture of 900 ml of deionized water and 100 ml of 1,2-propandiol. In the developer solution 4.0 g of hydrochinon is dissolved in a mixture of 1000 ml of a 0.1-*M* citrate buffer, pH 4.5, 0.1% formaldehyde, and 200 ml of 1,2-propandiol. Immediately before use, equal volumes of silver and developer solutions are mixed together in a clean vessel.

For silver enhancement of gold-labeled cell cultures, the silver enhancement mixture is added to the culture dishes. The dishes are best placed in a tumbling box at 16 to 18°C at moderate room illumination. After 15 to 20 min, the development is stopped by washing with deionized water. Where the amount of labeling is low, a second or even a third development is recommended, using freshly mixed developer. If necessary, development can be monitored in an inverted light microscope and stopped at the latest when background precipitation in the control specimen is visible.

As shown in the following section, our silver enhancement procedure may be applied with success to both light and electron microscopic investigations using gold-conjugated immunomarkers and gold-conjugated ligands in cell cultures.

B. Detection of Silver-Intensified Gold Probes and Immunogold Markers in Light and Electron Micrographs of Cell Cultures

Sites of immunogold binding marking endogenous proteoglycans, thrombospondin, and fibronectin, as well as depositions of exogenously applied gold-conjugated probes, are very easily detected by the silver enhancement method. For example, endogenous fibronectin is revealed deposited close to endothelial cells and smooth muscle cells in a fibrillar organization (Figure 9a), as has also been established by fluorescence microscopy. The silver enhancement technique, however, proves to be much more sensitive in the detection of antigens than does fluorescence labeling. This is clearly demonstrated by visualization of cell migration tracks containing thrombospondin, HS-proteoglycan, and fibronectin (Figure 9b).[33] On the cell surface a homogeneous distribution of thrombospondin-gold/silver marker is seen (Figure 9c). The density of the silver marker varies between individual cells, indicating different degrees of secretory activity in the cell culture. The distribution pattern corresponds to cell surface-associated thrombospondin/HS-proteoglycan granules, as seen in electron micrographs (Figure 6a). By positive staining of the cell cultures with fuchsin dye, the nucleus appears red, with less intensive staining of the cytoplasm. In black and white micrographs this staining adds useful background information and does not detract from the visibility of the black silver precipitate.

The experiments using 3-nm immunogold markers to detect thrombospondin (Figures 7a, b) provide a further example illustrating the sensitivity of silver enhancement.[89] By light microscopy, silver enhancement reveals endogenous thrombospondin in a network-like structure around the nucleus (Figure 7d). Electron microscopy of the preparations shows that this structure represents the endoplasmic reticulum (Figure 7c).

As ultrasmall-sized gold particles are hardly visible in the electron microscope at intermediate magnifications, a brief silver intensification step may be used to increase the size of the particulate marker to facilitate analysis of large areas at intermediate and low magnifications (cf. Figures 7a and 7b). For this purpose it can be advantageous to use silver lactate instead of silver nitrate. Although silver lactate is rather unstable and light sensitive, it produces spherical, rather than crystalline, depositions of metallic silver in the specimens.

The silver enhancement method has also proven invaluable for following cell-surface binding and internalization of thrombospondin-gold in cultured cells.[34] Even by standard light microscopy, intracellular accumulations of silver granules are observed after endocytotic uptake of thrombospondin-gold (Figure 9d). Electron microscopy confirms that these sites of accumulation represent lysosomes close to the nucleus. Loss of cell-surface-bound thrombospondin-

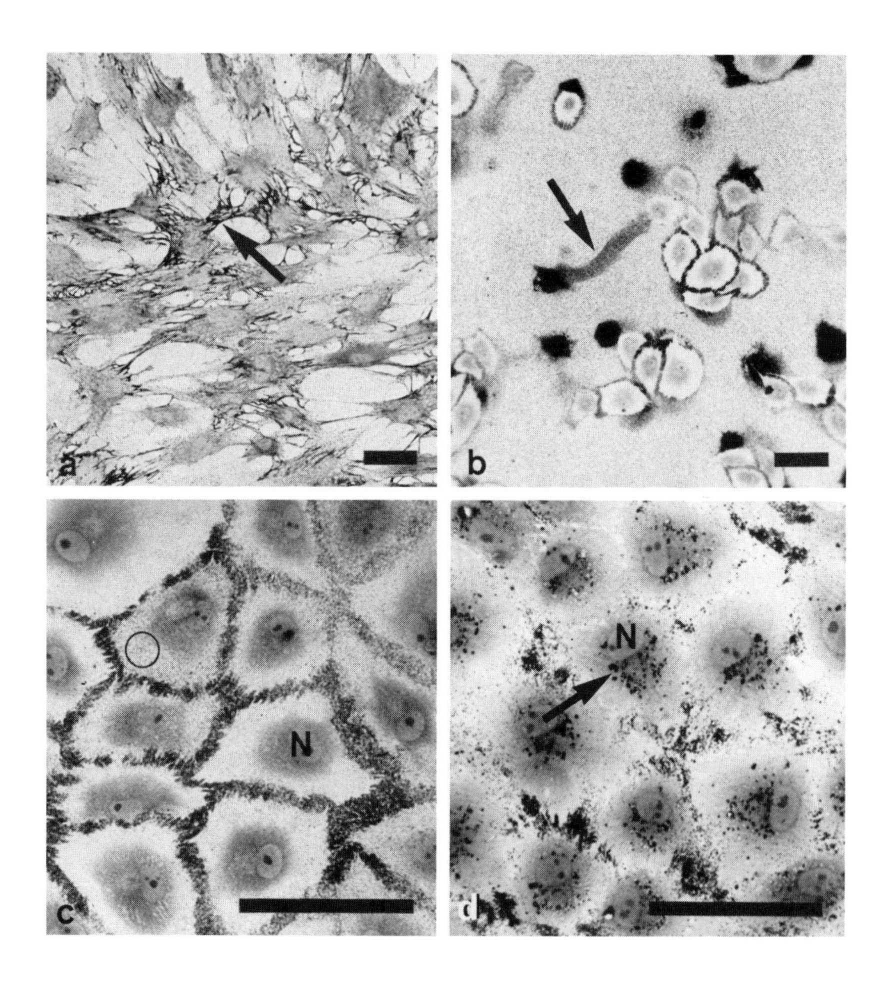

FIGURE 9. Silver enhancement of vascular cell cultures for light microscopic detection of immunogold markers and thrombospondin-gold probes. (a) Smooth muscle cell culture labeled with antibodies to fibronectin and a gold-conjugated secondary antibody. Dense silver precipitate reveals characteristic fibrils of fibronectin (arrow). (b) Subconfluent endothelial cells labeled with antibodies to thrombospondin and a gold-conjugated secondary antibody. Dense silver precipitate is seen in circular cell adhesion sites and migration tracks (arrow). (c) Confluent endothelial cell culture labeled with antithrombospondin. The cells show different amounts of homogeneously distributed thrombospondin-gold/silver marker (circle). (d) Internalized thrombospondin-gold is visualized by silver enhancement as large silver grains close to the nucleus (N). (Bars: 100 μm).

gold and intracellular translocation can be clearly followed over time. Monensin-induced inhibition of translocation of thrombospondin-gold probes can be readily demonstrated. Binding of gold-conjugated thrombospondin fragments can also be examined by light microscopy.

A few hours only are normally needed for colloidal gold conjugation, cell culture incubation, and silver enhancement for a preliminary characterization of ligand binding and uptake in cultured cells. The examples presented here clearly show that protein-gold-labeled cell cultures may be routinely examined after silver enhancement in a light microscope.

C. Summary and Discussion of Results Obtained with Silver Enhancement

Our studies have shown that thrombospondin is highly enriched, together with fibronectin and HS-proteoglycan, in cell attachment and in cell migration tracks on the bottom of the culture dish.[33] In cell migration tracks, HS-proteoglycan reveals a fibrillar pattern similar to that of fibronectin, while thrombospondin is deposited more homogeneously. This suggests involvement of these substances in cell adhesion and migration processes, both of which are important events in the early stages of atherosclerosis.

By using a combination of cuprolinic blue staining, immunogold labeling, and silver enhancement, we have further shown that on the surface of endothelial cells, thrombospondin is structurally organized, together with HS-proteoglycan, in granular structures.

In a recent report we have shown that exogenous thrombospondin-gold probes also bind to granular and fibrillar structures and to sites of cell-cell contacts on the cell surface; cell migration tracks are labeled as well.[34] Conjugation of intact thrombospondin and proteolytic fragments of thrombospondin with colloidal gold enables detection of binding sites and investigations of the mechanisms of uptake in endothelial cells. Binding of thrombospondin-gold probes is inhibited in the presence of heparin. Lack of binding and uptake is also observed with gold-conjugated thrombospondin fragments devoid of the HS-binding region. These studies provide strong evidence that HS-proteoglycans act as thrombospondin receptors or, at least, as specific mediators of receptor binding in coated pits of porcine endothelial cells.

V. CONCLUDING REMARKS

The work presented in this review has shown how modern cytochemical techniques — in particular cuprolinic blue staining, colloidal gold labeling, and silver enhancement techniques — can be applied to identify specific components of the extracellular matrix and investigate their interactions with vascular cells. These approaches have opened new opportunities for exploring the role of extracellular matrix components in atherosclerosis, which remain to be more fully exploited in the future.

ACKNOWLEDGMENTS

I wish to thank my dear colleagues and friends, Prof. Dr. Eckhart Buddecke, Dr. Annette Schmidt, Dr. Peter Vischer, and Dipl.-Biol. Verona Faber, for providing proteoglycans, thrombospondin, and antibodies that were used in our collaborative experiments for many years, as well as for many helpful discussions on the functions of these glycoconjugate molecules in the vascular wall. I am grateful to Prof. Dr. Horst Robenek for initiating studies using gold conjugates, from which much of my work has developed. I wish also to thank my technical assistant, Daniel Ziomek, for the excellent stainings and preparations of cell cultures and arterial tissues, and Marianne Opalka for preparing the photomicrographs. The experimental work described herein was supported by grants from the Deutsche Forschungsgemeinschaft (VO 399/1-1, VO 3991-2, SFB 310/B6).

REFERENCES

1. **Ross, R.,** The pathogenesis of atherosclerosis — an update, *N. Engl. J. Med.*, 314, 488, 1986.
2. **Salisbury, B.G.J. and Wagner, W.D.,** Isolation and preliminary characterization of proteoglycans dissociatively extracted from human aorta, *J. Biol. Chem.*, 256, 8050, 1981.
3. **Hollander, W.,** Unified concept of the role of acid mucopolysaccharides and connective tissue proteins in the accumulation of lipids, lipoproteins, and calcium in the atherosclerotic plaque, *Exp. Mol. Pathol.*, 25, 106, 1976.
4. **Berenson, G.S., Radhakrishnamurthy, B., Srinivasan, S.R., Vijayagopal, P., and Dalferes, E.R.,** Arterial wall injury and proteoglycan changes in atherosclerosis, *Arch. Pathol. Lab. Med.*, 112, 1002, 1988.
5. **Radhakrishnamurthy, B., Srinivasan, S.R., Vijayagopal, P., and Berenson, G.S.,** Arterial wall proteoglycans — biological properties related to pathogenesis of atherosclerosis, *Eur. Heart J.*, 11 (Suppl. E), 148, 1990.
6. **Wight, T.N., Kinsella, M.G., and Potter-Perigo, S.,** Proteoglycans synthesized and secreted by cultured vascular cells, in *Extracellular Matrix: Structure and Function*, Reddi, A.H., Ed., Alan R. Liss, New York, 1985, 321.
7. **Völker, W., Schmidt, A., and Buddecke, E.,** Cytochemical changes in a human arterial proteoglycan related to atherosclerosis, *Atherosclerosis*, 77, 117, 1989.
8. **Hollmann, J., Schmidt, A., von Bassewitz, D.-B., and Buddecke, E.,** Relationship of sulfated glycosaminoglycans and cholesterol content in normal and arteriosclerotic human aorta, *Arteriosclerosis*, 9, 154, 1989.
9. **Marcum, J.A., Reilly, C.F., and Rosenberg, R.D.,** Heparan sulfate species and blood vessel wall function, in *Biology of Proteoglycans*, Wight, T.N. and Mecham, R.P., Eds., Academic Press, San Diego, 1987, 301.
10. **Baenziger, N.L., Brodie, G.N., and Majerus, P.W.,** A thrombin-sensitive protein of human platelet membranes, *Proc. Natl. Acad. Sci. U.S.A.*, 68, 240, 1971.
11. **Mosher, D.F.,** Physiology of thrombospondin, *Ann. Rev. Med.*, 41, 85, 1990.
12. **McPherson, J., Sage, H., and Bornstein, P.,** Isolation and characterization of a glycoprotein secreted by aortic endothelial cells in culture: apparent identity with platelet thrombospondin, *J. Biol. Chem.*, 256, 11,330, 1981.
13. **Wight, T.N., Raugi, G.J., Mumby, S.M, and Bornstein, P.,** Light microscopic immunolocation of thrombospondin in human tissues, *J. Histochem. Cytochem*, 33, 295, 1985.
14. **Raugi, G.J., Mullen, J.S., Bark, D.H., Okada, T., and Mayberg, M.R.,** Thrombospondin deposition in rat carotid artery injury, *Am. J. Pathol.*, 137, 179, 1990.

15. **Majack, R.A., Cook, S.C., and Bornstein, P.,** Platelet-derived growth factor and heparin-like glycosaminoglycans regulate thrombospondin synthesis and deposition in the matrix by smooth muscle cells, *J. Cell Biol.*, 101, 1059, 1985.
16. **Majack, R.A., Goodman, L.V., and Dixit, V.M.,** Cell surface thrombospondin is functionally essential for vascular smooth muscle cell proliferation, *J. Cell Biol.*, 106, 415, 1988.
17. **Scott, J.E.,** Proteoglycan histochemistry — a valuable tool for connective tissue biochemists, *Coll. Relat. Res.*, 5, 541, 1985.
18. **Van Kuppevelt, T.H.M.S.M., Rutten, T.L.M., and Kuyper, C.M.A.,** Ultrastructural localization of proteoglycans in tissue using cuprolinic blue according to the critical electrolyte concentration method: comparison with biochemical data from literature, *Histochem. J.*, 19, 520, 1987.
19. **Huang, W., Richardson, M., Alavi, M.Z., and Moore, S.,** Proteoglycan distribution in rat aortic wall following indwelling catheter injury, *Atherosclerosis*, 51, 59, 1984.
20. **Huang, W. and Haust, M.D.,** Proteoglycans in human atherosclerotic lesions — a pilot qualitative and quantitative study by ruthenium red, *Histopathology*, 8, 835, 1984.
21. **Völker, W., Schmidt, A., Oortmann, W., Broszey, T., Faber, V., and Buddecke, E.,** Mapping of proteoglycans in atherosclerotic lesions, *Eur. Heart J.*, 11 (Suppl. E), 29, 1990.
22. **Scott, J.E., Kyffin, T.W., and Morris, G.A.,** Copper tetrapyridino 'phtalocyanin' (Cuprolinic blue) differs in shape from the palladium and platinum analogues, and this affects staining of polynucleotides, *Basic Appl. Histochem.*, 35, 7, 1991.
23. **Blumenfeld, O.O., Bienkowski, R.S., Schwartz, E., and Seifter, S.,** Extracellular matrix of smooth muscle cells, in *Biochemistry of Smooth Muscle*, Vol. II, Stephens, N.L., Ed., CRC Press, Boca Raton, FL,1983, 137.
24. **Wight, T.N., Lark, M.W., and Kinsella, M.G.,** Blood vessel proteoglycans, in *Biology of Proteoglycans*, Wight, T.N. and Mecham, R.P., Eds., Academic Press, San Diego, 1987, 267.
25. **Spooner, B. and Thompson-Pletscher, H.A.,** Matrix accumulation and the development of form: proteoglycans and branching morphogenesis, in *Regulation of Matrix Accumulation*, Mecham, R.P., Ed., Academic Press, San Diego, 1986, 399.
26. **Heinegard, D., Franzen, A., Hedbom, E., and Sommarin, Y.,** Common structures of the core proteins of interstitial proteoglycans, in *Functions of the Proteoglycans*, CIBA Foundation Symposium 124, John Wiley & Sons, Chichester, 1986, 69.
27. **Ruoslathi, E.,** Structure and biology of proteoglycans, *Ann. Rev. Cell Biol.*, 4, 1988, 229.
28. **Scott, J.E.,** Collagen-proteoglycan interactions. Localization of proteoglycans in tendon by electron microscopy, *Biochem. J.*, 187, 887, 1980.
29. **Scott, J.E. and Orford, C.R.,** Dermatan sulphate- rich proteoglycan associates with rat tail-tendon collagen at the d-band in the gap region, *Biochem. J.*, 197, 213, 1981.
30. **Seligman, A.M., Wasserkrug, H.L., and Hanker, J.S.,** A new staining method (OTO) for enhancing contrast of lipid-containing membranes and droplets in osmium tetroxide-fixed tissue with osmiophilic thiocarbohydrazide (TCH), *J. Cell Biol.*, 30, 424, 1966.
31. **Guyton, J.R., Bocan, T.M.A., and Schifani, T.A.,** Quantitative ultrastructural analysis of perifibrous lipid and its association with elastin in nonatherosclerotic human aorta, *Arteriosclerosis*, 5, 644, 1985.
32. **Guyton, J.R. and Klemp, K.F.,** Ultrastructural discrimination of lipid droplets and vesicles in atherosclerosis: value of osmium-thiocarbohydrazide-osmium and tannic acid-paraphenylenediamine techniques, *J. Histochem. Cytochem.*, 36, 1319, 1988.
33. **Vischer, P., Völker, W., Schmidt, A., and Sinclair, N.,** Association of thrombospondin of endothelial cells with other matrix proteins and cell attachment sites and migration tracks, *Eur. J. Cell Biol.*, 47, 36, 1988.
34. **Völker, W., Schön, P., and Vischer, P.,** Binding and endocytosis of thrombospondin and thrombospondin fragments in endothelial cell cultures analyzed by cuprolinic blue staining, colloidal gold labeling, and silver enhancement techniques, *J. Histochem. Cytochem.*, 39, 1385, 1991.

35. **Völker, W., Schmidt, A., and Buddecke, E.,** Compartmentation and characterization of different proteoglycans in bovine arterial wall, *J. Histochem. Cytochem.*, 34, 1293, 1986.
36. **Völker, W., Schmidt, A., and Buddecke, E.,** Mapping of proteoglycans in human arterial tissue, *Eur. J. Cell Biol.*, 45, 72, 1987.
37. **Camejo, G., Lopez, A., Lopez, F., and Quinones, J.,** Interaction of low density lipoproteins with arterial proteoglycans. The role of charge and sialic acid content, *Atherosclerosis*, 55, 93, 1985.
38. **Wagner, W.D., Salisbury, B.G.J., and Rowe, H.A.,** A proposed structure of chondroitin 6-sulfate proteoglycan of human normal and adjacent atherosclerotic plaque, *Arteriosclerosis*, 6, 407, 1986.
39. **Dalferes, E.R., Radhakrishnamurthy, B., Ruiz, H.A., and Berenson, G.S.,** Composition of proteoglycans from human atherosclerotic lesions, *Exp. Molec. Pathol.*, 47, 363, 1987.
40. **Vijayagopal, P., Srinivasan, S.R., Dalferes, E.R., Radhakrishnamurthy, B., and Berenson, G.S.,** Effect of low-density lipoproteins on the synthesis and secretion of proteoglycans by human endothelial cells in culture, *Biochem. J.*, 255, 639, 1988.
41. **Srinivasan, S.R., Vijayagopal, P., Eberle, K., Radhakrishnamurthy, B., and Berenson, G.,** Low-density lipoprotein binding affinity of arterial wall proteoglycans: characteristics of a chondroitin sulfate proteoglycan subfraction, *Biochim. Biophys. Acta*, 1006, 159, 1989.
42. **Camejo, G., Linden, T., Olsson, U., Wiklund, O., Lopez, F., and Bondjers, G.,** Binding parameters and concentration modulate formation of complexes between LDL and arterial proteoglycans in serum, *Atherosclerosis*, 79, 121, 1989.
43. **Sambandam, T., Baker, J.R., Christner, J.E., and Ekborg, S.L.,** Specificity of the low density lipoprotein-glycosaminoglycan interaction, *Arterioscler. Thromb.*, 11, 561, 1991.
44. **Ylä-Herttuala, S., Sumuvuori, H., Karkola, K., Möttönen, M., and Nikkari, T.,** Glycosaminoglycans in normal and atherosclerotic human coronary arteries, *Lab. Invest.*, 54, 402, 1986.
45. **Ylä-Herttuala, S., Solakivi, T., Hirvonen, J., Laaksonen, H., Möttönen, M., Pesonen, E., Raekallio, J., Akerblom, H.K., and Nikkari, T.,** Glycosaminoglycans and apolipoproteins B and A-1 in human aortas. Chemical and immunological analysis of lesion-free aortas from children and adults, *Arteriosclerosis*, 7, 333, 1987.
46. **Srinivasan, S.R., Yost, C., Bhandaru, R.R., Radhakrishnamurthy, B., and Berenson, G.S.,** Lipoprotein-glycosaminoglycan interactions in aortas of rabbits fed atherogenic diets containing different fats, *Atherosclerosis*, 43, 289, 1982.
47. **Alavi, M. and Moore, S.,** Kinetics of low density lipoprotein interactions with rabbit aortic wall following balloon catheter deendothelialization, *Arteriosclerosis*, 4, 395, 1984.
48. **Wagner, W.,** Proteoglycan structure and function as related to atherosclerosis, *Ann. N.Y. Acad. Sci.*, 454, 52, 1985.
49. **Hoff, H.F. and Wagner, W.D.,** Plasma low density lipoprotein accumulation in aortas of hypercholesterolemic swine correlates with modifications in aortic glycosaminoglycan composition, *Atherosclerosis*, 61, 231, 1986.
50. **Alavi, M.Z., Richardson, M., and Moore, S.,** The *in vitro* interactions between serum lipoproteins and proteoglycans of the neointima of rabbit aorta after a single balloon catheter injury, *Am. J. Pathol.*, 134, 287, 1989.
51. **Roth, J.,** The colloidal gold marker system for light and electron microscopic cytochemistry, in *Techniques in Immunocytochemistry*, Vol. 2, Bullock, G.R. and Petrusz, P., Eds., Academic Press, London, 1983, 217.
52. **Hayat, M.A.,** *Colloidal Gold. Principles, Methods and Applications*, Academic Press, Inc., San Diego, 1989, 1.
53. **Beesley, J.E.,** *Colloidal Gold: A New Perspective for Cytochemical Marking, Microscopy Handbooks*, Royal Microscopical Society, Oxford, 1989, 1.
54. **Stow, J.L., Kjellen, L., Unger, E., Höök, M., and Farquhar, M.G.,** Heparan sulfate proteoglycans are concentrated on the sinusoidal plasmalemmal domain and in intracellular organelles of hepatocytes, *J. Cell Biol.*, 100, 975, 1985.

55. **Lark, M.W., Yeo, T.-K., Mar, H., Lara, S., Hellström, I., Hellström, K.-E., and Wight, T.N.,** Arterial chondroitin sulfate proteoglycan: localization with a monoclonal antibody, *J. Histochem. Cytochem.*, 36, 1211, 1988.
56. **Mannweiler, K., Hohenberg, H., Bohn, W., and Rutter, G.,** Protein-A gold particles as markers in replica immunocytochemistry: high resolution electron microscope investigations of plasma membrane surfaces, *J. Microsc.*, 126, 145, 1982.
57. **Ratcliff, A., Fryer, P.R., and Hardingham, T.E.,** Proteoglycan biosynthesis in chondrocytes: protein A-gold localization of proteoglycan protein core and chondroitin sulfate within Golgi subcompartments, *J. Cell Biol.*, 101, 2355, 1985.
58. **Van Berkel, J., Steup, M., Völker, W., Robenek, H., and Flügge, U.I.,** Polypeptides of the chloroplast envelope membranes as visualized by immunochemical techniques, *J. Histochem. Cytochem.*, 34, 577, 1986.
59. **Morris, R.E. and Saelinger, C.B.,** Problems in the production and use of 5 nm avidin-gold colloids, *J. Microsc.*, 143, 171, 1986.
60. **Nakajima, M., Ito, N., Nishi, K., Okamura, Y., and Hirota, T.,** Immunogold labeling of blood-group antigens in human salivary glands using monoclonal antibodies and the streptavidin-biotin technique, *Histochemistry*, 87, 539, 1987.
61. **Horisberger, M.,** Lectin cytochemistry, in *Immunolabelling for Electron Microscopy*, Polak, J. M. and Varndell, I. M., Eds., Elsevier, Amsterdam, 1984, 249.
62. **Roth, J.,** Postembedding cytochemistry with gold-labelled reagents: a review, *J. Microsc.*, 143, 125, 1986.
63. **Bendayan, M.,** Enzyme-gold electron microscopic cytochemistry: a new affinity approach for the ultrastructural localization of macromolecules, *J. Electron Microsc. Tech.*, 1, 349, 1984.
64. **Hutchinson, N.J.,** Hybridisation histochemistry: *in situ* hybridisation at the electron microscope level, in *Immunolabelling for Electron Microscopy*, Polak, J.M. and Varndell, I.M., Eds., Elsevier, Amsterdam, 1984, 341.
65. **Binder, M., Tourmente, S., Roth, J., Renaud, M., and Gehring, W.J.,** *In situ* hybridization at the electron microscope level: localization of transcripts on ultrathin sections of Lowicryl K4M-embedded tissue using biotinylated probes and protein A-gold complexes, *J. Cell Biol.*, 102, 1646, 1986.
66. **Robenek, H., Rassat, J., Hesz, A., and Grünwald, J.,** A correlative study on the topographical distribution of the receptors for low density lipoprotein (LDL) conjugated to colloidal gold in cultured human skin fibroblasts employing thin section, freeze fracture, deep-etching, and surface replication techniques, *Eur. J. Cell Biol.*, 27, 242, 1982.
67. **Robenek, H. and Severs, N.,** Double labeling of lipoprotein receptors in fibroblast cell surface replicas, *J. Ultrastruct. Res.*, 87, 149, 1984.
68. **Willingham, M.C. and Pastan,** I., Morphologic methods in the study of endocytosis in cultured cells, in *Endocytosis*, Pastan, I. and Willingham, M.C., Eds., Plenum Press, New York, 1985, 281.
69. **Smedrod, B., Malmgren, M., Ericsson, J., and Laurent, T.C.,** Morphological studies on endocytosis of chondroitin sulphate proteoglycan by rat liver endothelial cells, *Cell Tissue Res.*, 253, 39, 1988.
70. **Frens, G.,** Controlled nucleation for the regulation of the particle size in monodisperse gold suspensions, *Nature*, 241, 20, 1973.
71. **Mühlpfordt, H.,** The preparation of colloidal gold particles using tannic acid as an additional reducing agent, *Experientia*, 38, 1127, 1982.
72. **Baschong, W., Lucocq, J.M., and Roth, J.,** "Thiocyanate gold": small (2-3 nm) colloidal gold for affinity cytochemical labeling in electron microscopy, *Histochemistry*, 83, 409, 1985.
73. **Van Bergen en Henegouwen, P.M.P. and Leunissen, J.L.M.,** Controlled growth of colloidal gold particles and implications for labelling efficiency, *Histochemistry*, 85, 81, 1986.

74. **De Mey, J.R.,** The preparation of immunoglobulin gold conjugates (IGS reagents) and their use as markers for light and electron microscopic immunocytochemistry, in *Immunocytochemistry*, Cuello, A.C., Ed., John Wiley & Sons, Chichester, 1983, 347.
75. **Völker, W., Schmidt, A., Robenek, H., and Buddecke, E.,** Binding and degradation of proteoglycans by cultured arterial smooth muscle cells. I. Endocytosis and intracellular translocation of proteoglycan-gold conjugates, *Eur. J. Cell Biol.*, 34, 110, 1984.
76. **Völker, W., Schmidt, A., Buddecke, E., Themann, H., and Robenek, H.,** Binding and degradation of proteoglycans by cultured arterial smooth muscle cells. II. Binding sites of proteoglycans on the cell surface, *Eur. J. Cell Biol.*, 36, 58, 1985.
77. **Patterson, S. and Verduin, B.J.M.,** Applications of immunogold labelling in animal and plant virology, *Arch. Virol.*, 97, 1, 1987.
78. **Bastholm, L., Nielson, M.H., and Larsson, L.-I.,** Simultaneous demonstration of two antigens in ultrathin cryosections by a novel application of an immunogold staining method using primary antibodies from the same species, *Histochemistry*, 87, 229, 1987.
79. **Rutter, G., Bohn, W., Hohenberg, H., and Mannweiler, K.,** Demonstration of antigens at both sides of plasma membranes in one coincident electron microscopic image: a double-immunogold replica study of virus-infected cells, *J. Histochem. Cytochem.*, 36, 1015, 1988.
80. **Van Den Pol, A.N.,** Silver-intensified gold and peroxidase as dual ultrastructural immunolabels for pre- and postsynaptic neurotransmitters, *Science*, 228, 332, 1985.
81. **Danscher, G. and Norgaard, J.O.R.,** Ultrastructural autometallography: a method for silver amplification of catalytic metals, *J. Histochem. Cytochem.*, 33, 706, 1985.
82. **Birrell, G.B. and Hedberg, K.K.,** Immunogold labeling with small gold particles: silver enhancement provides increased detectability at low magnifications, *J. Electron Microsc. Tech.*, 5, 219, 1987.
83. **Lah, J.J., Hayes, D.M., and Burry, R.W.,** A neutral pH silver development method for the visualization of 1-nanometer gold particles in pre-embedding electron microscopic immunocytochemistry, *J. Histochem. Cytochem.*, 38, 503, 1990.
84. **Stierhof, Y.-D., Humbel, B.M., and Schwarz, H.,** Suitability of different silver enhancement methods applied to 1 nm colloidal gold particles: an immunoelectron microscopic study, *J. Electron Microsc. Tech.*, 17, 336, 1991.
85. **Schmidt, A. and Buddecke, E.,** High-uptake and low-uptake forms of proteoglycans secreted by arterial smooth muscle cells, *Eur. J. Biochem.*, 153, 269, 1985.
86. **Schmidt, A., Bunte, A., and Buddecke, E.,** Proliferation-dependent changes of proteoglycan metabolism in arterial smooth muscle cells, *Biol. Chem. Hoppe Seyler*, 368, 277, 1987.
87. **Danscher, G. and Norgaard, J.O.R.,** Light microscopic visualization of colloidal gold on resin-embedded tissue, *J. Histochem. Cytochem.*, 31, 1394, 1983.
88. **Hacker, G.W.,** Silver-enhanced colloidal gold for light microscopy, in *Colloidal Gold. Principles, Methods, and Applications,* Hayat, M.A., Ed., Academic Press, Inc., San Diego, 1989, 297.
89. **Völker, W. and Faber, V.,** A two component silver intensification kit for the detection of colloidal gold markers in cultured cells, in *Procedures in Electron Microscopy*, Wilson, A.J. and Beesley, J., Eds., John Wiley & Sons, Chichester, in press.

Chapter 8

ROLE OF MODIFIED LIPOPROTEINS IN ATHEROSCLEROSIS

Wulf Palinski

TABLE OF CONTENTS

ISBN 0-8493-5505-2

I. INTRODUCTION

The last decade has led to significant advances in the understanding of the pathogenesis of atherosclerosis. The correlation between hypercholesterolemia and coronary heart disease has been established by numerous studies, and clinical intervention trials have proven that lowering of plasma cholesterol reduces cardiovascular mortality.[1-4] These findings have given new weight to the lipid infiltration theory of atherosclerosis, but despite a wealth of new knowledge about the cellular and molecular processes within the vascular wall, the mechanism by which low-density lipoproteins (LDL) contribute to the formation of the lesion are not fully understood. Increasing evidence indicates that the pathogenic potential of lipoproteins is enhanced by a number of postsecretory modifications,[5,6] and recent experiments have demonstrated that such modifications occur within atherosclerotic lesions.[7-10] The present chapter, based in large part on work from the Specialized Center of Research on Arteriosclerosis in La Jolla, CA, reviews some of the evidence for the occurrence of modified lipoproteins *in vivo* and discusses their potential role in the initiation and progression of atherosclerosis.

The slow progression and the complexity of the various forms of lesions constitute formidable obstacles in the search for the pathological mechanisms leading to atherosclerosis. From a clinical point of view, advanced atherosclerotic lesions have the greatest relevance, as they may result in ischemia and are the underlying cause of myocardial infarction and stroke. Fractures of plaques, associated with bleeding into the lesion and thrombus formation, have been found in the vast majority of cases of death from cardiovascular causes.[11,12] Subsequently, much emphasis has been placed on the composition and mechanical properties of advanced lesions that may predispose them to rupture,[13] and much attention has been focused on the cellular and molecular interactions in the later stages of lesion formation. Unfortunately, the advanced lesion may not be the best place to look for the primary factors responsible for atherogenesis. The advanced atherosclerotic lesion is a heterogeneous and very complex structure, comprising lipid deposits, areas of cellular proliferation, and increased connective tissue, as well as areas of necrosis.[11,14-16] Intense interactions are likely to occur between inflammatory and proliferative processes involving all cell types present in the lesion, the connective tissue matrix, and plasma components.[17-19] These complex interactions might offer multiple targets for intervention, but effectively prevent the differentiation between pathogenic causes and effects. The early stages of atherosclerosis, on the other hand, are far less complex and therefore more likely to reveal the factors primarily involved in the initiation of atherosclerosis.

II. THE INITIAL LESION: LIPOPROTEINS AND MACROPHAGES IN THE FATTY STREAK

The fatty streak is considered to be the earliest morphologically detectable stage of atherosclerosis. Although fatty streaks may be reversible, their occurrence at predilection sites typical of later stages of atherosclerotic lesions, as well as the natural progression of fatty streaks to more advanced lesions observed in animal models of experimentally induced atherosclerosis, suggest that they are indeed the precursor of transitional lesions and fibrous plaques.[14,20-22] The fatty streak is characterized by two elements: the accumulation of lipids in the subintima, and the presence of large numbers of cholesterol-loaded foam cells. Whether these two events occur simultaneously, or whether one is dependent upon the other, has yet to be established, at least for the human fatty streak. In cholesterol-fed rabbits, focal accumulations of LDL precede monocyte penetration.[23] Endothelial damage has long been held responsible for initiating lesion formation, although the endothelial coverage of fatty streaks generally appears intact. Subtle forms of endothelial damage affecting its barrier function for LDL could, of course, be responsible for increased inflow of lipids into the intima.[17] The focal nature of fatty streaks suggests that local factors, such as specific hemodynamic conditions, could cause subtle localized damage that may act synergistically with hypercholesterolemia. Accumulation of cholesterol and apoprotein in the arterial wall could, in theory, also result from changes in the cellular lipoprotein metabolism or from increased retention of lipoproteins in the extracellular spaces, e.g., as a result of complexing with glycosaminoglycans. Retention may indeed be the principal cause for lipid accumulation, at least in cholesterol-fed rabbits where retention of LDL in the extracellular space of lesion-susceptible areas of the aorta is the earliest identifiable event.[24]

The second characteristic element of the early fatty streak is the accumulation of foam cells derived almost exclusively from circulating monocytes.[25-27] In animal models of atherosclerosis, increased monocyte adhesion at lesion-susceptible areas can be induced by hyperlipidemia.[28] The adhesion of monocytes is likely to be mediated by specific endothelial leukocyte adhesion molecules (ELAMs).[29-31] Chemotactic factors within the lesion or secreted by endothelial cells probably play a central role in the initial attraction of monocytes and in their subsequent penetration of the intima.[32,33] Once in the intima, the monocytes undergo phenotypic changes to resident macrophages and foam cells.

III. MECHANISMS OF LIPOPROTEIN UPTAKE BY MACROPHAGES: WHAT LEADS TO FOAM CELL FORMATION?

Occasional observations of monocytes penetrating the vascular wall provide no indication that they have taken up significant amounts of cholesterol during their circulation in plasma, prior to penetration.[25,28] It can therefore be assumed that the excessive uptake of cholesterol by macrophages, and their resulting transformation into foam cells, occurs within the vascular wall. This points to an apparent paradox: although we know that the cholesterol accumulated in the fatty streak originates primarily from plasma LDL, it is not possible to transform macrophages into foam cells by incubation with high concentrations of LDL, *in vitro*. Although macrophages express a small number of the classical B/E receptors, the uptake of LDL by these receptors is regulated by the intracellular LDL concentration.[34] Furthermore, foam cell formation and atherogenesis are greatly enhanced in patients with homozygous familial hypercholesterolemia and in Watanabe Heritable Hyperlipidemic (WHHL) rabbits, both of whom have no functional LDL receptors on macrophages (or any other cell).[35-37]

In 1979 Goldstein et al. described a receptor on macrophages that mediated rapid uptake and degradation of acetylated LDL.[35] Uptake of acetylated LDL and other forms of chemically modified LDL by this "scavenger" receptor is not downregulated by high LDL concentrations and is several-fold greater than the uptake of native LDL by the B/E receptor. *In vitro* uptake of chemically modified lipoproteins leads to rapid foam cell formation. The scavenger receptor, found only on monocyte/macrophages, endothelial cells, Kupffer cells, and sinusoidal liver cells,[38] recognizes derivatized ε-amino groups of lysine residues of the apolipoprotein. Derivatizations of lysines profoundly affect the biological properties of proteins,[39,40] and the removal of modified lipoproteins could therefore be expected to be a primary function of macrophages.[35] Acetylation of lysine residues of the apoprotein provides a model of lipoprotein modification, but the physiologic occurrence of acetylation remains doubtful. The search for naturally occurring forms of modified lipoproteins was greatly enhanced when Henriksen et al. discovered that LDL incubated with cultured endothelial cells undergoes extensive structural modification, leading to its recognition by the scavenger receptor.[41]

IV. THE OXIDATIVE MODIFICATION AND ITS PATHOGENIC POTENTIAL

The chemical nature of this "endothelial-cell-modified" LDL (EC-LDL) has been extensively studied by our group. Compared to native LDL, EC-LDL is characterized by increased density, increased negative charge, greater electrophoretic mobility, decreased polyunsaturated fatty acids, increased lysolecithin content, and extensive fragmentation of apoprotein B-100.[5,41,42] The EC modification is clearly an oxidative process. It is presumed to be initiated by the peroxidation of polyunsaturated fatty acids in the phospholipid fraction of LDL and can be completely inhibited by antioxidants, such as butylated hydroxytoluene (BHT) and vitamin E, or by addition of plasma (plasma proteins may act as free radical scavengers, and may provide a variety of antioxidants). Lipid peroxidation results in the formation of highly reactive short chain-length aldehyde fragments, such as malondialdehyde (MDA) or 4-hydroxynonenal (4-HNE).[43,44] These aldehydes can then react with free amino-lysine residues of the apoprotein, leading to the formation of various adducts recognized by the scavenger receptor.[39,45] The EC modification is dependent on the presence of metal ions such as Cu^{++}. The addition of metal chelators, such as EDTA, completely prevents the modification of LDL by endothelial cells. Incubation of

LDL with 5 m*M* Cu^{++} ions in the absence of cells generates an oxidized form of LDL that shares many of the properties of EC-LDL.[45-47] A phospholipase A_2 enzymatic activity intrinsic to the LDL particle contributes to the extensive breakdown of lecithin to lysolecithin, and to the formation of lipid peroxides during EC modification.[48,49] Incubation of LDL with (snake venom) phospholipase A_2 and lipoxygenase, in the absence of cells, can again mimic EC-induced modification *in vitro*, whereas addition of phospholipase A_2 inhibitors prevents the oxidative modification by endothelial cells.[48,50]

The oxidative modification of the apoprotein leads to the formation of aldehyde-lysine adducts recognized by the scavenger receptor and mediates the rapid uptake of the modified lipoprotein by macrophages, but it may contribute to atherogenesis by other mechanisms, too. The cytotoxicity of high concentrations of LDL for cultured endothelial cells has long been known.[51,52] It is due to lipid peroxidation products generated during the modification of LDL by the endothelial cells; in other words, the cytotoxic effects are not an inherent property of native LDL, but of oxidized LDL. It is conceivable that oxidatively modified LDL formed in the atherosclerotic lesion could contribute to functional damage of the endothelium or even to endothelial denudation, which would in turn lead to further influx of lipoproteins into the vascular wall. Furthermore, Quinn et al. have demonstrated that the lipid fraction of oxidatively modified LDL has chemotactic activity for circulating monocytes, whereas it inhibits the motility of tissue macrophages.[53,54] Thus, oxidized LDL could increase the number of macrophages in the fatty streak by promoting the recruitment of circulating monocytes and could contribute to their accumulation by reducing their motility once they have reached the subintima. The degree of oxidative modification required to obtain chemotactic effects is probably very limited. Minimally modified LDL, i.e., LDL that has been oxidized to a degree that does not lead to significant uptake via the scavenger receptor, has been shown to stimulate monocyte-endothelial interactions and to induce the expression of the chemotactic factor MCP-1 and of leukocyte-colony-stimulating factors by endothelial cells.[55] Last but not least, oxidized LDL has been reported to inhibit the release of endothelium-derived relaxing factor (EDRF), which may result in localized vasoconstriction in coronary arteries.[56]

Endothelial cells are not the only cells capable of inducing oxidative modification of LDL. Cultured macrophages and smooth muscle cells also generate oxidized LDL, at least *in vitro*,[52,57-60] and we have recently shown that macrophages and macrophage-derived foam cells extracted from atherosclerotic lesions of WHHL rabbits are capable of promoting such oxidation.[61] Thus, all three major cell types found in the vascular lesion could contribute to the modification of LDL accumulated in the intima.

Based on these observations, a hypothesis was proposed (by our group[5]) that integrates pathogenetic mechanisms suggested by the lipid infiltration theory and the response to injury hypothesis. According to this hypothesis, hypercholesterolemia would induce adherence and penetration of monocytes, concomitant with lipid accumulation in the intima. Part of the lipid would be oxidatively modified by endothelial cells and macrophages, within the vessel. In turn, oxidized lipids would increase monocyte adhesion, penetration, and accumulation by the mechanisms described above. Oxidized lipids could then contribute to foam cell formation and, by causing endothelial injury, lead to further influx of lipids. Once the endothelial denudation occurs, the mechanisms proposed by the original response to injury hypothesis could come into action. Platelets may adhere to the exposed subendothelium and release growth factors. Platelet-derived growth factors, together with growth factors released by macrophages and other cells present within the initial lesion, may then account for cell proliferation and the progression of the fatty streak to more advanced stages of lesion. As this hypothesis was based in large part on *in vitro* observations, the demonstration that lipoprotein oxidation occurs *in vivo* was essential for its verification.

V. EVIDENCE FOR THE OCCURRENCE OF OXIDATIVE MODIFICATION *IN VIVO*

Two basic problems confront any experimental approach to demonstrate the presence of oxidized lipoproteins: first, the ubiquitous nature of oxidative processes places great emphasis on the prevention of artifacts during any procedure; second, and more important, detection of "oxidized LDL" is complicated by the complex nature of the oxidative modification. Western blots of oxidized lipoproteins show numerous apoprotein fragments.[9] *A priori*, it was not known which of these fragments might contain epitopes specific to the oxidative modification, and what the chemical nature of the adducts formed would be. To overcome these difficulties, we developed a panel of antibodies against different "model" epitopes that are formed during the oxidation of LDL, and used these antibodies for immunocytochemistry and other immunological methods, postulating that the simultaneous presence of a variety of such adducts in tissue sections or arterial lipoprotein extracts would demonstrate the occurrence of lipoprotein oxidation. In the following the development of these antibodies will be briefly described, and several mutually supportive lines of evidence will be presented for the occurrence of oxidative modification of lipoproteins *in vivo*. These include (1) the immunocytochemical demonstration of oxidation-specific epitopes in atherosclerotic lesions of WHHL rabbits and humans, (2) the demonstration that lipoproteins gently extracted from lesions contain oxidation-specific epitopes and share compositional and biological properties with LDL oxidized *in vitro*, (3) the presence of circulating autoantibodies against at least one form of oxidatively modified lipoprotein, and (4) the observation that treatment of WHHL rabbits with probucol, a potent lipophilic antioxidant, reduces atherogenesis.

TABLE 1
Antibodies Against Epitopes Generated during the Oxidative Modification

Antibody	Type	Immunogen	Specificity
MAL-2	Guinea pig antiserum	Guinea pig MDA-LDL	MDA-lysine
MDA2	Mouse monoclonal	Mouse MDA-LDL	MDA-lysine
HNE-6	Guinea pig antiserum	Guinea pig 4-HNE-LDL (reduced)	4-HNE-lysine
NA59	Mouse monoclonal	Mouse 4-HNE-LDL (reduced)	4-HNE-lysine
OLF4-3C10	Mouse monoclonal	Apoprotein fragments of Cu^{++}-oxidized mouse LDL	?

A. Development of Oxidation-Specific Antibodies

The oxidative modification of LDL is accompanied by extensive degradation of its polyunsaturated fatty acids, leading to the generation of a number of highly reactive short-chain fragments.[43,62] Malondialdehyde (MDA) and 4-hydroxynonenal (4-HNE) are two such reactive aldehydes.[44,63,64] During the oxidative modification of LDL, lipid peroxidation products can be conjugated to the apoprotein predominantly by covalent linkage with ε-amino groups of lysine residues.[44,45] Apoprotein B-100 contains about 360 lysine residues,[65,66] and the close proximity between apoprotein and lipid in the LDL particle makes it likely that a number of different aldehydes formed by the peroxidation of polyunsaturated lipids will be conjugated with these lysines. Thus, the simultaneous presence of different lysine-aldehyde conjugates on the apoprotein would provide additional evidence for the occurrence of oxidative modification of LDL. We therefore generated a panel of monospecific antibodies against a variety of lysine derivatives (Table 1).

The immunization of animals with homologous LDL, modified by conjugation of lysine groups with short-chain aldehydes such as formaldehyde or glucose, induces formation of antibodies against the lysine adduct, i.e., methyl-lysine or glucitol-lysine.[67-69] The specificity of such antibodies tends to be for a narrow epitope of the lysine adduct. In other words, the modified lysine will be recognized not only on modified LDL, but also on a variety of other similarly modified proteins. Polyvalent antisera and corresponding monoclonal antibodies against compounds formed during the oxidative modification were generated in a similar fashion by immunizing guinea pigs and mice with homologous LDL conjugated *in vitro* with MDA or 4-HNE. The modification and immunization procedures, as well as the determination of antibody titers and their characterization by competitive solid-phase radioimmunoassays, are described in detail in Reference 9. The specificity of the monoclonal antibodies closely resembled that of the polyvalent antisera. As expected, the antibodies generated against MDA-LDL recognized MDA-lysine epitopes, independently of the modified protein, whereas native LDL and conjugates of LDL with other aldehydes were not recognized. The MDA-lysine-specific antibodies also recognized some material in Cu^{++}-oxidized LDL; i.e., the oxidation of LDL by Cu^{++} ions in the absence of exogenous MDA also led to the generation of some MDA-lysine epitopes. The extent of binding of the antiserum MAL-2 (specific for the MDA-lysine epitope) to MDA-modified LDL increased with increasing degree of modification of its lysine residues. Although these antibodies recognize a specific epitope on a variety of proteins, it is essential to keep in mind that binding is not simply a linear function of the number of specific epitopes on any given protein, but that the nature of the protein, and in particular its tertiary structure, will, to a significant extent, determine the access of the antibody to these epitopes. Their application for immunocytochemistry or other immunochemical methods will therefore provide only semiquantitative information.

Antisera and monoclonal antibodies generated against reduced 4-HNE-LDL showed analogous specificity for the 4-HNE-lysine epitope. The reduced form of 4-HNE-LDL had been chosen as immunogen, as nonreduced 4-HNE-LDL had resulted in weak antibody formation. However, the 4-HNE-lysine epitope was recognized both on reduced and nonreduced proteins. No crossreactivity was observed between MDA- and 4-HNE-lysine-specific antibodies.

Although MDA-lysine and 4-HNE-lysine conjugates are known to be formed during the oxidation of LDL *in vitro*, no assumptions could be made *a priori* in regard to their quantitative occurrence *in vivo*. In addition to these two well-characterized "model epitopes", a large variety of other aldehyde-lysine adducts are likely to be formed as a result of lipid peroxidation. In order to detect such unknown adducts, we also generated monoclonal antibodies against LDL oxidized with Cu^{++} ions, which causes the breakdown of apoprotein B-100 into multiple apoprotein fragments and constitutes a good model of cell-induced lipoprotein oxidation. To increase the probability of generating an immune response against oxidation-specific epitopes on the apoprotein and to expose epitopes possibly masked by lipids, we immunized with a mixture of delipidated apoprotein fragments from murine LDL oxidized for 4 and 24 h.[9,70] Hybridoma supernates were screened against the immunizing agents, as well as intact copper-modified human LDL and native human LDL, and clones reacting with the oxidized forms were selected. One of the resulting monoclonal antibodies (OLF4-3C10) was characterized in greater detail.[9] Not surprisingly, it recognized lipoproteins oxidized to various extent, but not native proteins or lipoproteins. Recognition of oxidized lipoproteins was always enhanced by delipidation. Although the exact epitope recognized remains unknown, the fact that these antibodies recognized apoprotein fragments of LDL oxidized to various extent, as well as fragments of oxidized HDL, indicates that it is most likely a lipid-protein adduct similar to those above described.

B. Presence of Oxidation-Specific Epitopes in Atherosclerotic Lesions

1. Methods

a. Tissue Preparation for Lipoprotein Extraction and Immunostaining

Watanabe Heritable Hyperlipemic rabbits between 4 months and 4 years were anesthetized with Ketamine/Xylazine (35 mg/kg i.m. and 5 mg/kg i.m., respectively), and their aortic tree was perfusion fixed with formol-sucrose (4% paraformaldehyde plus 5% sucrose, pH 7.4) containing 50 μ*M* BHT and 1m*M* EDTA by way of a cannula inserted into the left ventricle. Tissue segments for immunostaining were rinsed overnight in 0.1 *M* sodium phosphate buffer, pH 7.4, containing 4% sucrose, 50 μ*M* BHT, and 1 m*M* EDTA at 4°C to remove remaining fixative; dehydrated through a graded series of ethanol concentrations, and embedded in paraffin. In some animals used for both immunostaining and extraction of intimal LDL, the aorta was perfused only with buffer containing BHT and EDTA, and segments for immunostaining were subsequently immersion fixed overnight in either formal-sucrose or Omnifix (Xenetics Biomedical, Inc., Irvine, CA) containing antioxidants.

b. Immunocytochemistry

Serial 5 to 7 μm-thick sections were cut from the paraffin-embedded aorta segments (arch, thoracic, and abdominal segments adjacent to major branch points), rehydrated, and immunostained using an avidin-biotin-alkaline phosphatase system (Vector Labs, Burlingame, CA). Alkaline phosphatase was selected, instead of horseradish peroxidase, to avoid the need to preblock endogenous peroxidase by exposure to hydrogen peroxide and thus potentially induce oxidation of arterial proteins.[71] Serial sections were immunostained for 1 h at room temperature with the antisera MAL-2 (anti MDA-lysine), HNE-6 (anti-4-HNE-lysine), the monoclonal antibodies OLF4-3C10 (anti-Cu^{++}-oxidized apo B fragments), MDA2 (anti MDA-lysine), and NA59 (anti-4-HNE-lysine), as well as with MB47, a monoclonal antibody against the receptor binding site of native apoprotein B-100,[72] and YE-1, a guinea pig antiserum against native WHHL-LDL.[10] In addition, RAM-11, a monoclonal antibody that specifically recognizes rabbit macrophages,[73] and HHF-35,[74] a monoclonal antibody against actin, were used to detect macrophages and smooth muscle cells. The optimum dilution of each primary antibody was established empirically. Controls included nonspecific antibodies (horse and goat sera) and oxidation-specific antisera preincubated with their respective antigen to absorb specific antibodies. For example, MAL-2 was preincubated successively in cell culture dishes that had been coated with MDA-LDL until antibody binding studies indicated that the preabsorbed MAL-2 antisera had no titer against MDA-LDL. Additional competition controls were performed by adding excess immunogen to the primary antibody prior to immunostaining. To exclude oxidative artifacts or unspecific binding of antibodies or staining reagents, appropriate controls were performed in which each step in the immunostaining procedure was omitted, one at a time.

2. *Distribution of Oxidation-Specific Epitopes*

Immunocytochemistry with all of the antibodies against oxidation-specific epitopes demonstrated that these epitopes are present in atherosclerotic lesions, ranging from early fatty streaks to very advanced fibrous plaques.[10] Other lipid-rich tissues treated in an identical fashion to the atherosclerotic artery, such as the adrenal gland, lung, or liver tissues, were generally free of staining, showing that the oxidation-specific epitopes were not generated during dehydration, embedding, rehydration, and staining procedures.

In early fatty streaks, staining was predominantly macrophage associated, and very little extracellular staining was observed. By contrast, immunocytochemistry with antibodies recognizing native apoprotein B exhibited diffuse, predominantly extracellular staining in the areas beneath the subendothelial macrophages and along the internal elastic lamina.

In transitional and early fibrous lesions, the macrophage-associated staining still predominated, but we also observed limited particulate staining of intimal smooth muscle cells and light diffuse staining of the extracellular matrix, especially in the core and cap regions. At higher magnification, the cellular staining of intimal macrophages and foam cells showed punctate and circular patterns (Figures 1A, B). These probably represent epitopes associated with membranes delimiting intracellular lipid accumulations (the lipid itself having been extracted by the ethanol gradient during embedding and immunostaining procedures). The annular staining pattern is reminiscent of ceroid staining described in human atherosclerotic lesions.[75] Ceroid, an inert and very stable lipid-protein conjugate that cannot be extracted with ethanol-xylene and stains with lipophilic stains, has been suggested to be an oxidation product because its formation can be influenced by oxidizing and reducing conditions.[75] Staining of transitional and early fibrous lesions with apo-B-specific monoclonal antibodies was diffuse and limited to the intima (Figure 1C).

The distribution of the different epitopes recognized by our panel of oxidation-specific antibodies was surprisingly similar. Staining was usually most intense in the shoulder area of lesions, coinciding with dense accumulations of macrophages (Figure 2). In this particular lesion the staining for oxidation-specific epitopes was clearly complementary to that for native apoprotein (Panels A–D). Preabsorbed controls, as well as competition and nonimmune controls, were entirely devoid of staining, again demonstrating the specificity of the antibodies.

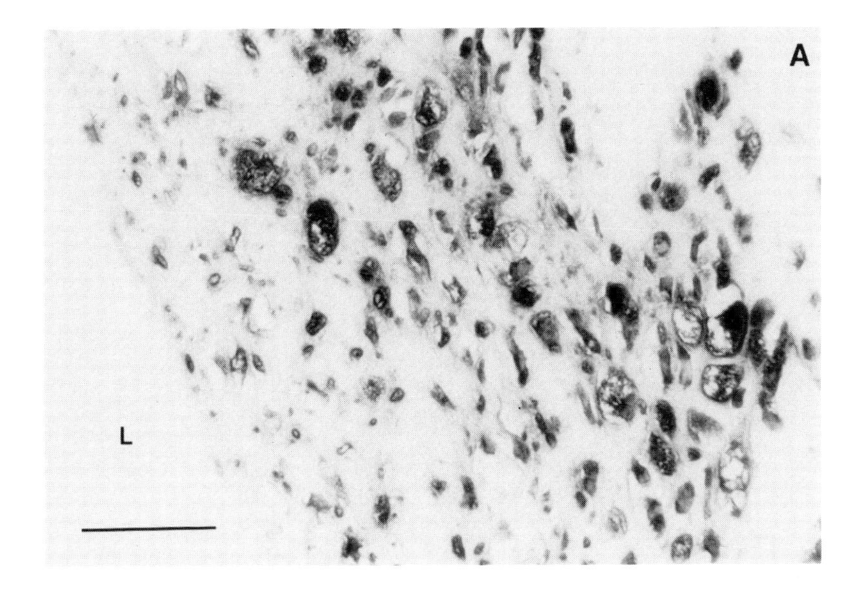

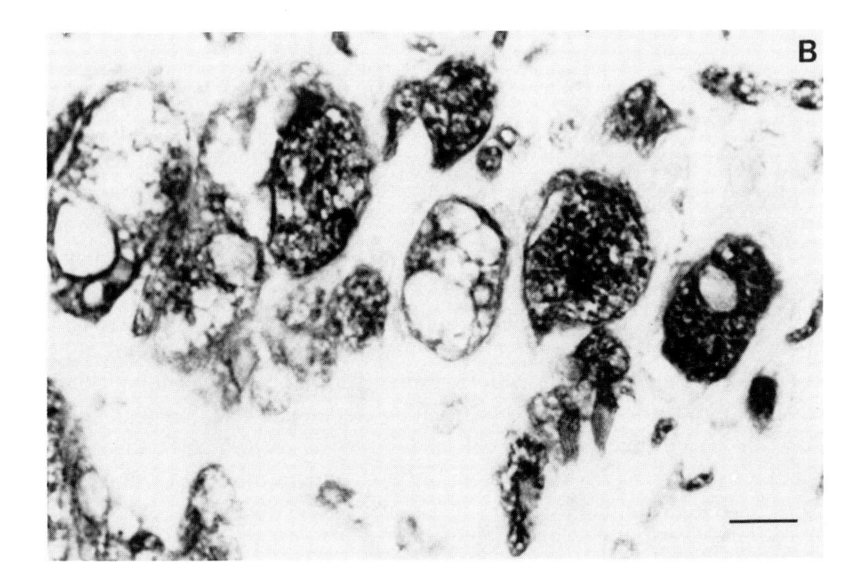

FIGURE 1. Immunostaining of a transitional lesion of the thoracic aorta of a WHHL rabbit with the avidin-biotin-alkaline phosphatase method. Panel A shows a macrophage-rich area of the lesion, stained with a 1:1000 dilution of the monoclonal antibody OLF4-3C10, which recognizes oxidized apoprotein fragments. As in all early lesions, staining is predominantly cell associated. At higher magnification (Panel B) the staining of macrophages and macrophage-derived foam cells reveals both particulate and annular patterns. By contrast, staining with a 1:250 dilution of MB 47, a monoclonal antibody against the receptor-binding region of native apoprotein B-100, is predominantly diffuse and extracellular (Panel C). Staining with MB 47 (or YE-1, an antibody against native WHHL-LDL) was often found in areas underneath subendothelial macrophage accumulations and in the proximity to the internal elastic lamina (IEL). (L = lumen). (Bar: 50 μm in Panel A; bar: 10 μm in Panels B and C).

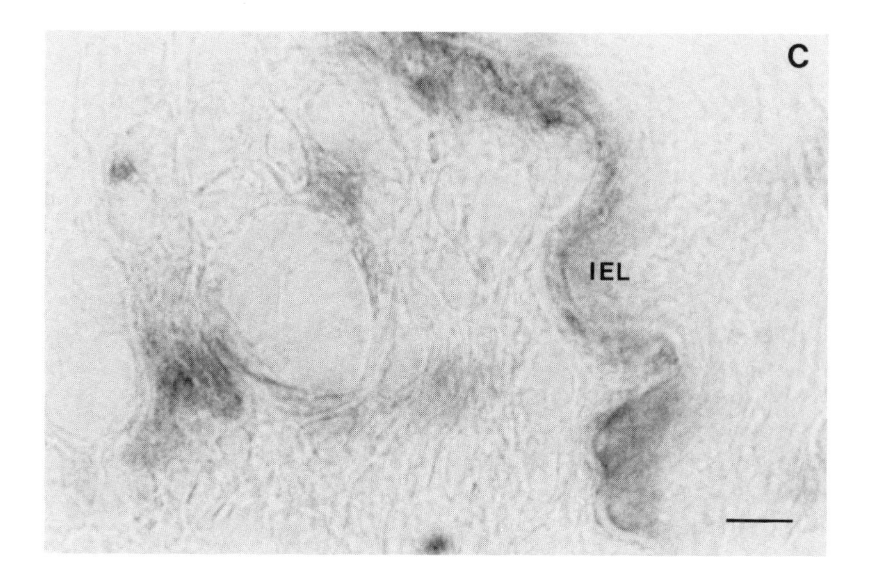

FIGURE 1 (continued).

In advanced lesions, extensive cell-associated staining was still observed, in particular, in the macrophage-rich shoulder and subendothelial regions. However, in areas with fewer cells, there was increased extracellular, diffuse staining with the antibodies against oxidation-specific epitopes and colocalization with intact apoprotein B. In addition, the necrotic core showed some annular staining patterns reminiscent of ceroid staining (Figure 3), and increased staining of intimal and medial smooth muscle cells. Figure 4 is an example of an advanced lesion that simultaneously exhibits both diffuse and cell-associated staining patterns. In the relatively cellular area immediately beneath the endothelium, both cell-associated and extracellular staining are found. Beneath the cellular area a band of diffuse extracellular staining can be detected above an area of widely dispersed smooth muscle cells exhibiting only cell-associated staining. At the base of the lesion in the necrotic core is a large area of entirely extracellular diffuse staining. Similar examples of simultaneous cellular and extracellular staining of complex lesions were obtained with all antibodies against oxidation-specific epitopes. Some staining of smooth muscle cells by the oxidation-specific antibodies was also frequently observed in advanced lesions.

Extensive *in vitro* evidence has proven that endothelial cells, macrophages, and smooth muscle cells are capable of effecting the oxidation of LDL.[41,52,57,59,60] The immunocytochemical demonstration of oxidation-specific epitopes within macrophages therefore does not allow one to differentiate whether these epitopes result from intracellular oxidation or from uptake of extracellular oxidized lipoproteins, or to determine the extent of extracellular oxidation induced by macrophages. In fatty streaks and transitional lesions, oxidation-specific epitopes were predominantly macrophage associated, whereas native LDL was found primarily extracellularly. This observation suggests that any extracellular oxidized LDL is rapidly taken up by the macrophages.

The extensive particulate staining observed in macrophages probably represents an accumulation of oxidized epitopes in intracellular organelles, possibly lysosomes. Several explanations could account for the intracellular accumulation of oxidized lipoproteins. The presence of aldehyde-modified lysine residues on apo B could render apo B relatively resistant to lysosomal degradation.[76] Crosslinking of apoprotein fragments with lysosomal membranes could also contribute to the marked accumulation of such epitopes in the macrophages. Current studies using electron microscopic immunocytochemistry should determine the exact intracellular location of these epitopes. Immunostaining of serial sections of a transitional lesion (Figure 2) provides evidence for the contribution of macrophages to the oxidative modification of LDL in the lesion; an acellular fibrous streak in the middle of an otherwise macrophage-rich cellular fatty streak stains with MB47, specific for the native apoprotein B, but is devoid of oxidation epitopes. By contrast, the adjacent macrophage-rich regions stain extensively with antibodies against the various oxidation-specific epitopes. The absence of oxidation in cell-free areas and the predominance of cell-associated staining in the fatty streak and transitional lesion would also be consistent with the hypothesis proposed by Steinberg et al.[5] that oxidative modification of LDL occurs in microdomains adjacent to cells, where prooxidant conditions may be sufficient to initiate the oxidative modification of LDL.

In more advanced lesions that contain relatively few macrophages, increasing diffuse, extracellular staining of oxidation-specific epitopes and a greater degree of colocalization with native LDL were found. This may indicate that lipoproteins are trapped in the extracellular spaces by complexing with matrix proteins[20,77] and have become inaccessible to cells capable of taking them up. Formation of complexes with matrix components may, on the other hand, increase the residence time of LDL and subject it to spontaneous or cell-induced lipid peroxidation. The intense extracellular staining observed in the necrotic core is likely to represent, at least in part, oxidation products released from dead and decaying cells.

In order to determine whether oxidation-specific epitopes also occur in nonatherosclerotic arteries, sections of New Zealand White rabbit aorta and sections of aorta from WHHL rabbits that did not exhibit any lesions were examined. All of these sections were devoid of specific staining, with the exception of the aortic adventitia, which stained independently of the presence of atherosclerotic lesion. The reason for this adventitial staining is unclear, but it is noteworthy that the adventitia contains both collagen and proteoglycans that may be responsible for LDL trapping, as well as a significant number of macrophages capable of modifying it. Furthermore, the oxygen concentration in the adventitia relative to the inner media or intima is high.[78]

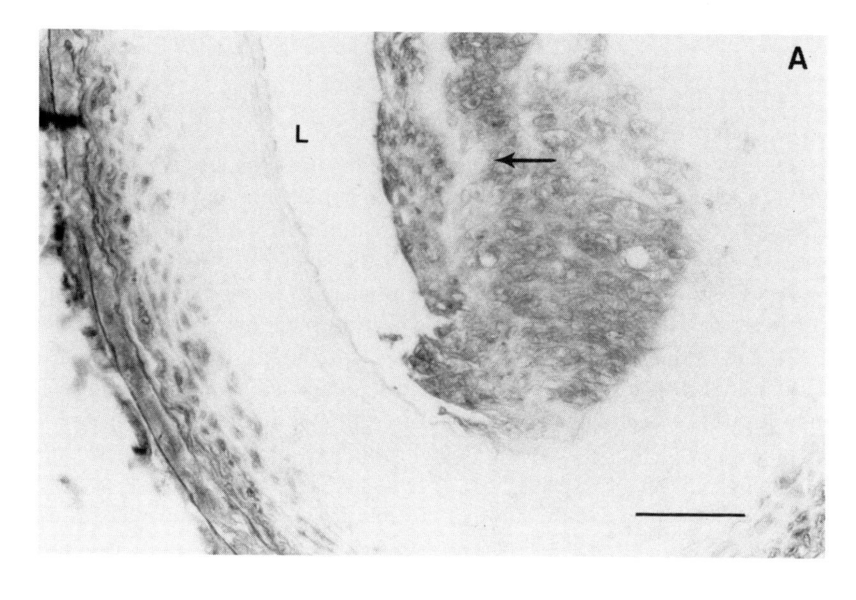

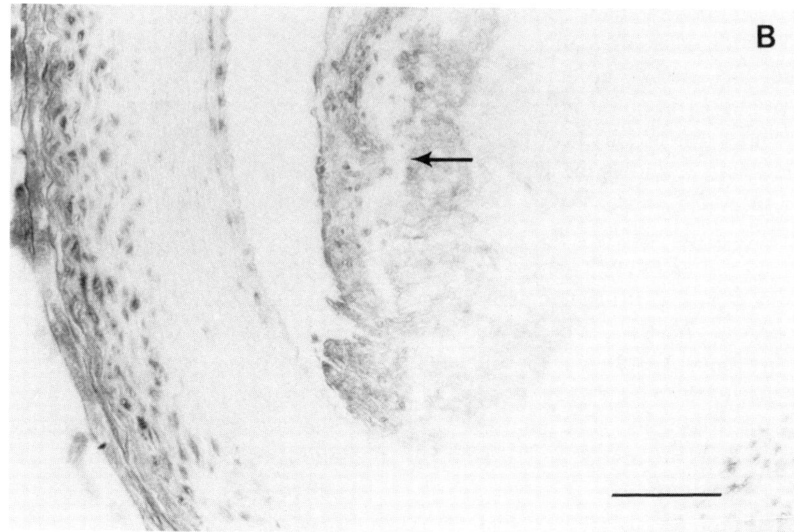

FIGURE 2. Comparative immunostaining of serial sections of a transitional lesion with the monoclonal antibody MDA2 (dilution 1:50) (Panel A) and the guinea pig antiserum HNE-6 (1:2000) (Panel B). Staining patterns obtained with these, and all other, oxidation-specific antibodies was very similar. Oxidation-specific epitopes were found predominantly in areas rich in macrophages, as seen in Panel D, stained with the macrophage-specific monoclonal antibody RAM-11 (1:1000). By contrast, an area relatively devoid of cells (arrow) stained with MB 47, a monoclonal antibody against native LDL (1:250) (Panel C). (L = lumen). (Bars: 100 μm).

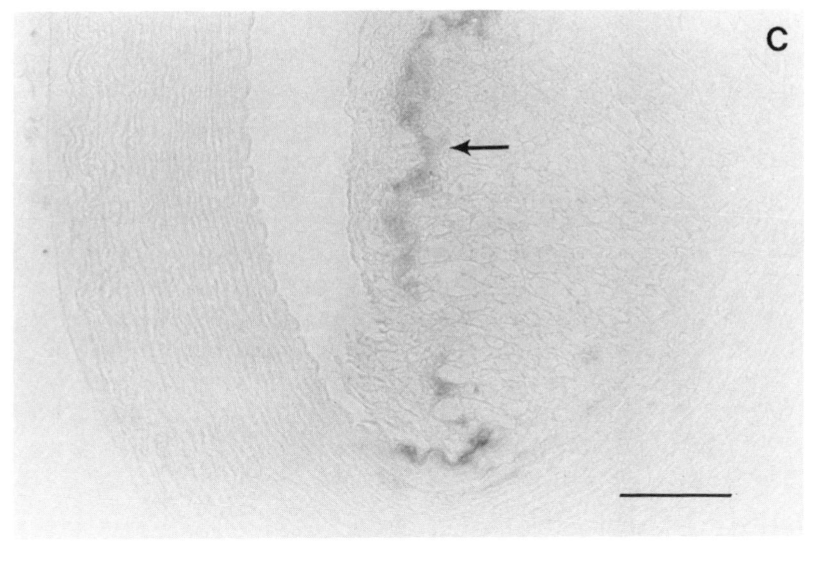

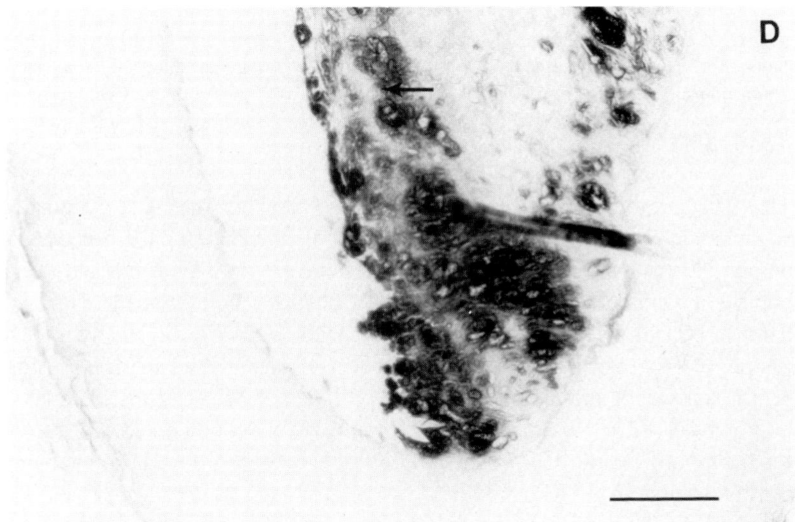

FIGURE 2 (continued).

The immunocytochemical application of five different antisera and monoclonal antibodies on serial sections of atherosclerotic lesions has enabled us to demonstrate the occurrence of oxidative modification in atherosclerotic lesions. Additional evidence for the *in vivo* presence of individual oxidative epitopes has been recently presented by other groups. Mowri et al. raised an antibody against homogenized atherosclerotic plaque that, in turn, bound to *in vitro* oxidized

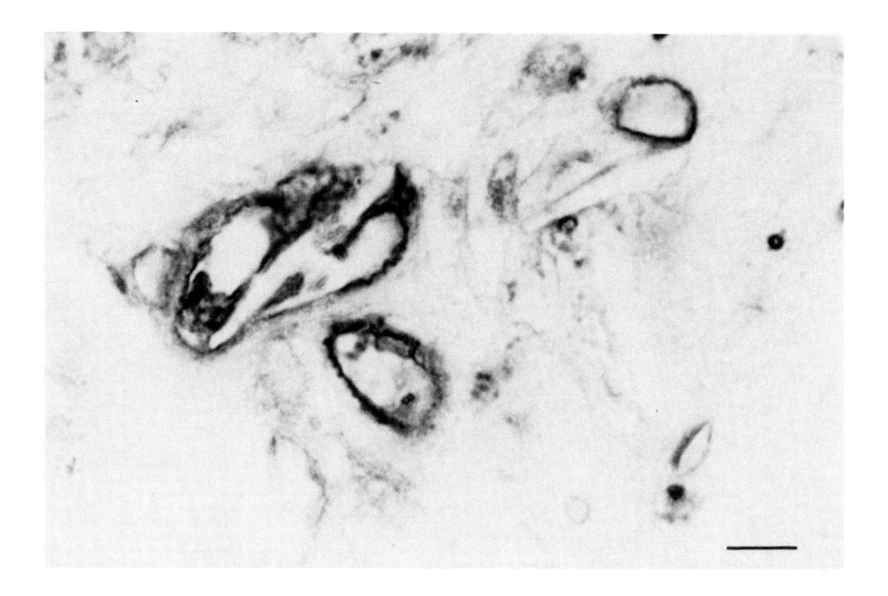

FIGURE 3. Necrotic core of an advanced lesion of a WHHL rabbit aorta, stained with 1:1000 dilution of the monoclonal antibody OLF4-3C10. These annular staining patterns, reminiscent of ceroid staining, were also observed with other oxidation-specific antibodies. Although the macrophage-associated staining observed in early stages of atherosclerosis showed similar annular patterns, extracellular ring-like staining was only found in advanced lesions. (Bar: 10 μm).

LDL.[79] Haberland et al. first demonstrated the presence of MDA-conjugated proteins in lesions. The staining obtained with their monoclonal antibody was comparable to what we found in advanced lesions; i.e., MDA-lysines and native apoprotein were colocalized. However, they did not report any cell-associated staining, typical of the earlier lesion, and their antibody did not recognize epitopes in Cu^{++}-oxidized LDL.[80] Salmon et al. also generated an MDA-lysine-specific antiserum and demonstrated the presence of MDA-lysine residues in apoprotein B of Cu^{++}-oxidized LDL.[81] Finally, Boyd et al. generated a monoclonal antibody against Cu^{++}-oxidized LDL that recognized material in atherosclerotic lesions of WHHL rabbits.[82]

Although the immunocytochemical application of five different antisera and monoclonal antibodies on serial sections of atherosclerotic lesions has enabled us to establish the occurrence of oxidative modification *in vivo*, we have to keep in mind that the antibodies recognize oxidation-specific antibodies not only on LDL, but also on a variety of other lipoproteins and proteins. In order to demonstrate that the oxidation-specific epitopes in atherosclerotic lesions represent, at least in part, oxidized LDL, we extracted LDL from the same lesions used for immunocytochemistry and showed that it contains these epitopes.

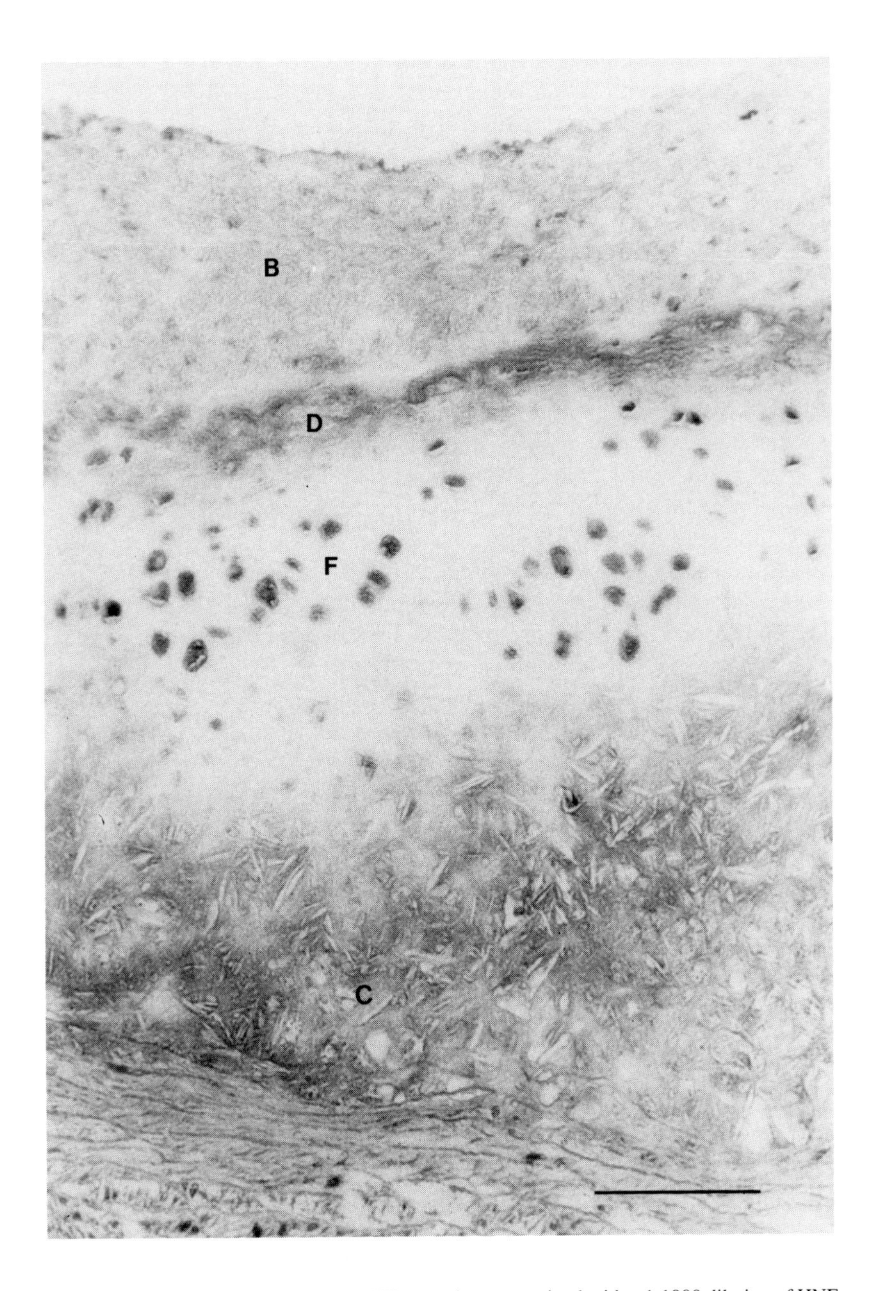

FIGURE 4. Advanced lesion of a WHHL aorta immunostained with a 1:1000 dilution of HNE-6 (anti-4-HNE-lysine), showing predominantly diffuse, extracellular staining in the necrotic core (C) and in a relatively acellular area (D), an area of virtually exclusive cell-associated staining (F), and an area of both cellular and extracellular staining (B) adjacent to the lumen. (Bar: 100 μm).

C. Lipoproteins Extracted from Lesions

Although lipoproteins can be extracted from arterial walls by a variety of methods, the direct demonstration of oxidatively modified LDL in atherosclerotic lesions is complicated by a number of problems that are likely to occur during the extraction and purification procedures.[83] Gentle extraction methods may not yield a representative spectrum of lipoproteins present in the lesion. Intact LDL is found extracellularly in advanced atherosclerotic lesions and normal artery, as demonstrated by numerous studies.[84,85] However, only small amounts of native apoprotein B are present extracellularly in the early fatty streaks. Furthermore, oxidation-specific epitopes are found predominantly to be cell associated, especially in the early lesions, and may therefore be difficult to extract by gentle methods. The theoretical caveats regarding the composition of extracted lipoproteins are confirmed by the observation of Smith[86] that little LDL can be extracted from human fatty streaks consisting primarily of foam cells. More aggressive extraction methods, on the other hand, are more likely to result in oxidative and other artifacts, such as lipoprotein aggregation. The problems that occurr during the subsequent purification and/or concentration of the arterial extracts include loss of apoproteins or apoprotein fragments during ultracentrifugation, insufficient differentiation between proteins and apoproteins by immunoprecipitation or immunoaffinity chromatography, and lipoprotein aggregation during ultracentrifugation.[83] Several groups have reported altered properties of lipoproteins isolated from aortas, but whether these represent LDL oxidatively modified *in vivo* remains uncertain.[87-90]

We have used the method of Ylä-Herttuala et al.,[87] designed to avoid lipoprotein denaturation or aggregation, to determine whether LDL extracted from the same aortas used for our immunocytochemical studies contained oxidatively modified LDL or other forms of modifications that may lead to enhanced uptake by macrophages.[7,8] Three lines of evidence indicated that LDL gently extracted from rabbit and human atherosclerotic lesions ("lesion LDL") had been oxidatively modified:[8]

1. Lesion LDL and *in vitro* Cu^{++}-oxidized LDL showed similar physical and chemical properties. Compared to plasma LDL, they both had higher electrophoretic mobility, higher density, higher free-cholesterol content, and higher proportions of sphingomyelin and lysophosphatidylcholine in their phospholipid fraction. The apoprotein of lesion LDL showed lower-molecular-weight fragments similar to those found in Cu^{++}-oxidized LDL.
2. In Western blots, both the intact apoprotein B band and some of the apo B fragments of lesion LDL were recognized by our antisera against MDA-lysine and 4-hydroxynonenal lysine adducts, whereas plasma LDL and LDL from normal human intima were not recognized.
3. Lesion LDL and Cu^{++}-oxidized LDL also shared biological properties. Compared to plasma LDL, lesion LDL induced much greater stimulation of cholesterol esterification and was degraded more rapidly by macroph-

ages. Degradation of radiolabeled lesion LDL was competitively inhibited by unlabeled lesion LDL, Cu^{++}-oxidized LDL, polyinosinic acid, and MDA-LDL, but not by native LDL, indicating uptake by the scavenger receptor. Finally, lesion LDL showed chemotactic activity for monocytes, which indicates lipid peroxidation.[53,54]

These studies show that at least some of the oxidation-specific epitopes immunocytochemically detected in atherosclerotic lesions result from the oxidative modification of LDL. Together, immunocytochemistry and LDL extraction provide strong evidence for the presence of oxidized LDL in atherosclerotic lesions.

D. Circulating Autoantibodies Recognize Modified Lipoproteins in Lesions

Relatively minor modifications of LDL render it highly immunogenic even in homologous species. We have made use of this immunogenicity to generate the oxidation-specific antisera and monoclonal antibodies described above. The occurrence of oxidative modifications *in vivo* and the presence of several forms of modified lysines in the arterial wall could therefore also lead to the formation of autoantibodies. When we screened sera of both rabbits and humans, we indeed found autoantibodies recognizing MDA-LDL in every serum tested, with titers ranging from 512 to greater than 4,096.[7] Autoantibodies in human serum, respectively in the IgG fraction of human serum, were characterized by competitive radioimmunoassays. Although the immunogen that induced their formation is *a priori* unknown, they recognize primarily the MDA-lysine epitope on MDA-LDL or other MDA-modified proteins and thus resemble the induced antiserum MAL-2 and the monoclonal antibody MDA2.

In order to test whether circulating autoantibodies would recognize modified lysines in atherosclerotic lesions, we used human sera or protein-A purified immunoglobulin fractions as primary antibodies for immunocytochemistry.[10] Figure 5 shows the cell-associated staining in the shoulder region of a transitional lesion with the IgG fraction of a human serum that had a high titer of autoantibodies. Similar staining was obtained with whole sera. Staining patterns resembled those observed with MAL-2, OLF4-3C10, HNE-6, or NA59. Human serum preabsorbed with MDA-LDL showed no staining, again demonstrating the specificity of the autoantibodies for MDA-lysines.

Vice versa, immunoglobulins extracted from atherosclerotic lesions of WHHL rabbits recognized bands of MDA-LDL in Western blots.[91] Thus, it seems that at least a fraction of the immunoglobulins found in atherosclerotic lesions have properties similar to those of the circulating autoantibodies.

E. Effect of Antioxidant Drugs on Atherogenesis

The occurrence of oxidative modification in the vascular wall suggests that it may contribute to atherogenesis by several of the mechanisms described in

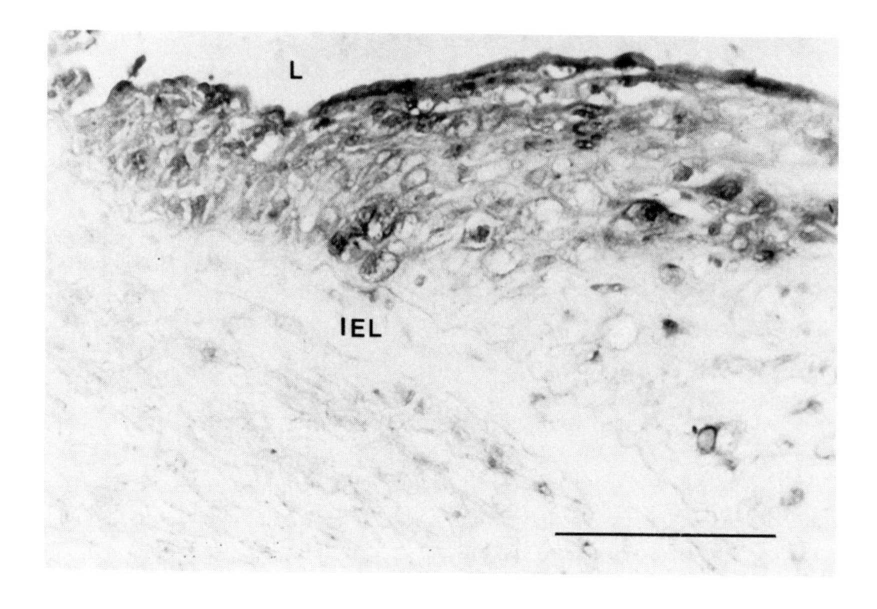

FIGURE 5. Staining of a transitional lesion of a WHHL aorta with a 1:20 dilution of an immunoglubulin G fraction isolated by protein A affinity column from a human serum with high titer of autoantibodies against MDA-LDL. The staining pattern obtained with autoantibodies was predominantly cell associated, showed greatest intensity in macrophage-rich areas of lesions, and generally resembled the staining patterns found with induced antibodies against MDA-lysines and other oxidation-specific epitopes. (Bar: 100 μm).

Section IV, but the presence of oxidized LDL per se does not prove its pathogenic role. On the other hand, the hypothesis that it contributes to atherogenesis implies that interventions designed to reduce intravascular lipoprotein oxidation should lead to reduced lesion formation. Carew et al. provided evidence that this is indeed the case, by showing that WHHL rabbits treated with probucol developed significantly less fatty-streak lesions than did controls.[92] Probucol, commonly used as a lipid-lowering drug, is a powerful inhibitor of the oxidative modification of LDL *in vitro*.[93] This antioxidative effect is probably enhanced by the fact that, in plasma, probucol is carried predominantly in LDL and VLDL, owing to its highly lipophilic properties. The antiatherogenic effect of probucol was clearly unrelated to its hypocholesterolemic action, as the plasma cholesterol level of the probucol-treated rabbits was not lower than that of rabbits treated with lovastatin, a lipid-lowering drug that lacks antioxidant properties. Degradation of the "protected" LDL in macrophage-rich fatty streaks of the probucol-treated WHHL rabbits was about half of that in fatty streaks of controls, whereas the rate of degradation was not different in nonlesioned areas of the same aortas. These results sustain the hypothesis that LDL undergoes oxidative modification and is subsequently taken up by the scavenger receptor of macrophages.

VI. OTHER FORMS OF LIPOPROTEIN MODIFICATION THAT MAY CONTRIBUTE TO FOAM CELL FORMATION

As we have shown, oxidative modification of LDL occurs *in vivo* and can induce foam cell formation by way of its recognition and rapid uptake by the scavenger receptor. However, a number of other modifications recognized by scavenger receptors, as well as receptor-independent uptake mechanisms, could account for lipid accumulation in macrophages.

The common element of cell-induced, enzymatic, or chemical modification of LDL that leads to recognition by the scavenger receptor(s) is the derivatization of the free amino groups of lysine residues in apoprotein B. Recent experimental evidence suggests that a number of different high-affinity receptors may exist for different forms of modification.[94-97] Whether these represent genuinely different receptors or variations of the very complex scavenger receptor[98] remains to be seen.

Although the biological evidence accumulated so far has been centered on the oxidative modification, many different forms of chemically or biologically modified LDL are recognized by these receptors, and increasing evidence indicates that LDL modified by glycosylation may also be of pathophysiological relevance.

Proteins exposed to reducing sugars such as glucose can undergo nonenzymatic glycosylation. Initially, Schiff bases are formed at a rate reflecting the concentration of glucose. The labile Schiff bases then undergo a chemical rearrangement to form more stable Amadori products.[99] A well-known example of such an early glycosylation product is hemoglobin A_{1c}. The early glycosylation reaction is generally reversible, and the formation of Amadori products reaches an equilibrium after several weeks. However, on long-lived proteins, such as membrane and structural proteins, reactive dicarbonyl compounds can be formed that, in turn, propagate the reaction and result in the irreversible formation of heterocyclic protein adducts. These "advanced glycosylation end products" (AGE) are characterized by a brown fluorescence and extensive crosslinking. By interfering with functionally important lysine groups and amino terminal amines, nonenzymatic glycosylation and AGE formation have the potential to affect the biological and structural properties of proteins and may be, in part, responsible for the progression of many diabetic complications, including microvascular pathology and increased atherosclerosis (for reviews see References 100–102).

Nonenzymatic glycosylation and AGE formation could potentially contribute to atherogenesis by a number of pathways:

1. Nonenzymatic glycosylation of LDL and other lipoproteins occurs to a greater extent in hyperglycemic than in euglycemic subjects.[103] Glycosylation of lysine residues on apoprotein B therefore could alter

normal lipoprotein metabolism, as the regulation of plasma LDL depends on its recognition by the LDL receptor.[104]

2. Glycosylated proteins may contribute to foam cell formation, either by rapid receptor-mediated uptake or by phagocytosis of heavily crosslinked particles. Proteins modified by advanced glycosylation are indeed specifically recognized by scavenger-like receptors.[96,97,105]
3. Advanced glycosylation products could also crosslink LDL to structural proteins in the vascular wall, prolonging its half-life.[106] Increased exposure to prooxidant conditions may thus enhance the generation of oxidized LDL. Several *in vitro* observations suggest that lipid peroxidation and formation of Amadori products and AGE may be linked. Products of lipid peroxidation increase the crosslinking of nonenzymatically glycosylated collagen, and conversely, glucose and glycosylated collagen increase free radical production and catalyze lipid peroxidation.[107-109] The putative reciprocal enhancement of oxidation and glycosylation may be of particular interest in other tissues subject to diabetic complications. For example, activated leukocytes, known to release oxidation-promoting agents, have been demonstrated to play an important pathogenic role in diabetic retinopathy.[110] They are responsible for capillary occlusions and are found in large numbers in areas of neovascularization and in areas rich in lipids. It is therefore conceivable that these lipids undergo peroxidation and contribute to the progression of diabetic retinopathy.
4. Finally, we found, in sera of normal and diabetic human subjects, autoantibodies that recognize Amadori products[111] and epitopes of AGE.[112] These autoantibodies may just represent markers of AGE formation, but may also contribute to atherogenesis by forming immune complexes with their respective antigens, within the arterial wall.[113,114]

Although scavenger receptors may account for the enhanced uptake of oxidized or glycosylated lipoproteins by macrophages, receptor-independent uptake mechanisms could also lead to lipid accumulation in macrophages.

The role of the LDL receptor (or "B/E receptor") in foam cell formation has been questioned because of the observation that atherosclerosis is enhanced in receptor-deficient humans and animals and that incubation of macrophages with LDL cannot induce foam cell formation. Nevertheless, this receptor could contribute to the cholesterol enrichment of macrophages by at least two mechanisms β: first, its affinity for β-VLDL is much greater than that for LDL, and the uptake of β-VLDL is not downregulated by intracellular cholesterol; second, the receptor is involved in the enhanced uptake of aggregated LDL by phagocytosis.[115] Recent evidence suggests that aggregated LDL may actually occur *in vivo*.[116] LDL aggregates resulting from self aggregation, modification by neutrophils,[117] or complexing with collagen, proteoglycans, or fibronectin would be potential substrates for phagocytosis.

Finally, the presence of circulating autoantibodies against glycosylated forms of LDL[111] and against at least one form of oxidatively modified LDL,[7] as well as the circumstance that they recognize their specific antigens in atherosclerotic lesions, suggest that immune complexes of modified LDL with these autoantibodies may form in the lesion. Such immune complexes could contribute to atherogenesis by several mechanisms (reviewed in Reference 113) and might be taken up by macrophages via the Fc receptor.

VII. CONCLUSIONS

It is now well established that a number of postsecretory modifications of lipoproteins increase their atherogenic potential. The direct demonstration of the occurrence of modified lipoproteins *in vivo* strongly supports the "modified lipoprotein hypothesis" of atherogenesis.[5] The common pathogenic element of different forms of modification is their rapid uptake by macrophages in the vascular wall. The uptake of large quantities of modified lipoproteins may be mediated by receptors, such as the scavenger receptor(s), which recognize oxidized lipid-protein adducts, scavenger-like receptors that recognize advanced glycosylation end products, or the Fc receptor, which may be responsible for the uptake of immune complexes potentially formed by autoantibodies and modified proteins. Receptors may also play a contributing role in the uptake of aggregated LDL by phagocytosis. The different forms of modification are probably intricately linked. Examples of such interactions are the reciprocal enhancement of oxidative modification and glycosylation, and the crosslinking of lipoproteins, structural proteins, and immunoglobulins by AGE or lipid peroxidation products such as MDA.

The understanding of these atherogenic mechanisms may indicate potential pathways for intervention. If we assume that the elimination of detrimental material is the primary function of macrophages and that the rapid uptake of modified lipoproteins has, at least initially, a beneficial effect, the prevention of lipoprotein modification would be desirable. Several mechanisms could contribute to reduce the formation of oxidized LDL. Most importantly, the reduction of plasma LDL levels may result in less substrate for oxidative modification by vascular cells. The composition of the LDL particle may offer another approach. Parthasarathy et al. recently showed that the substitution of polyunsaturated fatty acids by monounsaturated fatty acids (by a diet rich in oleic acid) protects LDL from oxidative modification, probably by interfering with the propagation of lipid peroxidation, which normally occurs in the presence of polyunsaturated fatty acids.[118] Similar data are now available from human studies.[119] The level of antioxidant protection, e.g., by natural antioxidants such as vitamin E or carotinoids, is likely to influence the propagation of oxidative processes.[62] Lipophilic antioxidants may increase the resistance of LDL to oxidation, and agents such as probucol and butylated hydroxytoluene have been demonstrated to reduce atherogenesis.[92,120] Intervention designed to reduce the formation of AGE could potentially be of great significance for the prevention of diabetic complications and, in particular, could reduce structural damage resulting from crosslinking in the arterial wall. Aminoguanidine, an aldehyde scavenger with a higher affinity for the reactive intermediate products of glycosylation than the lysine residues, has been proposed to reduce AGE formation,[121] but its effectiveness for atherogenesis requires further investigation.

The role of oxidized lipoproteins has been studied primarily in view of its recognition by specific high-affinity receptors. Recent evidence suggests that it may also cause or mediate a multitude of other atherogenic processes.[122] These include the reduced endothelium-mediated vasodilation in response to stimuli such as acetylcholine, serotonin, or bradykinin,[123-125] as well as activation of immune responses and inflammatory processes.[113] In particular, immune complexes between circulating autoantibodies and modified lipoproteins present in the artery wall may contribute to the activation of immune cells. Finally, oxidized lipoproteins stimulate platelet aggregation and interact with coagulation factors.[126] These proaggregatory effects may contribute to plaque fissuring and thrombosis in the late stages of atherosclerosis.

ACKNOWLEDGMENTS

The author would like to thank Dr. Joseph L. Witztum for his continuous support and critical review of this chapter. This work was supported by NHLBI Grant HL-14197 (Specialized Center of Research on Arteriosclerosis).

REFERENCES

1. **Goldstein, J.L. and Brown, M.S.,** The low density lipoprotein pathway and its relation to atherosclerosis, *Ann. Rev. Biochem.*, 46, 897–930, 1977.
2. Consensus conference, Lowering blood cholesterol to prevent coronary heart disease, *JAMA*, 153, 2080–2086, 1985.
3. **Tyroler, H.A.,** Lowering plasma cholesterol levels decreases risk of coronary heart disease: an overview of clinical trials, in *Hypercholesterolemia and Atherosclerosis*, Steinberg, D. and Olefsky, J.M., Eds., Churchill Livingstone, New York, 1987, 99–116.
4. **Steinberg D.,** The cholesterol controversy is over. Why did it take so long? *Circulation*, 80 (4), 1070–1078, 1989.
5. **Steinberg D., Parthasarathy, S., Carew, T.E., Khoo, J.C., and Witztum, J.L.,** Beyond cholesterol. Modifications of low density lipoprotein that increase its atherogenicity, *N. Engl. J. Med.*, 320, 915–924, 1989.
6. **Steinberg, D. and Witztum, J.L.,** Lipoproteins and atherogenesis. Current concepts., *JAMA*, 264 (23), 3047–3052, 1990.
7. **Palinski, W., Rosenfeld, M.E., Ylä-Herttuala, S., Gurtner, C., Socher, S.A., Butler, S., Parthasarathy, S., Carew, T.E., Steinberg, D., and Witztum, J.L.,** Low density lipoprotein undergoes oxidative modification *in vivo*, *Proc. Natl. Acad. Sci. U.S.A.*, 86 (4), 1372–1376, 1989.

8. **Ylä-Herttuala, S., Palinski, W., Rosenfeld, M.E., Parthasarathy, S., Carew, T.E., Butler, S., Witztum, J.L., and Steinberg, D.,** Evidence for the presence of oxidatively modified low density lipoprotein in atherosclerotic lesions of rabbit and man, *J. Clin. Invest.*, 84 (4), 1086–1095, 1989.
9. **Palinski, W., Ylä-Herttuala, S., Rosenfeld, M.E., Butler, S., Socher, S.A., Parthasarathy, S., Curtiss, L.K., and Witztum, J.L.,** Antisera and monoclonal antibodies specific for epitopes generated during the oxidative modification of low density lipoprotein. *Arteriosclerosis*, 10 (3), 325–335, 1990.
10. **Rosenfeld, M.E., Palinski, W., Ylä-Herttuala, S., Butler, S., and Witztum, J.L.,** Distribution of oxidation-specific lipid-protein adducts and apolipoprotein B in atherosclerotic lesions of varying severity from WHHL rabbits, *Arteriosclerosis*, 10 (3), 336–349, 1990.
11. **Davies, M.J. and Thomas, A.C.,** Plaque fissuring — the cause of acute myocardial infarction, sudden ischemic death, and crescendo angina, *Br. Heart J.*, 53, 363–373, 1985.
12. **Davies, M.J.,** A macro and micro view of coronary vascular insult in ischemic heart disease, *Circulation* (Suppl. II), 82 (3), 38–48, 1990.
13. **Richardson, P.D., Davies, M.J., and Born, G.V.R.,** Influence of plaque configuration and stress distribution on fissuring of coronary atherosclerotic plaques, *Lancet*, (II), 941–944, 1989.
14. **Small, D.M.,** Progression and regression of atherosclerotic lesions. Insights from lipid physical biochemistry, *Arteriosclerosis*, 8, 103–129, 1988.
15. **Ross, R., Wight, T.N., Strandness, E., and Thiele, B.,** Human atherosclerosis. I. Cell constitution and characteristics of advanced lesions of the superficial femoral artery, *Am. J. Pathol.*, 114, 79–93, 1984.
16. **Gown, A.M., Tsukada, T., and Ross, R.,** Human Atherosclerosis. II. Immunocytochemical analysis of the cellular composition of human atherosclerotic lesions, *Am. J. Pathol.*, 125, 191–207, 1986.
17. **Ross, R.,** The pathogenesis of atherosclerosis — an update. *N. Engl. J. Med.*, 314, 488–500, 1986.
18. **Munro, J.M. and Cotran, R.S.,** The pathogenesis of atherosclerosis: atherogenesis and inflammation, *Lab. Invest.*, 58, 249–261, 1988.
19. **Camejo, G.,** The interaction of lipids and lipoproteins with the intercellular matrix of arterial tissue: its possible role in atherogenesis, *Adv. Lipid Res.*, 19, 1–53, 1982.
20. **Ross, R. and Glomset, J.A.,** The pathogenesis of arteriosclerosis, *N. Engl. J. Med.*, 295, 369–377, 1976.
21. **Poole, J.C.F. and Florey, H.W.,** Changes in the endothelium of the aorta and the behaviour of macrophages in experimental atheroma of rabbits, *J. Pathol. Bacteriol.*, 75, 245–251, 1958.
22. **Stary, H.C. and Malinow, M.R.,** Ultrastructure of experimental coronary artery atherosclerosis in cynomolgus macaques: a comparison with lesions of other primates, *Atherosclerosis*, 43, 151–175, 1982.
23. **Schwenke, D.C. and Carew, T.E.,** Initiation of atherosclerotic lesions in cholesterol-fed rabbits. I. Focal increases in arterial LDL concentration precede development of fatty streak lesions, *Arteriosclerosis*, 9, 895–907, 1989.
24. **Schwenke, D.C. and Carew, T.E.,** Initiation of atherosclerotic lesions in cholesterol-fed rabbits. II. Selective retention of LDL vs. selective increases in LDL permeability in susceptible sites of arteries, *Arteriosclerosis*, 9, 908-918, 1989.
25. **Fowler, S., Shio, H., and Haley, W.J.,** Characterization of lipid-laden aortic cells from cholesterol-fed rabbits. IV. Investigation of macrophage-like properties of aortic cell populations, *Lab. Invest.*, 41, 372–378, 1979.
26. **Gerrity, R.G.,** The role of the monocyte in atherogenesis. I. Transition of blood-borne monocytes into foam cells in fatty lesions, *Am. J. Pathol.*, 103, 181–190, 1981.
27. **Aqel, N.M., Ball, R.Y., Waldman, H., and Mitchinson, M.J.,** Monocytic origin of foam cells in human atherosclerotic plaques, *Atherosclerosis*, 53, 265–271, 1984.

28. **Gerrity, R.G., Naito, H.K., Richardson, M., and Schwartz, C.J.,** Dietary induced atherogenesis in swine, *Am. J. Pathol.*, 95, 775–793, 1979.
29. **Bevilacqua, M., Pober, J.S., Wheeler, M.E., Cotran, R.S., and Gimbrone, M.A.,** Interleukin 1 acts on cultured human vascular endothelium to increase the adhesion of polymorphonuclear leukocytes, monocytes, and related leukocyte cell lines, *J. Clin. Invest.*, 76, 2003–2011, 1985.
30. **Bevilacqua, M.P., Pober, J.S., Mendrick, D.L., Cotran, R.S., and Gimbrone, M.A.,** Identification of an inducible endothelial-leukocyte adhesion molecule, *Proc. Natl. Acad. Sci. U.S.A.*, 84, 9238–9242, 1987.
31. **Cybulsky, M.I. and Gimbrone, M.A.,** Endothelial expression of a mononuclear leukocyte adhesion molecule during atherogenesis, *Science*, 251, 788–910, 1991.
32. **Berliner, J.A., Territo, M., Almada, L., Carter, A., Shafonsky, E., and Fogelman, A. M.,** Monocyte chemotactic factor produced by large vessel endothelial cells *in vitro*, *Arteriosclerosis*, 6, 254–258, 1986.
33. **Quinn, M.T., Parthasarathy, S., and Steinberg, D.,** Lysophosphatidylcholine: a new chemotactic factor for human monocytes and its potential role in atherogenesis, *Proc. Natl. Acad. Sci. U.S.A.*, 85, 2805–2809, 1988.
34. **Brown, M.S. and Goldstein, J.L.,** Regulation of plasma cholesterol by lipoprotein receptors, *Science*, 212, 628–635, 1981.
35. **Goldstein, J.L., Ho, Y.K., Basu, S.K., and Brown, M.S.,**. Binding site on macrophages that mediates uptake and degradation of acetylated low density lipoprotein, producing massive cholesterol deposition, *Proc. Natl. Acad. Sci. U.S.A.*, 76, 333–337, 1979.
36. **Buja, L.M., Kita, T., Goldstein, J.L., Watanabe, Y., and Brown, M.S.,** Cellular pathology of progressive atherosclerosis in the WHHL rabbit: an animal model of familial hypercholesterolemia, *Arteriosclerosis*, 3, 87–101, 1983.
37. **Rosenfeld, M.E., Tsukada, T., Gown, A.M., and Ross, R.,** Fatty streak initiation in the WHHL and comparably hypercholesterolemic fat-fed rabbits, *Arteriosclerosis*, 1, 9–23, 1987.
38. **Nagelkerke, J.F., Barto, K.P., and Van Berkel, T.J.,** *In vivo* and *in vitro* uptake and degradation of acetylated low density lipoprotein by rat liver endothelial, Kupffer, and parenchymal cells, *J. Biol. Chem.*, 258, 12,221–12,227, 1983.
39. **Weisgraber, K.H., Innerarity, T.L., and Mahley, R.W.,** Role of the lysine residues of plasma lipoproteins in high affinity binding to cell surface receptors on human fibroblasts, *J. Biol. Chem.*, 253, 9053–9062, 1978.
40. **Mahley, R.W., Innerarity, T.L., Weisgraber, K.H., and Oh, S.Y.,** Altered metabolism (*in vivo* and *vitro*) of plasma lipoprotein after selective chemical modification of lysine residues of the apoproteins, *J. Clin. Invest.*, 64, 743–750, 1979.
41. **Henriksen, T., Mahoney, E.M., and Steinberg, D.,** Enhanced macrophage degradation of low density lipoprotein previously incubated with cultured endothelial cells: Recognition by the receptor for acetylated low density lipoproteins, *Proc. Natl. Acad. Sci. U.S.A.*, 78, 6499–6503, 1981.
42. **Steinbrecher, U.P., Parthasarathy, S., Leake, D.S., Witztum, J.L., and Steinberg, D.,** Modification of low density lipoprotein by endothelial cells involves lipid peroxidation and degradation of low density lipoprotein phospholipids, *Proc. Natl. Acad. Sci. U.S.A.*, 81, 3883–3887, 1984.
43. **Esterbauer, H., Cheeseman, K.H., Dianzani, M.U., Poli, G., and Slather, T.F.,** Separation and characterization of the aldehydic products of lipid peroxidation stimulated by ADP-Fe^{2++} in rat liver microsomes, *Biochem. J.*, 208, 129–140, 1982.
44. **Fogelman, A.M., Schechter, J.S., Hokom, M., Child, J.S., and Edwards, P.A.,** Malondialdehyde alteration of low density lipoproteins leads to cholesteryl ester accumulation in human monocyte-macrophages, *Proc. Natl. Acad. Sci. U.S.A.*, 77, 2214–2218, 1980.
45. **Steinbrecher, U.P., Witztum, J.L., Parthasarathy, S., and Steinberg, D.,** Decrease in reactive amino groups during oxidation or endothelial cell modification of LDL. Correlation with changes in receptor-mediated catabolism, *Arteriosclerosis*, 7, 135–143, 1987.

46. **Parthasarathy, S., Fong, L., Otero, D., and Steinberg, D.,** Recognition of solubilized apoproteins from delipidated, oxidized low density lipoprotein (LDL) by the acetyl-LDL receptor, *Proc. Natl. Acad. Sci. U.S.A.*, 84, 537–540, 1987.
47. **Parthasarathy, S., Fong, L.G., Quinn, M.T., and Steinberg, D.,** Oxidative modification of LDL: comparison between cell-mediated and copper-mediated modification, *Eur. Heart J.*, 11(E), 83–87, 1990.
48. **Parthasarathy, S., Steinbrecher, U.P., Barnett, J., Witztum, J.L., and Steinberg, D.,** The essential role of phospholipase A_2 activity in endothelial cell-induced modification of low-density lipoprotein, *Proc. Natl. Acad. Sci. U.S.A.*, 82, 3000–3004, 1985.
49. **Parthasarathy, S. and Barnett, J.,** Phospholipase A_2 activity of low density lipoprotein, evidence for an intrinsic phospholipase A_2 activity of apoprotein B-100, *Proc. Natl. Acad. Sci. U.S.A.*, 87, 9741–9745, 1990.
50. **Sparrow, C.P., Parthasarathy, S., and Steinberg, D.,** Enzymatic modification of low density lipoprotein by purified lipoxygenase plus phospholipase A_2 mimics cell-mediated oxidative modification, *J. Lipid Res.*, 29, 745–753, 1988.
51. **Hessler, J.R., Robertson, A.L., Jr., and Chisolm, G.M.,** LDL-induced cytotoxicity and its inhibition by HDL in human vascular smooth muscle and endothelial cells in culture, *Atherosclerosis*, 32, 213–229, 1979.
52. **Morel, D.W., DiCorleto, P.E., and Chisolm, G.M.,** Endothelial and smooth muscle cells alter low density lipoprotein *in vitro* by free radical oxidation, *Arteriosclerosis*, 4, 357–364, 1984.
53. **Quinn, M.T., Parthasarathy, S., Fong, L.G, and Steinberg, D.,** Oxidatively modified low density lipoproteins: a potential role in recruitment and retention of monocyte/macrophages during atherogenesis, *Proc. Natl. Acad. Sci. U.S.A.*, 84, 2995–2998, 1987.
54. **Quinn, M.T., Parthasarathy S., and Steinberg, D.,** Lysophosphatidylcholine: a chemotactic factor for human monocytes and its potential role in atherogenesis, *Proc. Natl. Acad. Sci. U.S.A.*, 85, 2805–2809, 1988.
55. **Berliner, J.A., Territo, M.C., Sevanian, A., Ramin, S., Kim, J.A., Bamshad, B., Esterson, M., and Fogelman, A.M.,** Minimally modified low density lipoprotein stimulates monocyte endothelial interactions, *J. Clin. Invest.*, 85, 1260–1266, 1990.
56. **Kugiyama, K., Kerns, S.A., Morrisett, J.D., Roberts, R., and Henry, P.D.,** Impairment of endothelium-dependent arterial relaxation by lysolecithin in modified low density lipoproteins, *Nature*, 344, 160–162, 1990.
57. **Parthasarathy, S., Printz, D.J., Boyd, D., Joy, L., and Steinberg, D.,** Macrophage oxidation of low density lipoprotein generates a modified form recognized by the scavenger receptor, *Arteriosclerosis*, 6, 505–510, 1986.
58. **Hiramatsu, K., Rosen, H., Heinecke, J.W., Wolfbauer, G., and Chait, A.,** Superoxide initiates oxidation of low density lipoprotein by human monocytes, *Arteriosclerosis*, 7, 55–60, 1987.
59. **Cathcart, M.K., Morel, D.W., and Chisolm, G.M.,** Monocytes and neutrophils oxidize low density lipoprotein making it cytotoxic, *J. Leukoc. Biol.*, 38, 341–350, 1985.
60. **Heinecke, J.W., Rosen, H., and Chait, A.,** Iron and copper promote modification of low density lipoprotein by human arterial smooth muscle cells in culture, *J. Clin. Invest.*, 74, 1890–1894, 1984.
61. **Rosenfeld, M.E., Khoo, J.C., Miller, E., Parthasarathy, S., Palinski, W., and Witztum, J.L.,** Macrophage-derived foam cells freshly isolated from rabbit atherosclerotic lesions degrade modified lipoproteins, promote oxidation of LDL, and contain oxidation specific lipid-protein adducts, *J. Clin. Invest.*, 87, 90–99, 1991.
62. **Esterbauer, H., Jürgens, G., Quehenberger, O., and Keller, E.,** Autooxidation of human low density lipoprotein, loss of polyunsaturated fatty acids and vitamin E and generation of aldehyde, *J. Lipid Res.*, 28, 495–509, 1987.
63. **Benedetti, A., Comporti, M., and Esterbauer, H.,** Identification of 4–hydroxynonenal as a cytotoxic product originating from the peroxidation of liver microsomal lipids, *Biochim. Biophys. Acta*, 620, 281–296, 1980.

64. **Jürgens, G., Lang, J., and Esterbauer, H.,** Modification of human low-density lipoprotein by the lipid peroxidation product 4–hydroxynonenal, *Biochim. Biophys. Acta*, 875, 103–114, 1986.
65. **Law, S.W., Grant, S.M., Higuchi, K., Hospattanker, A., Lackner, K., Lee, N., and Brewer, H.B.,** Human liver apolipoprotein B-100 cDNA: complete nucleic acid and derived amino acid sequence, *Proc. Natl. Acad. Sci. U.S.A.*, 83, 8142–8146, 1986.
66. **Knott, J., Pease, R.J., Powell, L.M., Wallis, S.C., Rall, S.C., Jr., Innerarity, T.L., Taylor, Y.L., Marcel, Y.L., Milne, R., Johnson, D., Fuller, M., Lusis, A.J., McCarthy, B.J., Mahley, R.W., Levy-Wilson, B., and Scott, J.,** Complete protein sequence and identification of structural domains of human apolipoprotein B, *Nature*, 323, 734–738, 1986.
67. **Witztum, J.L., Steinbrecher, U.P., Fisher, M., and Kesaniemi, A.,** Nonenzymatic glucosylation of homologous LDL and albumin render them immunogenic in the guinea-pig, *Proc. Natl. Acad. Sci. U.S.A.*, 80, 2757–2761, 1983.
68. **Steinbrecher, U.P., Fisher, M., Witztum, J.L., and Curtiss, L.K.,** Immunogenicity of homologous low density lipoprotein after methylation, ethylation, acetylation or carbamylation: generation of antibodies specific for derivatized lysine, *J. Lipid Res.*, 25, 1109–1116, 1984.
69. **Curtiss, L.K. and Witztum, J.L.,** A novel method for generating region specific monoclonal antibodies to modified proteins: application to the identification of human glucosylated low density lipoproteins, *J. Clin. Invest.*, 72, 1427–1438, 1983.
70. **Parthasarathy, S., Fong, L.G., Otero, D., and Steinberg, D.,** Recognition of resolubilized apoproteins from delipidated, oxidatively modified low density lipoprotein (LDL) by the acetyl-LDL receptor, *Proc. Natl. Acad. Sci. U.S.A.*, 84, 537–540, 1987.
71. **Wieland, E., Parthasarathy, S., Walli, A.K., Seidel, D., and Steinberg, D.,** Modification of low density lipoprotein by H_2O_2 plus horseradish peroxidase (HRP), in *8th International Symposium on Atherosclerosis, Poster Session Abstract Book*, CIC Edizioni Internazionali, Rome, 1988, 1025.
72. **Young, S.G., Witztum, J.L., Casal, D.C., Curtiss, L.K., and Bernstein, S.,** Conservation of the low density lipoprotein receptor-binding domain of apoprotein B: demonstration by a new monoclonal antibody, MB 47, *Arteriosclerosis*, 6, 178–188, 1986.
73. **Tsukada, T., Rosenfeld, M.E., Ross, R., and Gown, A.M.,** Immunocytochemical analysis of cellular components in atherosclerotic lesions. Use of monoclonal antibodies with the Watanabe and fat-fed rabbit, *Arteriosclerosis*, 6, 601–613, 1986.
74. **Tsukada, T., Tippens, D., Mar, H., Gordon, D., Ross, R., and Gown, A.M.,** HHF-35: a muscle-actin-specific monoclonal antibody, *Am. J. Pathol.* 126, 51–60, 1986.
75. **Mitchinson, M.J., Hothersall, D.C., Brooks, P.N., and DeBurbure, C.Y.,** The distribution of ceroid in human atherosclerosis, *J. Pathol.*, 145, 177–183, 1985.
76. **Lougheed, M., Zhang, H.F., and Steinbrecher, U.P.,** Oxidized low density lipoprotein is resistant to cathepsins and accumulates within macrophages, *J. Biol. Chem.*, 266 (22), 14,519–14,525, 1991.
77. **Srinivasan, S.R., Yost, C., Bhandaru, R.R., Radhakrishnamurthy, B., and Berenson, G.S.,** Lipoprotein-glycosaminoglycan interactions in aortas of rabbits fed atherogenic diets containing different fats, *Atherosclerosis*, 43, 289–301, 1982.
78. **Niinikoski, J., Heughan, C., and Hunt, T.K.,** Oxygen tensions in the aortic wall of normal rabbits, *Atherosclerosis*, 17, 353–359, 1973.
79. **Mowri, H., Ohkuma, S., and Takano, T.,** Monoclonal $DLRI_a$/104G antibody recognizing peroxidized lipoproteins in atherosclerotic lesions, *Biochim. Biophys. Acta*, 963, 208–214, 1988.
80. **Haberland, M.E., Fong, D., and Cheng, L.,** Malondialdehyde-altered protein occurs in atheroma of Watanabe heritable hyperlipidemic rabbits, *Science*, 241, 215–241, 1988.
81. **Salmon, S., Maziere, C., Theron, L., Beucler, I., Ayrautt-Jarrier, M., Goldstein, S., and Polonovski, J.,** Immunological detection of low density lipoproteins modified by malondialdehyde *in vitro* or *in vivo*, *Biochim. Biophys. Acta*, 920, 215–220, 1987.

82. **Boyd, H.C., Gown, A.M., Wolfbauer, G., and Chait, A.,** Direct evidence for a protein recognized by a monoclonal antibody against oxidatively modified LDL in atherosclerotic lesions from a Watanabe Heritable Hyperlipidemic rabbit, *Am. J. Pathol.* 135 (5), 815–826, 1989.
83. **Ylä-Herttuala, S., Palinski, W., Rosenfeld, M.E., Witztum, J.L., and Steinberg, D.,** Lipoproteins in normal and atherosclerotic aorta, *Eur. Heart J.*, 11 (Suppl E), 88–99, 1990.
84. **Kao, V.C.Y. and Wissler, R.W.,** A study of the immunohistochemical localization of serum lipoproteins and other plasma proteins in human atherosclerotic lesions, *Exp. Molec. Pathol.*, 4, 465–479, 1965.
85. **Hoff, H.F., Lie, J.T., Titus, J.L., Bajardo, R.J., Jackson, R.L., DeBakey, M.E., and Gotto, A.M.,** Lipoproteins in atherosclerotic lesions: localization by immunofluorescence of apo-low density lipoproteins in human atherosclerotic arteries from normal and hyperlipoproteinemics, *Arch. Pathol.*, 99, 253–258, 1975.
86. **Smith, E.B.,** Plasma and tissue lipids in atherosclerosis, *Adv. Lipid Res.*, 12, 1–49, 1974.
87. **Ylä-Herttuala, S., Jaakkola, O., Ehnholm, C., Tikkanen, M.J., Solakivi, T., Sarkioja, T., and Nikkari, T.,** Characterization of two lipoproteins containing apolipoproteins B and E from lesion-free human aortic intima, *J. Lipid Res.*, 29, 563–572, 1988.
88. **Goldstein, J.L., Hoff, H.F., Ho, Y.K., Basu, S.K., and Brown, M.S.,** Stimulation of cholesteryl ester synthesis in macrophages by extracts of atherosclerotic human aortas and complexes of albumin/cholesterol, *Arteriosclerosis*, 1, 210–226, 1981.
89. **Morton, R.E., West, G.A., and Hoff, H.F.,** A low density lipoprotein-sized particle isolated from human atherosclerotic lesions is internalized by macrophages via a non-scavenger-receptor mechanism, *J. Lipid Res.*, 27, 1124–1134, 1986.
90. **Shaikh, M., Martini, S., Quiney, J.R., Baskerville, P., LaVille, A.E., Browse, N.L., Duffield, R., Turner, P.R., and Lewis, B.,** Modified plasma-derived lipoproteins in human atherosclerotic plaques, *Atherosclerosis*, 69, 165–172, 1988.
91. **Ylä-Herttuala, S., Butler, S., Picard, S., Palinski, W., Steinberg, D., and Witztum, J.L.,** Rabbit and human atherosclerotic lesions contain IgG that recognizes MDA-LDL and Copper-oxidized LDL, *Arterioscler. Thromb.*, 11 (5), 1426A, 1991.
92. **Carew, T.E., Schwenke, D.C., and Steinberg, D.,** Antiatherogenic effect of probucol unrelated to its hypocholesterolemic effect: evidence that antioxidants *in vivo* can selectively inhibit low density lipoprotein degradation in macrophage-rich fatty streaks slowing the progression of atherosclerosis in the WHHL rabbit, *Proc. Natl. Acad. Sci. U.S.A.*, 84, 7725–7729, 1987.
93. **Parthasarathy, S., Young, S.G., Witztum, J.L., Pittman, R.C., and Steinberg, D.,** Probucol inhibits oxidative modification of low density lipoprotein, *J. Clin. Invest.*, 77, 641–644, 1986.
94. **Sparrow, C.P., Parthasarathy, S., and Steinberg, D.,** A macrophage receptor that recognizes oxidized low density lipoprotein but not acetylated low density lipoprotein, *J. Biol. Chem.*, 264, 2599–2604, 1989.
95. **Arai, H.T., Kita, M., Yokode, S., Narumiya, S., and Kawai, C.,** Multiple receptors for modified low density lipoproteins in mouse peritoneal macrophages: different uptake mechanisms for acetylated and oxidized low density lipoproteins, *Biochem. Biophys. Res. Commun.*, 159, 1375–1382, 1989.
96. **Vlassara, H., Brownlee, M., and Cerami, A.,** High-affinity receptor-mediated uptake and degradation of glucose-modified proteins: a potential mechanism for the removal of senescent macromolecules, *Proc. Natl. Acad. Sci. U.S.A.*, 82, 5588–5592, 1985.
97. **Radoff, S., Vlassara, H., and Cerami, A.,** Characterization of a solubilized cell surface binding protein on macrophages specific for proteins modified non-enzymatically by advanced glycosylation endproducts, *Arch. Biochem. Biophys.*, 263, 418–423, 1988.
98. **Kodama, T., Freeman, M., Rohrer, L., Zabrecky, T., Matsudaira, P., and Krieger, M.,** Type I macrophage scavenger receptor contains alpha-helical and collagen-like coiled coils, *Nature*, 343, 531–535, 1990.

99. **Baynes, J.W., Watkins, N.G., Fisher, C.I., Hull, C.J., Patrick, J.S., Ahmed, M.U., Dunn, J.A., and Thorpe, S.R.,** The Amadori product on protein: structure and reactions, *Prog. Clin. Biol. Res.*, 304, 43–67, 1989.
100. **Brownlee, M., Vlassara, H., and Cerami, A.,** Nonenzymatic glycosylation and the pathogenesis of diabetic complications, *Ann. Intern. Med.*, 101, 527–537, 1984.
101. **Cerami, A., Vlassara, H., and Brownlee, M.,** Role of advanced glycosylation products in complications of diabetes, *Diabetes Care*, 11, 73–79, 1988.
102. **Njoroge, F.G. and Monnier, V.M.,** The chemistry of the Maillard reaction under physiological conditions — a review, *Prog. Clin. Biol. Res.*, 304, 85–107, 1989.
103. **Steinbrecher, U.P. and Witztum, J.L.,** Glucosylation of low density lipoproteins to an extent comparable to that seen in diabetes slows their catabolism, *Diabetes*, 33, 130–134, 1984.
104. **Brown, M.S. and Goldstein, J.L.,** A receptor-mediated pathway for cholesterol homeostasis, *Science*, 232, 34–47, 1986.
105. **Takata, K., Horiuchi, S., Araki, N., Shiga, M., Saitoh, M., and Morino, Y.,** Endocytic uptake of nonenzymatically glycosylated proteins is mediated by a scavenger receptor for aldehyde-modified proteins, *J. Biol. Chem.*, 263, 14,819–14,825, 1988.
106. **Brownlee, M., Vlassara, H., and Cerami A.,** Nonenzymatic glycosylation products of collagen covalently trap low-density lipoprotein, *Diabetes*, 34, 938–941, 1985.
107. **Mullarkey, C.J., Edelstein, D., and Brownlee, M.,** Free radical generation by early glycation products: a mechanism for accelerated atherogenesis in diabetes, *Biochem. Biophys. Res. Commun.*, 173 (3), 932–939, 1990.
108. **Hicks, M., Delbridge, L., Yue, D.K., and Reeve, T.S.,** Catalysis of lipid peroxidation by glucose and glycosylated collagen, *Biochem. Biophys. Res. Commun.*, 151, 649–655, 1988.
109. **Hicks, M., Delbridge, L., Yue, D.K., and Reeve, T.S.,** Increase in crosslinking of nonenzymatically glycosylated collagen induced by products of lipid peroxidation, *Arch. Biochem. Biophys.*, 268, 249–254, 1989.
110. **Schröder, S., Palinski, W., and Schmid-Schönbein, G.W.,** Activated monocytes and granulocytes, non-perfusion and neovascularization in experimental diabetic retinopathy, *Am. J. Pathol.,* 139 (1), 81–100, 1991.
111. **Witztum, J.L., Steinbrecher, U.P., Kesaniemi, Y.A., and Fisher, M.,** Autoantibodies to glucosylated proteins in the plasma of patients with diabetes mellitus, *Proc. Natl. Acad. Sci. U.S.A.*, 81, 3204–3208, 1984.
112. **Witztum, J.L. and Koschinsky, T.,** Metabolic and immunological consequences of glycation of low density lipoproteins. in *The Maillard Reaction in Aging, Diabetes and Nutrition*, Baynes, J.W. and Monnier, V.M., Eds., Alan R. Liss, New York, 1989, 219–234.
113. **Libby, P. and Hansson, G.K.,** Involvement of the immune system in human atherogenesis: current knowledge and unanswered questions, *Lab. Invest.*, 64 (1), 5–15, 1991.
114. **Brownlee, M., Pongor, S., and Cerami, A.,** Covalent attachment of soluble proteins by nonenzymatically glycosylated collagen: role of the *in situ* formation of immune complexes, *J. Exp. Med.*, 158, 1739–1744, 1983.
115. **Khoo, J.C., Miller, E., McLoughlin, P., and Steinberg, D.,** Enhanced macrophage uptake of low density lipoprotein after self-aggregation, *Arteriosclerosis*, 8, 348–358, 1988.
116. **Nievelstein, P.F.E.M., Fogelman, A.M., Mottino, G., and Frank, J.S.,** Lipid accumulation in rabbit aortic intima two hours after bolus infusion of low density lipoprotein. A deep-etch and immunolocalization study of ultra-rapidly frozen tissue, *Arterioscler. Thromb.*,11 (6), 1795–1805, 1991.
117. **Polacek, D., Byrne, R.E., and Scanu, A.M.,** Modification of low density lipoproteins by polymorphonuclear cell elastase leads to enhanced uptake by human monocyte-derived macrophages via the low density lipoprotein receptor pathway, *J. Lipid Res.*, 29, 797–808, 1988.

118. **Parthasarathy, S., Khoo, J.C., Miller, E., Barnett, J., Witztum, J.L., and Steinberg, D.,** Low density lipoprotein rich in oleic acid is protected against oxidative modification: implications for dietary prevention of atherosclerosis, *Proc. Natl. Acad. Sci. U.S.A.*, 87, 3894–3898, 1990.
119. **Reaven, P., Parthasarathy, S., Grasse, B.J., Miller, E., Alzman, F., Mattson, F.H., Khoo, J.C., Steinberg, D., and Witztum, J.L.,** Feasibility of using an oleate-rich diet to reduce the susceptibility of low-density lipoprotein to oxidative modification in humans, *Am. J. Clin. Nutr.*, 54, 701–706, 1991.
120. **Björkhem, I., Henriksson-Freyschuss, A., Breuer, O., Diczfalusy, U., Berglund, L., and Henriksson, P.,** The antioxidant butylated hydroxytoluene protects against atherosclerosis, *Arterioscler. Thromb.*, 11, 15–22, 1991.
121. **Brownlee, M., Vlassara, H., Kooney, A., and Cerami, A.,** Aminoguanidine prevents diabetes-induced arterial wall protein cross-linking, *Science*, 323 (4758), 1629–1632, 1986.
122. **Rosenfeld, M.E.,** Oxidized LDL affects multiple atherogenic cellular responses, *Circulation*, 83 (6), 2137–2140, 1991.
123. **Galle, J., Bassenge, E., and Busse, R.,** Oxidized low density lipoproteins potentiate vasoconstrictions to various agonists by direct interaction with vascular smooth muscle, *Circ. Res.*, 66, 1287–1293, 1990.
124. **Yokoyama, M., Hirata, K., Miyake, R., Akita, H., Ishikawa, Y., and Fukuzaki, H.,** Lysophospatidylcholine: essential role in the inhibition of endothelium-dependent vasorelaxation by oxidized low density lipoprotein, *Biochem. Biophys. Res. Commun.*, 168, 301–308, 1990.
125. **Aviram M.,** Modified forms of low density lipoprotein affect platelet aggregation *in vitro*, *Thromb. Res.*, 53, 561–567, 1989.
126. **Schuff-Werner, P., Claus, G., Armstrong, V.W., Kostering, H., and Seidel, D.,** Enhanced procoagulatory activity (PCA) of human monocytes/macrophages after *in vitro* stimulation with chemically modified LDL, *Atherosclerosis*, 78,109–112, 1989.

Chapter 9

CHARACTERIZATION OF LIPOPROTEIN METABOLISM IN CELLS ISOLATED FROM ATHEROSCLEROTIC ARTERIES

Olli Jaakkola

TABLE OF CONTENTS

ISBN 0-8493-5505-2

I. INTRODUCTION

Lipid-filled foam cells are a prominent feature of early atherosclerotic lesions, the fatty streaks, in both humans and experimental animal models.[1] The intracellular lipid occurs as droplets consisting predominantly of cholesteryl ester. The majority of foam cells are derived from circulating monocytes,[2-7] but as lesions progress, some smooth muscle cells also accumulate lipid inclusions, albeit to a lesser degree.[8] The cholesterol in the cells of the atherosclerotic lesion is derived mainly from plasma lipoproteins. In epidemiological studies the concentration of low-density lipoprotein (LDL) shows a significant association with the prevalence of atherosclerosis.[9,10] However, cell culture studies have shown that native LDL does not lead to cholesteryl ester accumulation in monocyte-macrophages[11] or in smooth muscle cells.[12] This review summarizes some of the cellular lipoprotein uptake mechanisms that may be of relevance to foam cell formation, in the light of results we have obtained in enzymatically isolated atherosclerotic lesion cells maintained in primary culture.[13-15]

II. MONOCYTE/MACROPHAGES IN ATHEROGENESIS

The earliest visible effect of hypercholesterolemia in experimental animals is the adherence of monocytes onto endothelial cells in the lesion-prone areas.[3,4,16] On the basis of ultrastructural studies, it is thought that circulating monocytes are able to penetrate through the endothelial cell lining into the intima. Early fatty streak lesions consist of accumulations of intimal monocyte-macrophages, identified as such by the use of specific monoclonal antibodies against monocyte epitopes both in humans[5,6,18,19] and in experimental animals.[7]

The entrance of monocytes into a developing atherosclerotic lesion is thought to be a biphasic process involving (1) adhesion of monocytes to the vascular endothelium in lesion-prone areas and (2) migration into the intima. The enhanced adhesion of monocytes onto the luminal surface probably requires changes both in the monocytes and in the endothelial cells. Hypercholesterolemia and the atherogenic lipoproteins LDL[20] and β-very low-density lipoprotein[21] (β-VLDL) have been shown to enhance the adhesivity of monocytes. On the other hand, endothelial cells are capable of expressing inducible adhesion molecules that mediate the adhesion of specific types of leukocytes.[22,23] Synthesis of these adhesion molecules is stimulated by several cytokines. Recently it was shown that, in hypercholesterolemic rabbits, adhesion molecules are expressed in the endothelial cells covering developing lesions, but not in those of the adjacent healthy vessel wall.[24] Atherosclerotic tissue has been shown to contain extractable chemotactic substances that stimulate the migration of monocytes through the endothelium into the intima.[25,26] Cultured endothelial cells[27,28] and smooth muscle cells[29,30] also synthesize such factors. One of the most important of the monocyte chemoattractants so far discovered is the newly characterized "monocyte chemotactic protein-1" (MCP-1), which is expressed in the macrophage-rich areas of atherosclerotic lesions.[31] Stimulation with cytokines[32-34] or with mildly oxidized LDL[35] enhances the production of MCP-1. On the other hand, oxidized LDL is itself chemotactic to monocytes.[27] Once a small number of monocytes has entered the intima, lesion progression may become accelerated by their presence, for example, by recruitment of more monocytes, oxidation of LDL present in the intimal extracellular space, and secretion of factors that stimulate migration and proliferation of smooth muscle cells.[36]

III. LIPOPROTEIN RECEPTORS AND FOAM CELL FORMATION

The lipoprotein metabolism of both monocyte-macrophages and smooth muscle cells has been extensively studied in culture. The smooth muscle cells used for such studies have usually been subcultures originating from arterial explants. Mouse peritoneal macrophages and human monocyte-derived macrophages have been used as models for lesion macrophages.[37-39] Although LDL is atherogenic, cell culture studies have shown that the apoB,E (LDL) receptor-mediated uptake of LDL is regulated by negative feedback mechanisms that prevent excess cholesterol uptake.[38] The mechanism of the accumulation of LDL-derived lipids in the arterial cells started to be unraveled after the discovery by Goldstein et al.[40] that chemical modification of LDL by acetylation enhances its uptake into macrophages. This uptake was subsequently found to be mediated by another receptor, the “acetyl-LDL receptor” or “scavenger receptor”. Reaction of LDL with several substances capable of modifying lysine residues in the apolipoprotein B (apoB) moeity of LDL leads to its recognition by the scavenger receptor. The scavenger receptor has a low specificity; several types of chemical modifications convert LDL to a ligand for this receptor, including acetoacetylation,[41] maleylation,[40] succinylation,[40] carbamylation,[42] and treatment with malondialdehyde.[43] In addition, the scavenger receptor recognizes many other negatively charged nonlipoprotein ligands.[11] The uptake of modified LDL via this pathway is not feedback regulated, and incubation of macrophages with acetylated or other modified forms of LDL leads to cholesteryl ester accumulation resembling that seen in atherosclerotic lesions.[11] The finding that malondialdehyde, which is liberated *in vivo* during lipid peroxidation, can convert LDL to a ligand recognized by the scavenger receptor raises the possibility that such modifications might also be plausible *in vivo*. It is, however, uncertain whether the concentration of malondialdehyde *in vivo* reaches levels that are sufficiently high to modify LDL, and so the relevance of this process in the pathogenesis of atherosclerosis remains doubtful.

Another ligand with more obvious *in vivo* relevance for the scavenger receptor emerged with the findings that endothelial cells,[44,45] as well as macrophages[46] and smooth muscle cells,[45] are capable of modifying LDL in such a way that it is rendered recognizable by the scavenger receptor. This modification was shown to involve oxidation of LDL lipids, fragmentation of apoB, and derivatization of its lysine residues by the aldehydes liberated by lipid oxidation.[47,48] Oxidatively modified LDL has been extracted from atherosclerotic lesions and shown to be metabolized via the scavenger receptor in macrophage cultures.[49]

β-VLDL is a lipoprotein found in the plasma of several species after cholesterol feeding[50-52] and in the plasma of patients with type III hyperlipoproteinemia.[52] In contrast to normal VLDL, β-VLDL has a core rich in cholesteryl ester, and its principal apoproteins are apoB and apoE. Receptor-mediated uptake of β-VLDL into macrophages leads to cholesterol accumulation.[53,54] The macrophage receptor mediating the uptake of β-VLDL appears to be a product of the LDL receptor gene,[55,56] and in macrophages it may mediate cholesterol accumulation.

IV. LIPOPROTEIN METABOLISM IN ATHEROSCLEROTIC LESION CELLS

Model cell cultures do not necessarily metabolize lipoproteins in the same way as cells of the atherosclerotic lesion, and in such systems, the relative contributions of the various cell types of the lesion to the pathogenesis of atherosclerosis cannot easily be determined. In an attempt to overcome this problem, one approach has been to use intact atherosclerotic arteries for studies of lipoprotein metabolism. When labeled atherogenic lipoproteins, such as LDL,[57-60] β-VLDL,[61] and acetylated LDL,[58] are injected into the circulation of an animal or added to the perfusion medium of the artery, they are preferentially taken up by lesion-bearing areas. However, this approach is limited because of the difficulty of differentiating between extracellular retention and cellular uptake of the lipoproteins, and the problem of discriminating the roles of the various cell types remains. Moreover, differences in the endothelial permeability in different areas and possible lipoprotein modifications in the intima may affect the results. A more specific approach to obtaining information about the mechanisms of cholesterol accumulation is provided by studies of lipoprotein and cholesterol metabolism in cells that have been isolated enzymatically from atherosclerotic lesions. This approach allows us to study the cells shortly after isolation, thus minimizing any changes in receptor activities that might otherwise occur, and forms the basis for the present study.

V. MATERIALS AND METHODS

A. Artery Specimens and Cell Isolation

In the studies discussed here, atherosclerosis was induced in New Zealand White rabbits by feeding them a commercial chow containing 1 to 2% cholesterol,[15,62] maintaining the serum cholesterol concentration above 40 mmol/l. This regime led to development of fatty streak-type lesions in 3 to 4 months. The control rabbits were fed a standard chow. The rabbits were killed, and the intima together with the inner media were dissected out from the aortas and used for cell isolations.

Human aortas were obtained from medicolegal autopsies within 6 h after death.[14] The aortas were removed and then transported in an ice bath to the laboratory. Fatty streak areas were visually identified, separated under a dissection microscope, and used for cell isolation.

Cells from rabbit and human aortas were isolated using a modification[13-15,62] of the collagenase-elastase digestion method.[63] The tissue was chopped into 0.5 × 0.5 mm pieces with a McIlwain-type tissue chopper, and the pieces were washed twice with Hepes-buffered Krebs-Ringer solution containing 0.2 m*M* $CaCl_2$ and 1 mg/ml soybean trypsin inhibitor. The enzymes, collagenase (500 units/ml) and elastase (100 units/ml), were dissolved in the same solution. The tissue pieces were digested sequentially with collagenase solution for 1 h, with elastase for 1 h, and finally with collagenase until a cell suspension was obtained. Similar results could be obtained by using collagenase and elastase simultaneously, but the first collagenase treatment in the sequential procedure provides an effective means for removing endothelial cells. Disintegration of loosely packed tissue pieces could be promoted by pipetting through a siliconized Pasteur pipette with a tip narrowed to 0.5 mm. The cell suspension was filtered through a 160 μm nylon net and washed twice with Hepes-buffered Krebs-Ringer solution containing 2 m*M* $CaCl_2$. The cells were counted in a hemocytometer, and viability was assessed by trypan blue exclusion. The cell yields varied between 3 to 9 × 10^6 cells/g wet tissue.

Aortic cells were seeded in tissue culture vials in Dulbecco's modification of Eagle's medium (DMEM) supplemented with 1% nonessential amino acids, 50 units/ml penicillin-streptomycin, and 10% fetal calf serum or rabbit serum. The cells were cultured according to standard procedures.

B. Isolation and Labeling of Lipoproteins

LDL (density = 1.019–1.063 g/ml) from human plasma and β-VLDL (density < 1.006 g/ml), intermediate density lipoprotein (IDL) (density = 1.006–1.019 g/ml) and LDL (density = 1.019–1.063 g/ml) from the plasma of cholesterol-fed rabbits were isolated by sequential ultracentrifugation using solid KBr to adjust the densities.[64] All solutions contained 5 mg/ml Na_2EDTA, 0.5 mg/ml GSH, and 1.3 mg/ml ε-aminocaproic acid to prevent oxidative and proteolytic damage.[65] The lipoproteins were dialyzed against several changes of a solution containing 0.15 *M* NaCl and 1 m*M* EDTA, pH 7.4. Lipoproteins were labeled with ^{125}I, using the iodine monochloride method.[66]

Lipoproteins were labeled with the fluorescent probe 3,3′-dioctadecylindocarbocyanine (DiI) according to Pitas et al.[67] LDL (1 to 2 mg) was mixed with

2 ml of lipoprotein-deficient serum (LPDS) prepared from calf serum. Then, while gently mixing, DiI stock solution (3 mg/ml in dimethyl sulfoxide) was added to a final concentration of 150 μg DiI/mg LDL. The solution was filter sterilized (0.45 μm for LDL; 0.8 μm for β-VLDL) and incubated at 37°C for 15 to 18 h. The density of the mixture was adjusted to 1.080 g/ml with KBr, and it was transferred to a 4.2 ml ultracentrifuge tube, overlayered with 1.063 g/ml KBr solution, and ultracentrifuged at 40,000 rpm and 10°C in a 6 × 4.2 ml swinging-bucket rotor (200,000 × g) for 20 h. The DiI-labeled lipoprotein at the top of the tube was collected by aspiration and dialyzed overnight against 0.15 mol/l NaCl containing 1 mmol/l Na_2EDTA, pH 7.4.

C. Microscopic Analysis of Lipoprotein Uptake

To analyze the cellular uptake of DiI-labeled lipoproteins, cell cultures on coverslips were incubated with the lipoproteins (10 μg/ml) for 4 to 5 h, washed with phosphate-buffered saline, fixed with 3% formaldehyde, and mounted on microscope slides. Fluorescence was observed by standard epifluorescence microscopy using a rhodamine filter.

Measurements of $[^3H]$oleate incorporation into cholesteryl esters and cellular binding and degradation of ^{125}I-labeled lipoproteins were done as described previously.[13-15]

D. Identification of Monocyte/Macrophages

Monocyte/macrophages in primary cultures from human lesions were identified with phycoerythrin-conjugated monoclonal antibody LeuM3.[14] In primary cultures of rabbit atherosclerotic lesions, macrophages were identified on the basis of their Fc receptors[13,15] or acid esterase activities.[13] Fc receptors were localized by rosetting with sheep erythrocytes coated with rabbit anti-sheep red blood cell IgG.[13,15]

Other methods referred to here are as described in the original publications.[13-15]

VI. RESULTS

A. Properties of Atherosclerotic Lesion Cells

The cultures obtained following enzymatic dispersal of cells from rabbit and human atherosclerotic lesions consist of macrophages and smooth muscle cells (Figures 1 and 2). In the rabbit cell cultures we identified macrophages on the basis of several markers, e.g., histochemical staining for acid lipase or nonspecific esterase[13] and the demonstration of Fc receptors.[13,15] In cell cultures from human fatty streak lesions, we used the monoclonal antibody LeuM3 for macrophage identification (Figure 3). In our studies the proportion of macrophages in rabbit cell cultures varied between 30 and 50%,[13] and in human cell cultures, between 40 and 70%.[14]

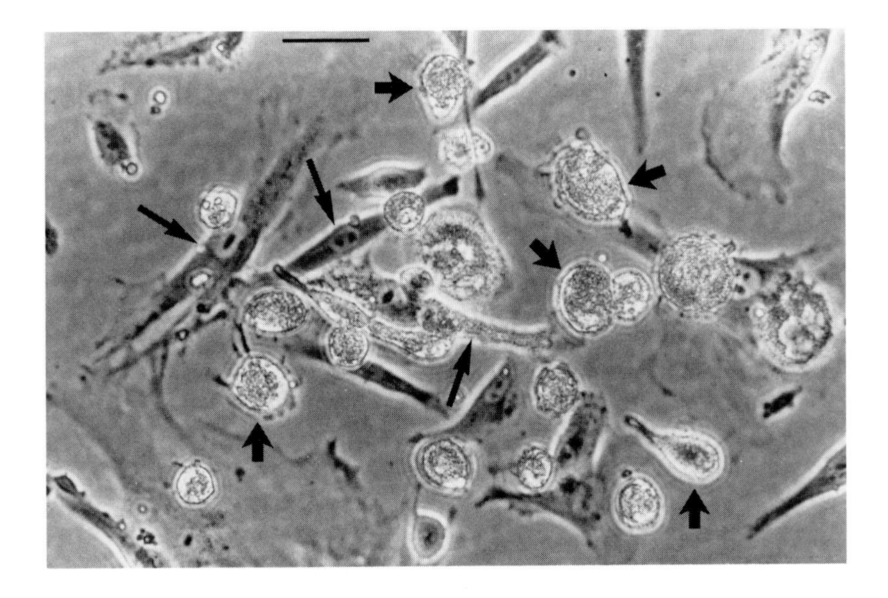

FIGURE 1. Enzymatically isolated cells from cholesterol-fed rabbit aorta in primary culture 2 d after isolation. Arrowheads indicate macrophages and arrows smooth muscle cells. (Bar: 20 μm).

The cells isolated from the aortas of cholesterol-fed rabbits contained 240 ± 74 μg (mean ± S.D.) free cholesterol and 1060 ± 445 μg of esterified cholesterol per milligram of cell protein.[15] The cholesteryl ester droplets in the atherosclerotic lesion cells could be localized on the basis of the ability of filipin to form fluorescent complexes with free cholesterol at the droplet margin.[13,14] In the freshly isolated lesion cell cultures from both rabbit and human lesions, most macrophages contained large amounts of brightly fluorescent cholesteryl ester droplets, whereas smooth muscle cells contained smaller amounts.

In cultures of enzyme-isolated atherosclerotic lesion cells, the smooth muscle cells start to proliferate a few days after seeding; in rabbit cell cultures this takes place on the second or third day,[13] and in human cell cultures, on the fifth or sixth day.[14] During proliferation, the cellular content of cholesteryl ester rapidly declines (Figure 4). The reduction is due both to extensive hydrolysis of cholesteryl esters and to rapid cell proliferation. The total mass of cholesterol per dish remains almost constant during the period.

B. Lipoprotein Uptake in Arterial Primary Cell Cultures

Lipoprotein metabolism in the cultures of cells from lesions of the atherosclerotic rabbits was studied after 2 d and 7 d, and those from human fatty streaks after 5 d and 12 d following isolation. In both cases the first time point represents a nonproliferating, macrophage-enriched phase before any major *in vitro* changes in the lipoprotein receptor activities have occurred. At the later time points, the

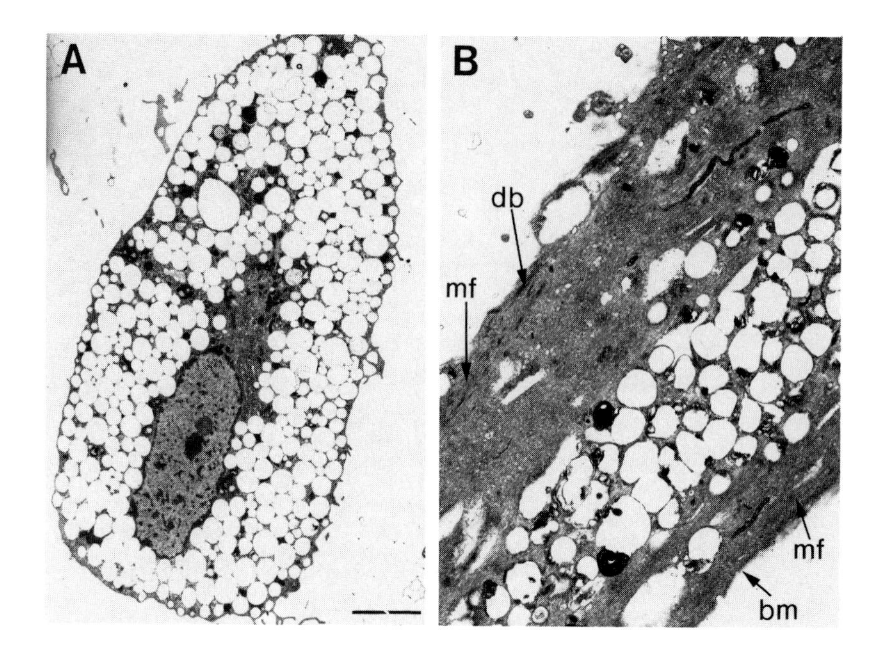

FIGURE 2. Electron microscopy of enzymatically isolated cells from human fatty streak lesions after 5 d in culture. A typical macrophage foam cell (A) and a smooth muscle cell (B) containing myofilaments (mf) and lipid inclusions. (db = dense bodies; bm = basal membrane). (Bar: 2 μm). (B is from Jaakkola, O., et al., *Atherosclerosis*, 79, 173, 1989, with permission.)

cultures are dominated by the rapidly proliferating smooth muscle cells, and changes in lipoprotein metabolism in the quiescent macrophages may also have occurred.

Lipoproteins fluorescently labeled with DiI were used to analyze the cellular uptake of lipoproteins. DiI-labeled acetyl-LDL, used as a model of modified LDL, is avidly internalized by macrophages both in rabbit (Figure 5A–C) and human lesion cell cultures (Figure 6A, B). In the freshly isolated cultures of rabbit atherosclerotic aortas, smooth muscle cells are also found to take up acetyl-LDL (Figure 5A, B), but this activity is almost completely lost in older cultures (Figure 5C). Abolition of fluorescence uptake from DiI-acetyl-LDL is observed in the presence of excess unlabeled acetyl-LDL, indicating the involvement of a limited number of binding sites in the uptake process. These findings are consistent with the expression of the scavenger receptor in the rabbit atherosclerotic lesions.

In the 2-day-old rabbit cell cultures, the lipoprotein that shows the highest rate of degradation is acetyl-LDL (Table 1). Acetyl-LDL induces more cholesterol esterification, as measured by [^{3}H]oleate incorporation, than do the other lipoproteins studied (Figure 7). Incubation with acetyl-LDL is also shown to lead

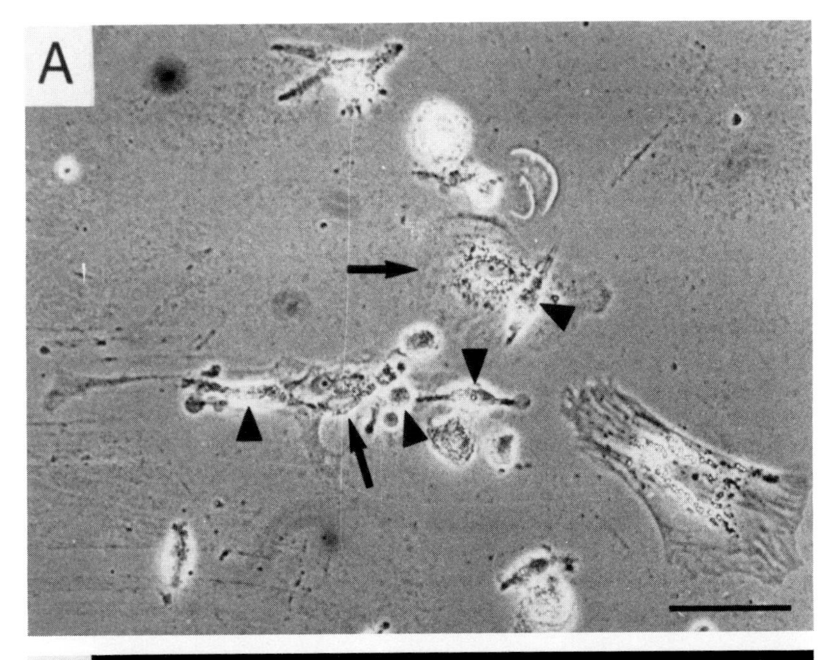

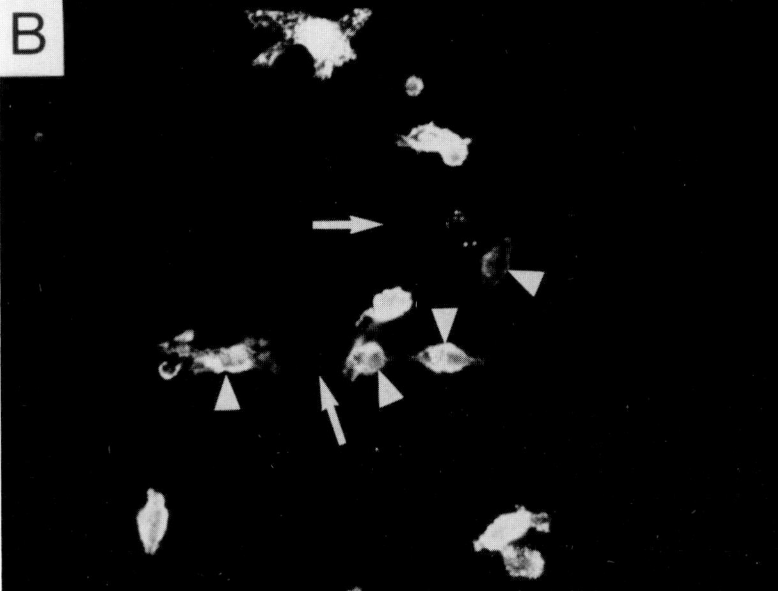

FIGURE 3. Identification of macrophages by fluorescence-labeled monoclonal antibody LeuM3 in 5-d-old human lesion cell culture. Phase contrast (A) and fluorescence views (B) of the same field are shown. (Bar: 20 μm). (From Jaakkola, O., et al., *Atherosclerosis*, 79, 173, 1989, with permission.)

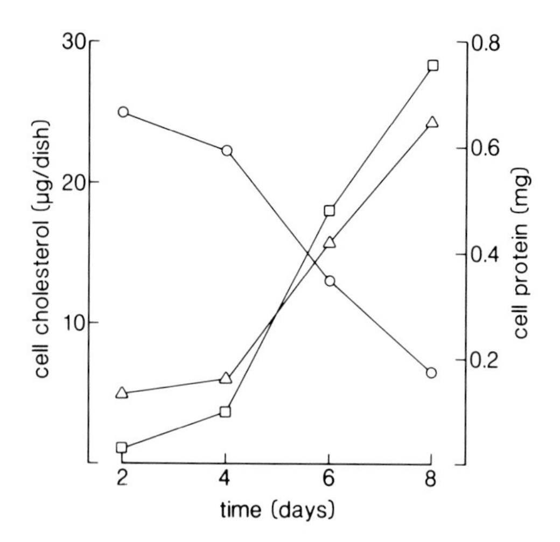

FIGURE 4. Change in cholesterol content in primary culture of cells isolated from cholesterol-fed rabbit aortas. (Δ = Free cholesterol; ○ = esterified cholesterol; □ = cell protein). (From Jaakkola, O. and Nikkari, T., *Am. J. Pathol.*, 137, 457, 1990, with permission.)

to a marked increase in the cellular cholesteryl ester content (Table 2). These findings indicate that cells in fatty streak lesions are capable of taking up modified lipoproteins in a manner leading to cholesteryl ester accumulation.

In the proliferating phase of the rabbit cell cultures (on day 7), the degradation of ^{125}I-acetyl-LDL (Table 1) and esterification of cholesterol with [^{3}H]oleate (Figure 7) are significantly decreased. When incubated with DiI-acetyl-LDL, most cells incorporate only low levels of fluorescence. However, in both human and rabbit cultures (Figure 5C), macrophages with bright DiI fluorescence are present, but they form only a small minority of cells.

Uptake of DiI-labeled β-VLDL into macrophages and smooth muscle cells can be demonstrated in primary cell cultures from rabbit (Figure 5D, E) and human arterial fatty streak lesions (Figure 6C, D). In the rabbit cell cultures, β-VLDL only slightly enhances the rate of cholesterol esterification both in 2-day- and 7-day-old cultures (Table 1). The uptake is not seen until some reduction in the cellular cholesteryl ester content has taken place, particularly in the rabbit fatty streak macrophages (Figure 5E). This is probably due to a suppression of the apoB, E (LDL) receptor activity in the presence of high cholesteryl ester content, as has been demonstrated in cultured peritoneal macrophages.[53] In both human and rabbit arterial cell cultures, DiI-β-VLDL is also internalized in smooth muscle cells. This uptake is stimulated when the apoB,E(LDL) receptor activity increases during cell proliferation. In the macrophage-enriched early cultures from both rabbits and humans, the uptake of LDL is very low when measured with any of the methods applied here (Tables 1, 2; Figure 7), whereas

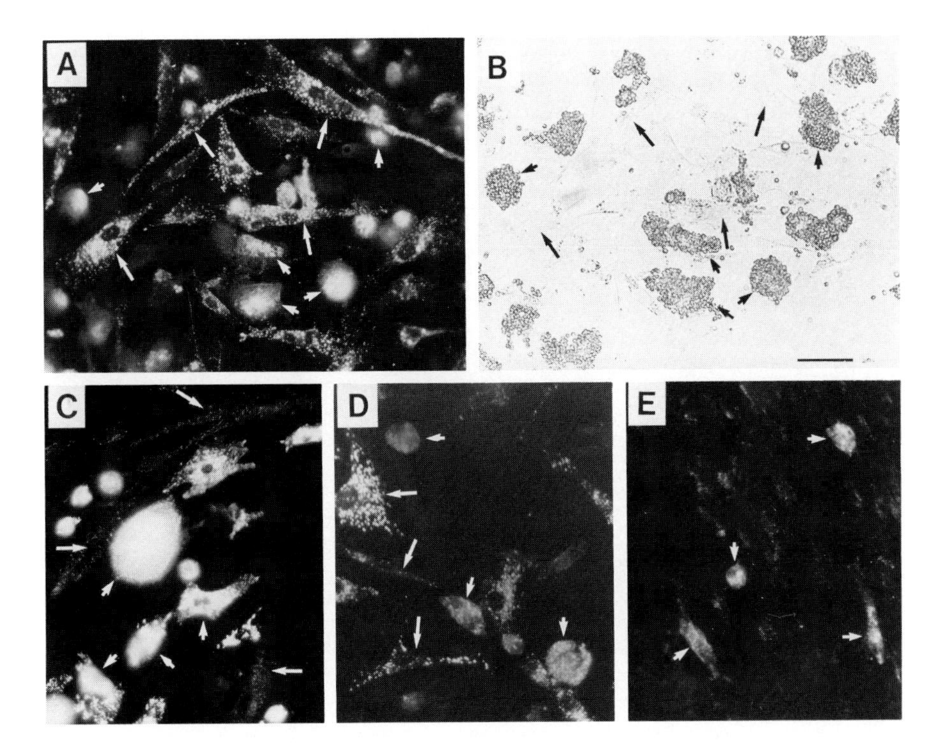

FIGURE 5. Uptake of DiI-labeled lipoproteins in cholesterol-fed rabbit aortic cells in primary culture. (A) Two-d-old culture incubated with (10 μg/ml) DiI-acetyl-LDL. (B) Phase contrast view of the same field as in (A) incubated with IgG-coated sheep red blood cells to identify Fc-receptor-positive macrophages. (C) Seven-d-old, confluent culture incubated with DiI-acetyl-LDL. Macrophages have accumulated bright fluorescence, whereas most smooth muscle cells have remained unlabeled. (D) Two-d-old culture incubated with (10 μg/ml) DiI-β-VLDL. Most macrophages show only low fluorescence, whereas occasional smooth muscle cells are brightly fluorescent. (E) Seven-d-old culture incubated with DiI-β-VLDL. On confluent areas the uptake of fluorescence into smooth muscle cells is low, and macrophages with more intense fluorescence can be distinguished. (Bar: 20 μm). (From Jaakkola, O. and Nikkari, T., *Am. J. Pathol.*, 137, 457, 1990, with permission.)

in the proliferating cultures, its uptake and degradation is considerably enhanced, and this is apparent in most smooth muscle cells. No significant uptake of native LDL into macrophages is seen.

VII. DISCUSSION

A. Characterization of the Isolated Arterial Cells

Cell isolation from arterial tissue using enzymes that degrade extracellular matrix components, most often collagenase and elastase, has been described for normal[63,68] and atherosclerotic[2,8,69,70] arterial tissue. The reported cell yields in

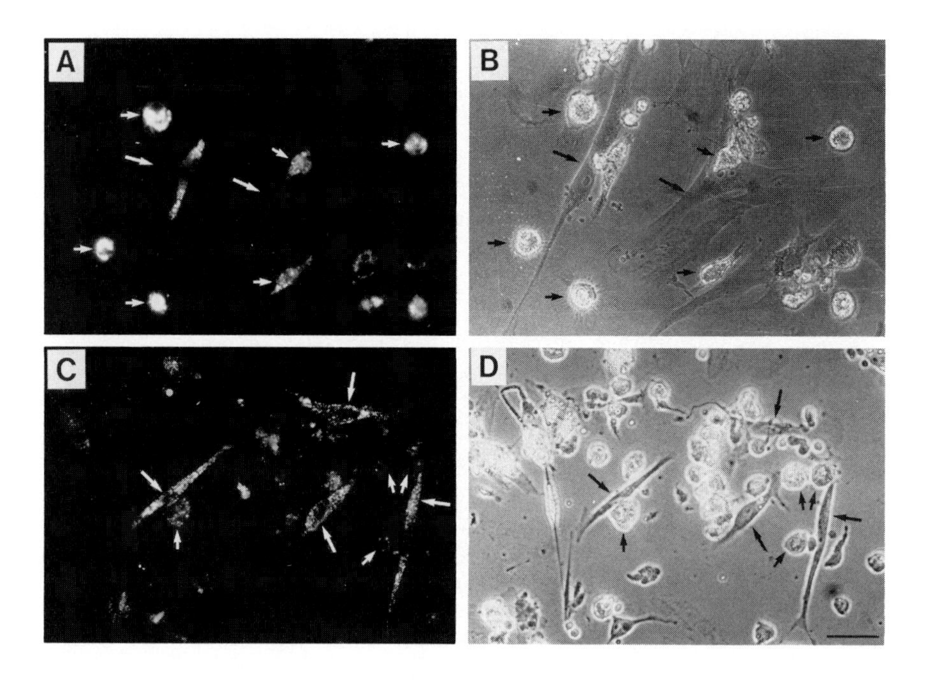

FIGURE 6. Uptake of DiI-labeled lipoproteins in human aortic fatty streak cells in 5-d-old primary cultures. The cultures were incubated with (10 µg/ml) DiI-acetyl-LDL (A and B) or DiI-β-VLDL (C and D). Fluorescence microscopic (A and C) and phase contrast (B and D) views of the same fields are shown. (Bar: 20 µm). (From Jaakkola, O., et al., *Atherosclerosis*, 79, 173, 1989, with permission.)

different studies have varied between 1×10^6 and 2.3×10^7 cells/g wet tissue, representing 20 to 70% of all cells measured as recovered DNA.[63,71] The cell recovery from atherosclerotic tissue has been reported to be somewhat lower than that from normal tissue.[71] The seeding efficiencies of enzymatically isolated arterial cells are reported to vary from 30 to 90%,[63,71] being slightly higher for cells from normal tissue. In the studies presented here, the cell yield was 3 to 6×10^6 cells/g wet tissue, and the seeding efficiencies varied between 40 and 60%. By these criteria, our isolation procedure gives results similar to those reported by others. Even though the cellular composition of cultures isolated enzymatically from atherosclerotic lesions is substantially more representative of the original tissue than that in explant-derived cultures, one has to bear in mind that it usually represents only a minority of the cells originally present in the tissue. Both the enzymatic digestion and the adherence to the culture surface may have selective effects, so that the cell composition of the cultures may not precisely match that of the original tissue.

TABLE 1
Degradation of ^{125}I-Labeled Lipoproteins (ng/mg) by Cells Cultured from Atherosclerotic Rabbit Aortas and by Mouse Peritoneal Macrophages[a]

	Cells from atherosclerotic aortas			
	Primary cultures			
Lipoprotein	2 d	7 d	Subcultures	Mouse peritoneal macrophages
β-VLDL	135.6 ± 54.2	113.1 ± 29.9	66.9 ± 36.1	478.4 ± 44.9
LDL	195.3 ± 19.8	353.8 ± 26.5	217.7 ± 22.2	378.2 ± 22.7
Acetyl-LDL	693.6 ± 14.1	169.1 ± 18.7	67.7 ± 9.20	865.2 ± 36.8

[a] Enzyme-isolated aortic cells were cultured for 2 d or 7 d or subcultured (first passage) for 2 d, and peritoneal macrophages were cultivated for 2 d. The cultures were incubated with ^{125}I-labeled lipoprotein (30 μg/ml), and after 8-h incubation, the degradation of the ^{125}I-labeled lipoprotein was determined. Values are means ± SD. (From Jaakkola, O. and Nikkari, T., *Am. J. Pathol.*, 137, 457, 1990, with permission.)

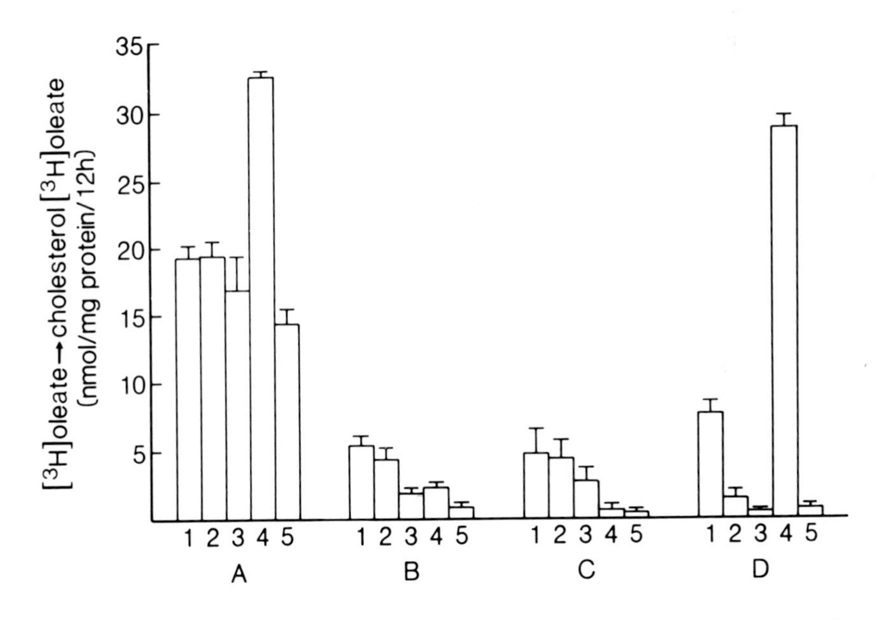

FIGURE 7. Incorporation of [^{3}H]oleate into cholesteryl esters in cultured cells. Rabbit aortic cells after 2 d (A) and 7 d (B) in culture, subcultured aortic cells (C), and mouse peritoneal macrophages (D) were incubated with the indicated lipoproteins (30 μg/ml) in the presence of [^{3}H]oleate, and after the incubation, the radioactivity incorporated into cholesteryl esters was determined. The lipoproteins were (1) β-VLDL, (2) IDL, (3) LDL, (4) acetyl-LDL, and (5) control without lipoprotein. (From Jaakkola, O., et al., *Atherosclerosis*, 79, 173, 1989, with permission).

TABLE 2
Effect of Incubation with Different Lipoproteins on the Cholesterol Content (μg/mg) of Rabbit Aortic Cells and of Mouse Peritoneal Macrophages[a]

		Cells from atherosclerotic aortas			
		Primary cultures			
Lipoprotein		2 d	7 d	Subcultures	Mouse peritoneal macrophages
β-VLDL	CE	258.4 ± 16.8	15.8 ± 0.6	5.9 ± 0.7	27.6 ± 8.8
	C	78.4 ± 4.3	35.3 ± 0.8	30.5 ± 1.8	52.6 ± 1.6
IDL	CE	291.6 ± 12.0	14.3 ± 1.6	5.1 ± 0.2	13.2 ± 9.7
	C	79.7 ± 2.6	34.5 ± 4.2	29.3 ± 0.6	43.5 ± 1.5
LDL	CE	279.7 ± 22.0	13.6 ± 0.9	3.9 ± 1.3	11.8 ± 4.4
	C	73.5 ± 3.7	31.3 ± 3.5	28.4 ± 1.7	36.3 ± 2.0
Acetyl-LDL	CE	331.1 ± 9.5	14.9 ± 1.0	3.6 ± 0.1	52.1 ± 8.0
	C	82.9 ± 3.4	29.2 ± 0.5	27.9 ± 1.1	58.3 ± 11.1
None	CE	262.3 ± 17.4	11.9 ± 1.6	1.8 ± 0.4	4.5 ± 1.7
	C	72.9 ± 2.3	27.3 ± 1.6	20.4 ± 3.7	28.1 ± 3.0

[a] Enzyme-isolated aortic cells from cholesterol-fed rabbits were cultured for 2 d or 7 d, subcultured aortic cells (first passage) and peritoneal macrophages for 2 d. The cultures were incubated for 12 h with the indicated lipoprotein (30 μg/ml). Lipids were extracted and fractionated, and free and esterified cholesterol (C, CE) were determined using gas chromatography. Values are means ± S.D. (From Jaakkola, O. and Nikkari, T., *Am. J. Pathol.*, 137, 457, 1990, with permission.)

B. Use of DiI Label in Lipoprotein Uptake Studies

DiI is a phospholipid analogue; its chemical structure includes two hydrocarbon chains and a polar, fluorescent head group. Labeling of lipoproteins with DiI is carried out in the presence of serum proteins[67] that assist the association of DiI with the lipoprotein phospholipid shell. The physicochemical properties of lipoproteins, such as density or electrophoretic mobility, did not appear to be significantly affected by the incorporation of the label. Our own experience and studies from several laboratories[67,72,73] indicate that, once incorporated, there is little passive exchange of DiI between lipoproteins. Comparative experiments[73] have shown that the uptake of DiI and [^{125}I] from labeled LDL occurs in a parallel manner, whereas DiI label slightly enhances acetyl-LDL uptake via the scavenger receptor, as compared with ^{125}I-labeled acetyl-LDL. DiI is a cumulative marker of lipoprotein uptake; i.e., when internalized into cells, it is preserved for several days without significant loss.[73,74] That the DiI label is transferred into cells only by binding or internalization of the intact labeled lipoprotein has been confirmed by experiments in which fixed, lipid-containing arterial cell cultures have been incubated with DiI-labeled LDL or acetyl-LDL. Thus, DiI-labeled lipoproteins are suitable for measurements of cellular lipoprotein uptake, and the fluorescence intensity is a quantitative marker of the extent of lipoprotein accumulation.

C. Lipoprotein Metabolism in Enzyme-Dispersed Arterial Cells

The results presented here show that acetyl-LDL, used as a model for modified LDL, is avidly metabolized by lesion-derived macrophages of both rabbit and human arteries. Similar findings have been reported by Pitas et al.[74] using explant-derived primary cultures, and by Rosenfeld et al.[75] using macrophage cultures isolated from aortas of cholesterol-fed rabbits. It has also been shown that in atherosclerotic rabbit aortas perfused with a solution containing labeled acetyl-LDL, lesion macrophages are the major cell type accumulating the label.[76] The findings suggest that the activity of the scavenger receptor pathway in the lesion macrophages is high enough to lead to cholesterol accumulation. The importance of this pathway is emphasized by the detection of oxidatively modified lipoproteins capable of binding to scavenger receptors, in human and animal atherosclerotic lesions.[49] A primary role for the scavenger receptor does not, however, rule out other possible mechanisms of macrophage foam cell formation, such as phagocytic uptake of LDL aggregates formed in a variety of ways.[77-81]

In freshly isolated rabbit lesion cell cultures, we also observed uptake of DiI-acetyl-LDL in some, but not all, smooth muscle cells. This implies a heterogeneity of the intimal smooth muscle cell population. The property was absent in older cultures, suggesting that the ability of acetyl-LDL uptake in smooth muscle cells is not preserved in culture. Recent data from our laboratory (Jaakkola et al., unpublished observation) and from others[82] suggest that rabbit arterial smooth muscle cells express scavenger receptor activity following stimulation of protein kinase C with phorbol ester, whereas human smooth muscle cells do not. The receptor may similarly be activated in the rabbit atherosclerotic lesions studied here, but if this is so, the factors or conditions contributing to this activation remain undefined. It is tempting to speculate that scavenger receptor-mediated uptake of modified lipoproteins might also be responsible for cholesterol accumulation in lesion smooth muscle cells, at least in rabbits.

Receptor-mediated uptake of β-VLDL in arterial macrophages has been considered to be one of the mechanisms of foam cell formation during atherogenesis,[54] and uptake of DiI-labeled β-VLDL by macrophages is apparent in primary cultures of both rabbit and human aortic cell cultures. However, particularly in rabbit macrophages, uptake of DiI-β-VLDL is not seen in the freshly isolated cells, but is stimulated during the primary culture period, simultaneously with the decrease in cellular cholesteryl ester content. This observation is in keeping with the finding that β-VLDL is taken up by macrophages via the apoB, E (LDL) receptor, which is downregulated by excess cholesterol.[53,54,83]

Although LDL is considered to be the main atherogenic lipoprotein, its uptake in freshly isolated lesion cells in culture appears to be very sparse. An increase in the metabolism of LDL is detected in the older, smooth muscle cell-enriched cultures, but this is related to cell proliferation and is not associated with cholesterol accumulation.[84,85]

VIII. CONCLUDING REMARKS

The results of lipoprotein-uptake studies with rabbit and human fatty streak lesion cells in primary culture indicate that modified LDL is avidly taken up by macrophages and results in cholesteryl ester accumulation. In rabbit cell cultures, smooth muscle cells also internalize modified LDL, but the role of this in cholesterol accumulation requires further study. By using cells isolated from early lesions, it has thus become possible to validate directly the roles of macrophages and smooth muscle cells derived initially from studies on standard cell cultures.

REFERENCES

1. **Ross, R.,** The pathogenesis of atherosclerosis — an update, *N. Engl. J. Med.*, 314, 488, 1986.
2. **Fowler, S., Shio, H., and Haley, N.J.,** Characterization of lipid-laden aortic cells from cholesterol-fed rabbits. IV. Investigation of macrophage-like properties of aortic cell populations, *Lab. Invest.*, 41, 372, 1979.
3. **Gerrity, R.G.,** The role of the monocyte in atherogenesis. I. Transition of blood-borne monocytes into foam cells in fatty lesions, *Am. J. Pathol.*, 103, 181, 1981.
4. **Faggiotto, A., Ross, R., and Harker, L.,** Studies on hypercholesterolemia in the nonhuman primate. I. Changes that lead to fatty streak formation, *Arteriosclerosis*, 4, 323, 1984.
5. **Aquel, N.M., Ball, R.Y., Waldmann, H., and Mitchinson, M.J.,** Monocytic origin of foam cells in human atherosclerotic plaques, *Atherosclerosis*, 53, 265, 1984.
6. **Gown, A.M., Tsukada, T., and Ross, R.,** Human atherosclerosis. II. Immunocytochemical analysis of the cellular composition of human atherosclerotic lesions, *Am. J. Pathol.*, 125, 191, 1986.
7. **Tsukada, T., Rosenfeld, M., Ross, R., and Gown, A.M.,** Immunocytochemical analysis of cellular components in atherosclerotic lesions. Use of monoclonal antibodies with the Watanabe and fat-fed rabbit, *Arteriosclerosis*, 6, 601, 1986.
8. **Haley, N.J., Shio, H., and Fowler, S.,** Characterization of lipid-laden aortic cells from cholesterol-fed rabbits. I. Resolution of aortic cell populations by metrizamide density gradient centrifugation, *Lab. Invest.*, 37, 287, 1977.
9. **Kannel, W.B., Castelli, W.P., Gordon, T., and McNamara, P.M.,** Serum cholesterol, lipoproteins and the risk of coronary heart disease. The Framingham study, *Ann. Intern. Med.*, 74, 1, 1971.
10. Lipid Research Clinics Program, The lipid research clinics coronary primary prevention trial results. I. Reduction in incidence of coronary heart disease, *JAMA*, 251, 351, 1984.
11. **Brown, M.S. and Goldstein, J.L.,** Lipoprotein metabolism in the macrophage: implications for cholesterol deposition in atherosclerosis, *Ann. Rev. Biochem.*, 52, 223, 1983.

12. **Goldstein, J.L., Anderson, R.G.W., Buja, L.M., Basu, S.K., and Brown, M.S.,** Overloading human aortic smooth muscle cells with low density lipoprotein-cholesteryl ester reproduces features of atherosclerosis *in vitro*, *J. Clin. Invest.*, 59, 1196, 1977.
13. **Jaakkola, O., Kallioniemi, O.-P., and Nikkari, T.,** Lipoprotein uptake in primary cell cultures of rabbit atherosclerotic lesions. A fluorescence microscopic and flow cytometric study, *Atherosclerosis*, 69, 257,1988.
14. **Jaakkola, O., Ylä-Herttuala, S., Särkioja, T., and Nikkari, T.,** Macrophage foam cells from human aortic fatty streaks take up β-VLDL and acetylated LDL in primary culture, *Atherosclerosis*, 79, 173, 1989.
15. **Jaakkola, O. and Nikkari, T.,** Lipoprotein degradation and cholesterol esterification in primary cell cultures of rabbit atherosclerotic lesions, *Am. J. Pathol.*, 137, 457, 1990.
16. **Rosenfeld, M.E., Tsukada, T., Gown, A.M., and Ross, R.,** Fatty streak initiation in Watanabe heritable hyperlipemic and comparably hypercholesterolemic fat-fed rabbits, *Arteriosclerosis*, 7, 9, 1987.
17. **Masuda, J. and Ross, R.,** Atherogenesis during low level hypercholesterolemia in the nonhuman primate. I. Fatty streak formation, *Arteriosclerosis*, 10, 164, 1990.
18. **Klurfeld, D.M.,** Identification of foam cells in human atherosclerotic lesions as macrophages using monoclonal antibodies, *Arch. Pathol. Lab. Med.*, 109, 445, 1985.
19. **Jonasson, L., Holm, J., Skalli, O., Bondjers, G., and Hansson, G.K.,** Regional accumulation of T cells, macrophages, and smooth muscle cells in the human atherosclerotic plaque, *Arteriosclerosis*, 6, 131, 1986.
20. **Alderson, L.M., Endemann, G., Lindsey, S., Pronczuk, A., Hoover, R.L., and Hayes, K.C.,** LDL enhances monocyte adhesion to endothelial cells *in vitro*, *Am. J. Pathol.*, 123, 334, 1986.
21. **Endemann, G., Pronzcuk, A., Friedman, G., Lindsey, S., Alderson, L., and Hayes, K.C.,** Monocyte adherence to endothelial cells *in vitro* is increased by β-VLDL, *Am. J. Pathol.*, 126, 1, 1987.
22. **Bevilacqua, M.P., Pober, J.S., Wheeler, M.E., Cotran, R.S., and Gimbrone, M.A., Jr.,** Interleukin 1 acts on cultured human vascular endothelium to increase the adhesion of polymorphonuclear leucocytes, monocytes and related leucocyte cell lines, *J. Clin. Invest.*, 76, 2003, 1985.
23. **Dustin, M.L., Rothlein, R., Dhan, A.K., Dinarello, C.A., and Springer, T.A.,** Induction by IL-1 and interferon-gamma: tissue distribution, biochemistry, and function of a natural adherence molecule (ICAM-1), *J. Immunol.*, 137,245, 1986.
24. **Cybulsky, M.I. and Gimbrone, M.A., Jr.,** Endothelial expression of a mononuclear leukocyte adhesion molecule during atherogenesis, *Science*, 251, 788, 1991.
25. **Gerrity, R.G., Goss, J.A., and Soby, L.,** Control of monocyte recruitment by chemotactic factor(s) in lesion-prone areas of swine aorta, *Arteriosclerosis*, 5, 55, 1985.
26. **Denholm, E.M. and Lewis, J.C.,** Monocyte chemoattractants in pigeon aortic atherosclerosis, *Am. J. Pathol.*, 126, 464, 1987.
27. **Quinn, M.T., Parthasarathy, S., Fong, L.G., and Steinberg, D.,** Oxidatively modified low density lipoproteins: a potential role in recruitment and retention of monocyte/macrophages during atherogenesis, *Proc. Natl. Acad. Sci. U.S.A.*, 84, 2995, 1987.
28. **Berliner, J.A., Territo, M., Almada, L., Carter, A., Shafonsky, E., and Fogelman, A.M.,** Monocyte chemotactic factor produced by large vessel endothelial cells *in vitro*, *Arteriosclerosis*, 6, 254, 1986.
29. **Jauchem, J.R., Lopez, M., Sprague, E.A., and Schwartz, C.J.,** Mononuclear cell chemoattractant activity from cultured arterial smooth muscle cells, *Exp. Mol. Pathol.*, 37, 166, 1982.
30. **Valente, A.J., Fowler, S.R., Sprague, E.A., Kelley, J.L., Suenram, C.A., and Schwartz, C.J.,** Initial characterization of a peripheral blood mononuclear cell chemoattractant derived from cultured arterial smooth muscle cells, *Am. J. Pathol.*, 117, 409, 1984.

31. **Ylä-Herttuala, S., Lipton, B.A., Rosenfeld, M.E., Särkioja, T., Yoshimura, T., Leonard, E.J., Witztum, J.L., and Steinberg, D.,** Expression of monocyte chemoattractant protein 1 in macrophage-rich areas of human and rabbit atherosclerotic lesions, *Proc. Natl. Acad. Sci. U.S.A.*, 88, 5252, 1991.
32. **Sica, A., Wang, J.M., Colotta, F., Dejana, E., Mantovani, A., Oppenheim, J.J., Larsen, C.G., Zachariae, C.O.C., and Matsushima, K.,** Monocyte chemotactic and activating factor gene expression induced in endothelial cells by IL-1 and tumor necrosis factor, *J. Immunol.*, 144, 3034, 1990.
33. **Rollins, B.J., Yoshimura, T., Leonard, E.J., and Pober, J.S.,** Cytokine-activated human endothelial cells synthesize and secrete a monocyte chemoattractant, MCP-1/JE, *Am. J. Pathol.*, 136, 1229, 1990.
34. **Wang, J.M., Sica, A., Peri, G., Walter, S., Padura, I.M., Libby, P., Ceska, M., Lindley, I., Colotta, F., and Mantovani, A.,** Expression of monocyte chemotactic protein and interleukin-8 by cytokine-activated human vascular smooth muscle cells, *Arterioscler. Thromb.*, 11, 1166, 1991.
35. **Cushing, S.D., Berliner, J.A., Valente, A.J., Territo, M.C., Navab, M., Parhami, F., Gerrity, R., Schwartz, C.J., and Fogelman, A.M.,** Minimally modified low density lipoprotein induces monocyte chemotactic protein 1 in human endothelial cells and smooth muscle cells, *Proc. Natl. Acad. Sci. U.S.A.*, 87, 5134, 1990.
36. **Mitchinson, M.J. and Ball, R.Y.,** Macrophages and atherosclerosis, *Lancet*, II, 146, 1987.
37. **Goldstein, J.L. and Brown, M.S.,** The low-density lipoprotein pathway and its relation to atherosclerosis, *Ann. Rev. Biochem.*, 46, 897, 1977.
38. **Mahley, R.W. and Innerarity, T.L.,** Lipoprotein receptors and cholesterol homeostasis, *Biochim. Biophys. Acta*, 737, 197, 1983.
39. **Steinberg, D.,** Lipoproteins and atherosclerosis. A look back and a look ahead, *Arteriosclerosis*, 3, 283, 1983.
40. **Goldstein, J.L., Ho, Y.K., Basu, S.K., and Brown, M.S.,** Binding site on macrophages that mediates uptake and degradation of acetylated low density lipoprotein, producing massive cholesterol deposition, *Proc. Natl. Acad. Sci. U.S.A.*, 76, 333, 1979.
41. **Mahley, R.W., Innerarity, T.L., Weisgraber, K.H., and Oh, S.Y.,** Altered metabolism (*in vivo* and *in vitro*) of plasma lipoproteins after selective chemical modification of lysine residues of the apoproteins, *J. Clin. Invest.*, 64, 743, 1979.
42. **Gonen, B., Cole, T., and Hahm, K.S.,** The interaction of carbamylated low-density lipoprotein with cultured cells. Studies with human fibroblasts, rat peritoneal macrophages and human monocyte-derived macrophages, *Biochim. Biophys. Acta*, 754, 201, 1983.
43. **Fogelman, A.M., Shechter, I., Seager, J., Hokom, M., Child, J.S., and Edwards, P.A.,** Malondialdehyde alteration of low density lipoproteins leads to cholesteryl ester accumulation in human monocyte-macrophages, *Proc. Natl. Acad. Sci. U.S.A.*, 77, 2214, 1980.
44. **Henriksen, T., Mahoney, E.M., and Steinberg, D.,** Enhanced macrophage degradation of biologically modified low density lipoprotein, *Arteriosclerosis*, 3, 149, 1983.
45. **Morel, D.W., DiCorleto, P.E., and Chisolm, G.M.,** Endothelial and smooth muscle cells alter low density lipoprotein *in vitro* by free radical oxidation, *Arteriosclerosis*, 4, 357, 1984.
46. **Parthasarathy, S., Prinz, D.J., Boyd, D., Joy, L., and Steinberg, D.,** Macrophage oxidation of low density lipoprotein generates a modified form recognized by the scavenger receptor, *Arteriosclerosis*, 6, 505, 1986.
47. **Steinbrecher, U.P., Parthasarathy, S., Leake, D., Witztum, J.L., and Steinberg, D.,** Modification of low density lipoprotein by endothelial cells involves lipid peroxidation and degradation of low density lipoprotein phospholipid, *Proc. Natl. Acad. Sci. U.S.A.*, 81, 3883, 1984.
48. **Steinberg, D., Parthasarathy, S., Carew, T.E., Khoo, J.C., and Witztum, J.L.,** Beyond cholesterol. Modifications of low-density lipoprotein that increase its atherogenicity, *N. Engl. J. Med.*, 320, 915, 1989.

49. **Ylä-Herttuala, S., Palinski, W., Rosenfeld, M.E., Parthasarathy, S., Carew, T.E., Butler, S., Witztum, J.L., and Steinberg, D.,** Evidence for the presence of oxidatively modified low density lipoprotein in atherosclerotic lesions of rabbit and man, *J. Clin. Invest.*, 84, 1086, 1989.
50. **Mahley, R.W., Innerarity, T.L., Weisgraber, K.H., and Fry, D.L.,** Canine hyperlipoproteinemia and atherosclerosis, *Am. J. Pathol.*, 87, 205, 1977.
51. **Kovanen, P.T., Brown, M.S., Basu, S.K., Billheimer, D.W., and Goldstein, J.L.,** Saturation and suppression of hepatic lipoprotein receptors: a mechanism for the hypercholesterolemia of cholesterol-fed rabbits, *Proc. Natl. Acad. Sci. U.S.A.*, 78, 1396, 1981.
52. **Fainaru, M., Mahley, R.W., Hamilton, R.L., and Innerarity, T.L.,** Structural and metabolic heterogeneity of β-very low density lipoproteins from cholesterol-fed dogs and humans with type III hyperlipoproteinemia, *J. Lipid Res.*, 23, 702, 1982.
53. **Goldstein, J.L., Ho, Y.K., Brown, M.S., Innerarity, T.L., and Mahley, R.W.,** Cholesteryl ester accumulation in macrophages resulting from receptor-mediated uptake and degradation of hypercholesterolemic canine β-very low density lipoproteins, *J. Biol. Chem.*, 255, 1839, 1980.
54. **Mahley, R.W., Innerarity, T.L., Brown, M.S., Ho, Y.K., and Goldstein, J.L.,** Cholesteryl ester synthesis in macrophages: stimulation by β-very low density lipoproteins from cholesterol-fed animals of several species, *J. Lipid Res.*, 21, 970, 1980.
55. **Koo, C., Wernette-Hammond, M.E., and Innerarity, T.L.,** Uptake of canine β-very low density lipoproteins by mouse peritoneal macrophages is mediated by a low density lipoprotein receptor, *J. Biol. Chem.*, 261, 11,194, 1986.
56. **Ellsworth, J.L., Kraemer, F.B., and Cooper, A.D.,** Transport of β-very low density lipoproteins and chylomicron remnants by macrophages is mediated by the low density lipoprotein receptor pathway, *J. Biol. Chem.*, 262, 2316, 1987.
57. **Slater, H.R., Shepherd, J., and Packard, C.J.,** Receptor-mediated catabolism and tissue uptake of human low density lipoprotein in the cholesterol-fed, atherosclerotic rabbit, *Biochim. Biophys. Acta*, 713, 435, 1982.
58. **Portman, O.W., O'Malley, J.P., and Alexander, M.,** Metabolism of native and acetylated low density lipoproteins in squirrel monkeys with emphasis on aortas with varying severities of atherosclerosis, *Atherosclerosis*, 66, 227, 1987.
59. **Carew, T.E., Pittman, R.C., Marchand, E.R., and Steinberg, D.,** Measurement *in vivo* of irreversible degradation of low density lipoprotein in the rabbit aorta. Predominance of intimal degradation, *Arteriosclerosis*, 4, 214, 1984.
60. **Bremmelgaard, A., Stender, S., Lorentzen, J., and Kjeldsen, K.,** *In vivo* flux of plasma cholesterol into human abdominal aorta with advanced atherosclerosis, *Arteriosclerosis*, 6, 442, 1986.
61. **Daugherty, A., Lange, L.G., Sobel, B.E., and Schonfeld, G.,** Aortic accumulation and plasma clearance of β-VLDL and HDL: effects of diet-induced hypercholesterolemia in rabbits, *J. Lipid Res.*, 26, 955, 1985.
62. **Kallioniemi, O.-P., Jaakkola, O., Nikkari, S.T., and Nikkari, T.,** Growth properties and composition of cytoskeletal and cytocontractile proteins in aortic cells isolated and cultured from normal and atherosclerotic rabbits, *Atherosclerosis*, 52, 13, 1984.
63. **Ives, H.E., Schultz, G.S., Galardy, R.E., and Jamieson, J.D.,** Preparation of functional smooth muscle cells from the rabbit aorta, *J. Exp. Med.*, 148, 1400, 1978.
64. **Hatch, F.T. and Lees, R.S.,** Practical methods for plasma lipoprotein analysis, *Adv. Lipid Res.*, 6, 1, 1968.
65. **Lee, D.M., Valente, A.J., Kuo, W.H., and Maeda, H.,** Properties of apolipoprotein B in urea and in aqueous buffers. The use of glutathione and nitrogen in its solubilization, *Biochim. Biophys. Acta*, 666, 133, 1981.

66. **Bilheimer, D.W., Eisenberg, S., and Levy, R.I.,** The metabolism of very low density lipoprotein proteins. I. Preliminary *in vitro* and *in vivo* observations, *Biochim. Biophys. Acta*, 260, 212, 1972.
67. **Pitas, R.E., Innerarity, T.L., Weinstein, J.N., and Mahley, R.W.,** Acetoacetylated lipoproteins used to distinguish fibroblasts from macrophages *in vitro* by fluorescence microscopy, *Arteriosclerosis*, 1, 177, 1981.
68. **Chamley, J.H., Campbell, G.R., McConnel, J.D., and Gröschel-Stewart, U.,** Comparison of vascular smooth muscle cells from adult human, monkey and rabbit in primary culture and in subculture, *Cell Tissue Res.*, 177, 503, 1977.
69. **Orekhov, A.N., Tertov, V.V., Novikov, I.D., Krushinsky, A.V., Andreeva, E.R., Lankin, V.Z., and Smirnov, V.N.,** Lipids in cells of atherosclerotic and uninvolved human aorta. I. Lipid composition of aortic tissue and enzyme-isolated and cultured cells, *Exp. Mol. Pathol.*, 42, 117, 1985.
70. **Naito, M., Kuzuya, M., Funaki, C., Nakayama, Y., Asai, K., and Kuzuya, F.,** Separation and characterization of macrophages and smooth muscle cells in rabbit atherosclerotic lesions, *Artery*, 14, 266, 1987.
71. **Orekhov, A.N., Krushinsky, A.V., Andreeva, E.R., Repin, V.S., and Smirnov, V.N.,** Adult human aortic cells in primary culture: heterogeneity in shape, *Heart Vessels*, 2, 193, 1986.
72. **Barak, L.S. and Webb, W.W.,** Fluorescent low density lipoprotein for observation of dynamics of individual receptor complexes on cultured human fibroblasts, *J. Cell Biol.*, 90, 595, 1981.
73. **Reynolds, G.D. and St. Clair, R.W.,** A comparative microscopic and biochemical study of the fluorescent and ^{125}I-labelled lipoproteins by skin fibroblasts, smooth muscle cells, and peritoneal macrophages in culture, *Am. J. Pathol.*, 121, 200, 1985.
74. **Pitas, R.E., Innerarity, T.L., and Mahley, R.W.,** Foam cells in explants of atherosclerotic rabbit aortas have receptors for β-very low density lipoproteins and modified low density lipoproteins, *Arteriosclerosis*, 3, 2, 1983.
75. **Rosenfeld, M.E., Khoo, J.C., Miller, E., Parthasarathy, S., Palinski, W, and Witztum, J.L.,** Macrophage-derived foam cells freshly isolated from rabbit atherosclerotic lesions degrade modified lipoproteins, promote oxidation of low-density lipoproteins, and contain oxidation-specific lipid-protein adducts, *J. Clin. Invest.*, 87, 90, 1991.
76. **Wiklund, O., Mattsson, L., Björnheden, T., Camejo, G., and Bondjers, G.,** Uptake and degradation of low density lipoproteins in atherosclerotic rabbit aorta: role of local LDL modification, *J. Lipid Res.*, 32, 55, 1991.
77. **Khoo, J.C., Miller, E., McLoughlin, P., and Steinberg, D.,** Enhanced macrophage uptake of low density lipoprotein after self-aggregation, *Arteriosclerosis*, 8, 348, 1988.
78. **Suits, A.G., Chait, A., Aviram, M., and Heinecke, J.W.,** Phagocytosis of aggregated lipoprotein by macrophages: low density lipoprotein receptor-dependent foam-cell formation, *Proc. Natl. Acad. Sci. U.S.A.*, 86, 2713, 1989.
79. **Ylä-Herttuala, S., Jaakkola, O., Solakivi, T., Kuivaniemi, H., and Nikkari, T.,** The effect of proteoglycans, collagen and lysyl oxidase on the metabolism of low density lipoprotein by macrophages, *Atherosclerosis*, 67, 73, 1986.
80. **Falcone, D.J., Mated, N., Shio, H., Minick, C.R., and Fowler, S.D.,** Lipoprotein-heparin-fibronectin-denatured collagen complexes enhance cholesteryl ester accumulation in macrophages, *J. Cell Biol.*, 99, 1266, 1984.
81. **Hoff, H.F. and O'Neil, J.,** Lesion-derived low density lipoprotein and oxidized low density lipoprotein share a lability for aggregation, leading to enhanced macrophage degradation, *Arterioscler. Thromb.*, 11, 1209, 1991.
82. **Pitas, R.E.,** Expression of the acetyl low density lipoprotein receptor by rabbit fibroblasts and smooth muscle cells. Upregulation by phorbol esters, *J. Biol. Chem.*, 265, 12,722, 1990.

83. **Koo, C., Wernette-Hammond, M.E., Garcia, Z., Malloy, M.J., Uauy, R., East, C., Bilheimer, D.W., Mahley, R.W., and Innerarity, T.L.,** Uptake of cholesterol-rich remnant lipoproteins by human monocyte-derived macrophages is mediated by low density lipoprotein receptors, *J. Clin. Invest.*, 81, 1332, 1988.
84. **Witte, L.D. and Cornicelli, J.A.,** Platelet-derived growth factor stimulates low density lipoprotein receptor activity in cultured human fibroblasts, *Proc. Natl. Acad. Sci. U.S.A.*, 77, 5962, 1980.
85. **Davies, P.F. and Kerr, C.,** Modification of low density lipoprotein metabolism by growth factors in cultured vascular cells and human skin fibroblasts. Dependence upon duration of exposure, *Biochim. Biophys. Acta*, 712, 26, 1982.

Chapter 10

ENDOCYTOSIS OF LIPOPROTEINS AND CHOLESTEROL HOMEOSTASIS

Horst Robenek and Nicholas J. Severs

TABLE OF CONTENTS

ISBN 0-8493-5505-2

I. INTRODUCTION

The discovery and characterization of a range of distinct lipoprotein receptors over the last 15 years stand as major milestones in the field of atherosclerosis research (for reviews see Goldstein and Brown,[1] Mahley and Innerarity,[2] Goldstein et al.,[3] Bradley and Gianturco,[4] and Hobbs et al.[5]). Lipoprotein receptors are integrated constituents of cell membranes that bind, and enable cells to internalize, specific classes of lipoproteins.[6-8] Lipoproteins are blood-borne complexes of protein and lipid that serve as vehicles for the transport of cholesterol and other lipids to and from cells. The primary control mechanisms for maintaining lipoprotein and cholesterol homeostasis center on lipoprotein receptors.[5] By virtue of their influence on the concentration of lipoproteins in the plasma and on the accumulation of lipoproteins in the arterial wall, lipoprotein receptors play a crucial part in the genesis and progression of atherosclerosis.[5]

Foam cells have long been recognized as a characteristic and major cellular constituent of the atherosclerotic lesion, and it is the interaction of these cells and their precursors with lipoproteins that arguably forms the most important link between cholesterol and atherosclerosis.[9-17] Foam cells are now known to originate predominantly from monocytes that, after entry into the intima, are transformed into macrophages. A prime thrust of our work has therefore been to elucidate the mechanisms of cholesterol influx and accumulation in macrophages.[18-25] Our studies, which use cultured cells as a model system, have helped delineate the receptor-mediated pathways by which macrophages interact with and take up lipoprotein-bound cholesterol. In particular, we have examined the responses of macrophages to β-very low density lipoproteins (β-VLDL) and chemically modified low density lipoproteins, both which are highly efficacious in converting macrophages into foam cells.[26-40]

The mechanisms of cholesterol regulation in macrophages differ from those of cells such as fibroblasts, which are equipped with a large population of low density lipoprotein (LDL) receptors. In fibroblasts, the activities of the receptor and the rate-limiting step of intracellular cholesterol synthesis are strictly controlled in tune with cellular needs. LDL receptor cells are consequently protected from excessive accumulation of cholesterol or cholesteryl esters; however high the external cholesterol concentration, they are not converted into foam cells. Foam cell formation occurs only via receptor processes that are not strictly regulated. In order to avoid accumulation of cholesterol, such cells, being unable to down-regulate their receptors, need to interact with potent cholesterol accepters. High density lipoproteins (HDL) fulfill this function; they stimulate removal of cholesterol from macrophages in culture, and *in vivo* have been implicated in the transport of cholesterol from peripheral tissues for disposal by the liver, a process called "reverse cholesterol transport".[21,25,41-43]

In our recent work, we have studied HDL interactions with macrophages, and have identified further processes that may contribute to reverse cholesterol transport. The first of these is an HDL-independent mechanism by which macrophages release excess cholesterol. Second, we have demonstrated *in vitro* that otherwise nonmotile lipid-laden foam cells can be reinduced to migrate in response to appropriate chemotactic stimuli. These findings have implications for the understanding of cholesterol efflux *in vivo* and for the concept of lesion regression. In this chapter, we will present these and related findings within the context of a more general survey of the mechanisms of cellular cholesterol influx and efflux, the receptors involved, and the modern labeling and ultrastructural techniques by which these processes can now be elucidated. We start by summarizing the background to lipoproteins and lipoprotein receptors.

A. Plasma Lipoproteins

Plasma lipoproteins are metabolically diverse spheroid or discoid macromolecules comprising complexes of lipids (triglycerides, cholesterol and phospholipids) and one or more apolipoproteins (apo) (for reviews see Goldstein and Brown,[1] Mahley and Innerarity,[2] Goldstein et al.,[3] Bradley and Gianturco,[4] and Assmann[44]). There are four major classes of lipoproteins defined by density and electrophoretic mobility; (1) chylomicrons, (2) very low-density lipoproteins (VLDL), (3) low density lipoproteins (LDL) and 4) high density lipoproteins (HDL). These major classes of lipoprotein are further divided into subclasses, and in addition, several less abundant but nevertheless important lipoproteins are present in normal plasma. Among these are lipoprotein Lp(a), a particle with a lipid composition similar to that of LDL, and intermediate-density lipoprotein (IDL), which has physical and chemical properties intermediate between those of VLDL and LDL.

There are four major groups of apolipoproteins, designated A, B, C, and E. Apart from binding and solubilizing lipids to facilitate their transport as stable complexes in the plasma, apolipoproteins act as ligands for the binding of lipoproteins to receptors and regulate the activities of enzymes involved in lipoprotein and lipid metabolism. With the exception of LDL, each class of lipoprotein contains a variety of apolipoproteins in differing proportions. Apolipoprotein A occurs in three main forms (A-I, A-II, and A-IV), two of which (A-I and A-II) form the major proteins of HDL. LDL contains only apolipoprotein B. Two forms of apolipoprotein B exist, apolipoprotein B-100 and apolipoprotein

B-48. The form of apolipoprotein B in LDL is apo B-100, and this apolipoprotein is also an obligate constituent of VLDL and IDL. In man, apolipoprotein B-48 is found in chylomicrons and chylomicron remnants. The lipoprotein Lp(a) is composed of an LDL-like particle to which apolipoprotein (a) is attached through disulfide bridging to apolipoprotein B-100.[45] Apolipoproteins C and E are present in various proportions in all lipoproteins except LDL. There are three genetic forms of apolipoprotein E; E2, E3, and E4. The relationship of HDL with apolipoprotein E is important, and it is useful to distinguish two variants of this lipoprotein; HDL with apolipoprotein E (cholesterol-rich, HDL_c or HDL1), and HDL without apolipoprotein E (cholesterol-poor).

In addition to these normal lipoproteins, a number of abnormal lipoprotein particles appear in the plasma during certain disease states and in experimental animals fed diets high in fat and cholesterol. LpX is found in obstructive liver disease and in familial lecithin:cholesterol acyl transferase (LCAT) deficiency. Patients with the genetic disorder Type III hyperlipoproteinemia possess apolipoprotein E only in the E2 isoform, and have high levels of β-migrating VLDL.[2,40] β-VLDL are cholesterol-enriched lipoproteins that contain apolipoproteins B and E as their major protein constituents. β-VLDL also appear as part of the altered plasma lipoprotein profile in experimental animals fed diets rich in fat and cholesterol. Another change occurring with these diets is an increase in HDL with apolipoprotein E, and a reduction of the usual form of HDL (without apo-E).[2] Other variants of particular importance are chemically modified lipoproteins. Oxidation of lipoproteins can occur *in vivo* during circulation in the plasma or perfusion through the arterial wall.[46] For studies on cultured cells, the chemical modifications of LDL that have been used include acetylation,[26,27] acetoacetylation,[2] malondialdehyde treatment,[35] oxidation,[31] and LDL-proteoglycan complex formation.[47,48]

B. Lipoprotein Receptors

By far the most extensively characterized receptor is the LDL (apo-B,E) receptor, the details of which were initially worked out on cultured fibroblasts in the pioneering studies of Goldstein and Brown.[49] Further lipoprotein receptor systems have since been characterized, with evidence for others that remain to be elucidated in detail. While most lipoprotein receptors function to provide a mechanism for receptor-mediated endocytosis — and thus catabolism of specific plasma lipoproteins — some receptors apparently have additional functions.

Lipoprotein receptors other than the LDL receptor have been documented in a number of cell types and tissues. In macrophages, receptors or binding sites for at least four different types of lipoproteins have been described: (1) normal LDL (LDL receptor), (2) the β-VLDL of patients with Type III hypercholesterolemia and the β-VLDL induced by high cholesterol diets (the "β-VLDL receptor"),[2] (3) negatively changed or oxidized LDL (scavenger receptor),[50,51] and (4) HDL (A-I or HDL receptor).[21] Recently, a receptor for apolipoprotein E-containing lipoproteins, the low density lipoprotein-related protein (LRP) has been identified in liver.[52,53] Lipoprotein receptors show a high degree of selectivity with respect to the apolipoprotein ligands they recognize. The occurrence of these lipoprotein receptors in different types of cell and other details of their properties are summarized in Table 1.

II. APPROACHES TO THE LOCALIZATION OF LIPOPROTEINS AND LIPOPROTEIN RECEPTORS

A. Localization at the Cell Surface

The realization of the central importance of plasma membrane receptors to cell function has sparked major efforts over the last decade to improve morphological methods for receptor localization at the electron microscopic level. The development of colloidal gold labeling, in particular, has revolutionized the way cell biologists approach the study of cell surface receptors. Colloidal gold particles offer the highest resolution of marking achievable by electron microscopy today, and not only permit precise ultrastructural localization but also offer opportunities for quantitative approaches to experimental investigations. The combination of gold labeling with platinum/carbon (Pt/C) replication techniques, which we played a part in developing and exploiting,[54] has become the method of choice for studying the surface distribution of lipoprotein receptors. This is because it presents unique "nonaveraged" information on receptor distribution and numbers as viewed from above the cell — information that cannot be obtained by other structural or biochemical analytical techniques.[54-58] The dense, particulate nature of the gold marker facilitates accurate identification of the labeled structure at high magnifications, and is equally useful at lower magnifications for rapid sampling and preliminary screening of experimental interventions. Moreover, by combining this approach with experiments designed to follow ligand-receptor interactions and receptor expression over time, a dynamic aspect to experimental investigations becomes possible.

1. Cytochemical Detection Systems

The cytochemical detection approaches we use for visualizing lipoprotein receptors at the cell surface can be classified into three categories. The first method uses lipoprotein-gold complexes presented to cells (Figure 1a–d) to permit identification of the sites to which they bind on the plasma membrane surface. Lipoproteins conjugated to colloidal gold particles are added to the culture medium of living cells, and the cells incubated at 4°C for various periods of time. The experiments are terminated by thorough washing with ice-cold buffer, followed by chemical fixation and processing for electron microscopy. In the second method, the approach is indirect (two-step) immunocytochemical labeling of receptors or receptor-bound lipoproteins; living cells are exposed to specific antibodies directed either against the receptor or against the lipoprotein

TABLE 1
Summary of the Properties of Lipoprotein Receptors

Receptor	Protein determinants for binding	Lipoproteins recognized	Function	Localization
LDL	apo-B, apo-E	LDL, VLDL, HDL1, Chylomicron remnants	Lipoprotein catabolism	Ubiquitous (esp. liver)
β-VLDL[a]	apo-E	β-VLDL (from Type III subjects and induced by cholesterol feeding animals)	Catabolism of abnormal VLDL: role in foam cell formation	Monocytes, Macrophages Foam cells Endothelial cells
LRP	apo-E	VLDL Chylomicron remnants	Catabolism of apo E-containing lipoproteins	Liver, Macrophages
Scavenger	Modified apo-B	Chemically modified lipoproteins (LDL, Lp(a), etc.)	Role in foam cell formation	Macrophages, Foam cells, Endothelia cells
HDL	apo-A-I	HDL, HDL-subclasses	Reverse cholesterol transport	Macrophages and other cell types

[a] A distinct β-VLDL receptor pathway has been described in the macrophage, although the literature on the type(s) of receptor responsible for β-VLDL uptake is confusing. Interpretation of experimental results is complicated by the presence in macrophages of an E-receptor (LRP equivalent) and the classical LDL receptor, both of which may participate in uptake of β-VLDL.

(prebound to the cell). This step is followed by incubation with protein A-gold complexes or peroxidase-/gold-labeled secondary antibodies at 4°C. Washing with ice-cold buffer, fixation, and processing for electron microscopy completes the procedure. The third method involves immunocytochemical labeling of receptor-bound lipoproteins in sectioned material. This can be done by incubating living cells with unlabeled lipoproteins at 4°C for 1 h, and after fixation, processing for cryo-ultrathin sectioning, or embedding in Lowicryl resin. Sections are treated sequentially with specific antibodies directed against apolipoproteins followed by protein A-gold or secondary antibody-gold complexes. All three procedures conducted at 4°C give information on the distribution of surface receptors captured in a state of "arrest" by the low temperature. Warming the cells to 37°C initiates dynamic changes that, with the aid of these detection systems, can be traced from the cell surface through to the intracellular pathways of the lipoprotein receptor systems.

2. *Surface Replication Techniques*

The preparation of surface replicas of dried cultured cells is a highly versatile technique for visualizing receptors at the cell surface. Further dimensions are added by the techniques of lysis-squirting combined with Pt/C replication, and the label-fracture technique. The methodologies involved in these techniques are summarized as follows:

Surface Replication
Label ➤ Fix ➤ Dry ➤ Pt/C replication ➤ Replica cleaning

Lysis-Squirting
Lysis ➤ Squirting ➤ Label ➤ Fix ➤ Dry ➤ Pt/C replication ➤ Replica cleaning

Label-Fracture
Label ➤ Fix ➤ Cryoprotect ➤ Freeze-fracture ➤ Pt/C replication ➤ Replica rinsed in water (no cleaning in bleach or acid)

In the now standard labeled-surface replica technique, cultured cells are gold-labeled using either the first or second method detailed above. They are then fixed, dried, and replicated with Pt/C. The drying step can be accomplished quickly by air drying after ethanol dehydration, or if desired, by critical point drying or freeze-drying. The gold particles remain trapped within the replica providing that the cleaning conditions used to remove the biological material from the replica are not too harsh. Upon examination in the electron microscope, the gold particles marking the positions of the receptors are clearly viewed from above the cell over extensive replica views of the external surface of the plasma membrane. Apart from the advantages inherent in the use of gold labeling, the particular advantages of the standard surface replica technique are that it is quick and straightforward to carry out, and extensive areas of the cell surfaces of multiple cells can be surveyed, thus enabling rapid and reliable assessment of results. This approach is particularly well-suited to detailed elucidation of the distribution and dynamics of receptors.

In the standard surface replication technique, the only aspect of the cell viewed is the external surface of the "dorsal" (i.e., uppermost) plasma membrane. The lysis-squirting and "sandwich" techniques enable the protoplasmic aspects of the plasma membranes of cultured cells to be exposed for gold labeling and subsequent visualization by surface replication.[59-62] In brief, lysis-squirting

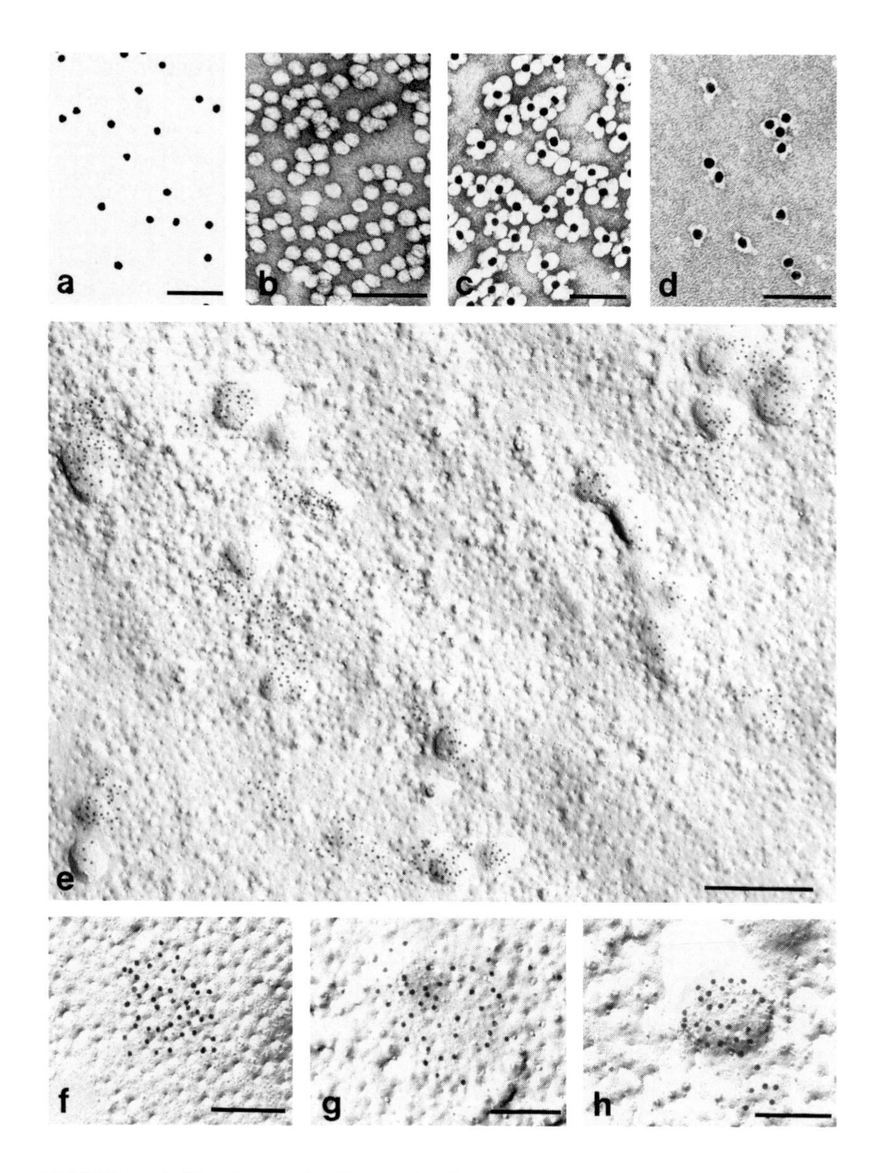

FIGURE 1. Gold markers, native lipoproteins, lipoprotein-gold complexes, and the labeling of LDL receptors in replicas, as viewed by transmission electron microscopy: (a) Example of a preparation of colloidal gold particles (average diameter 15 nm); (b) unconjugated (native) LDL molecules; (c) LDL-gold conjugate; (d) HDL-gold conjugate; (e) LDL receptor distribution studied by gold labeling and replicas (this example shows a label-fracture preparation from a fibroblast treated with LDL-gold conjugates like those in (c) for 1 h at 4°C. Note that all the gold particles, marking positions of LDL receptors, occur in clusters. Almost no single gold particles are present. Treatment with filipin (in glutaraldehyde) was given after the LDL-gold to discriminate between clathrin-coated and uncoated regions of the membrane, showing as unperturbed and pitted areas of

involves incubating cells attached to coverslips in hypotonic buffer, and then using a stream of buffer from a syringe or Pasteur pipette to shear away the bulk of the cell components. The protoplasmic surface of the "ventral" plasma membrane is thus exposed for immunolabeling. To reveal the protoplasmic aspect of the "dorsal" plasma membrane, a different approach (sandwich mounting) is required. This involves sandwiching a monolayer of cells between polylysine-coated coverslips, and separating the coverslips to break the cells apart. Once the sample has been labeled, drying is done by freeze-drying, critical point drying, or simply air-drying after ethanol dehydration, and replication carried out. The particular value of these approaches is that they permit investigation of the role of cytochemically identified cytoskeletal elements in receptor dynamics.

The label-fracture technique, developed by Pinto da Silva and Kan,[63] is one of a set of highly effective freeze-fracture cytochemical techniques developed over the last decade.[64] In label-fracture, label on the external surface of the plasma membrane is viewed superimposed upon a freeze-fracture replica view of the E-face of the plasma membrane. The procedure is to label the cells first, and then follow the routine freeze-fracture sequence up to and including the stage of Pt/C replication. Instead of cleaning the replicas in sodium hypochlorite (or another cleaning agent), however, they are washed extensively in distilled water and viewed directly in the electron microscope.[63] Where cells are fractured to reveal the E-face of the plasma membrane, the external half-membrane leaflet with still attached label remains stuck to the replica, allowing simultaneous visualization of surface label and E-face structural detail. To facilitate generation of the right type of fracture through the cultured cells, pieces of coverslip on which the cells have been grown are inverted onto drops of the mounting medium (polyvinyl alcohol), frozen, and then fractured by separating the coverslip from the frozen medium under high vacuum.[65]

All these labeling/replica techniques are limited to investigations of receptors on the surfaces of membranes that are directly accessible to the cytochemical reagents. Their application is thus restricted to the surfaces of cultured cell monolayers, cell suspensions, and the luminal membranes of surface epithelia. To localize lipoproteins or lipoprotein receptors within tissues or cells, rather different approaches have to be adopted. Sectioning techniques, and in particular pre-embedding and post-embedding immunocytochemical methodologies, are particularly appropriate.

---

the membrane, respectively); (f–h) Fibroblasts treated with LDL-gold conjugates under the same conditions as in (e). Groups and clusters of gold particles occur both over noncoated areas (f), mildly indented clathrin-coated areas (g) and clathrin-coated invaginations of the plasma membrane (h). The example in (f) represents more recently inserted receptors than those in (g). (b,c, and d are negatively stained preparations). (Bars: (a–d) 0.1 μm; (e) 0.5 μm; (f–h) 0.2 μm).

B. Localization within Tissues and Cells

1. Pre-Embedding Labeling Techniques

A standard approach we have used to follow the fate of internalized lipoproteins is to examine, by standard thin section electron microscopy, cell cultures that had been fixed at intervals after exposure to lipoprotein-gold complexes at 37°C. The disadvantage of this approach is that the intracellular pathway of the lipoprotein can often only be followed for a limited period owing to dissociation of lipoprotein and marker (e.g., upon lysosomal degradation of lipoprotein). Such aspects as receptor recycling and the initial synthesis and trafficking of receptors and lipoproteins cannot be investigated with this approach.

One technique that offers the possibility of investigating these aspects is pre-embedding cytochemistry. The principle of this technique is to make the cells permeable and allow penetration of cytochemical reagents before processing for thin section electron microscopy. We do this by fixing cells in McLean and Nakane's fixative[66] permeabilizing with saponin, and incubating with an appropriate dilution of primary antibodies against lipoproteins or lipoprotein receptors. Incubation with secondary peroxidase-conjugated antibodies follows, and peroxidase-activity is revealed by the diaminobenzidine method. Thin sections are examined without further staining.

2. Post-Embedding Labeling Techniques

The two low denaturing preparation procedures most widely used for intracellular localization of proteins are cryofixation followed by cryoultramicrotomy and low temperature embedding in hydrophilic acrylic resins such as Lowicryl followed by conventional sectioning.[67] In both procedures, the strategy is to carry out immunogold labeling of thin sections after they have been cut, rather than labeling specimens before embedding, as in the previous section.[68,69] In the post-embedding approach the cell does not itself present a barrier to reagent diffusion, and so permeabilization of membranes is not required.

In our hands, these post-embedding techniques have proved the most effective approaches for the detection of intracellular lipoproteins. For ultracryomicrotomy, formaldehyde-fixed cells or tissues are processed according to the method of Tokuyasu.[70] For embedding in Lowicryl, the cells are fixed in 4% formaldehyde or in a mixture of 0.5 to 3% glutaraldehyde and 4% formaldehyde, dehydrated in an ascending series of dimethylformamide, and embedded at low temperature in Lowicryl K4M following the protocol described by Völker et al.[71]

Immunocytochemical labeling is routinely done by floating grids serially, section side-down, on droplets of the following solutions; (1) buffer containing 0.2% glycine and/or 1% BSA, (2) an appropriate dilution of specific primary antibodies, (3) buffer wash, and (4) protein A-gold or secondary antibody-gold complexes. After the labeling procedure, sections are washed thoroughly with distilled water, and subsequently stained with uranyl acetate oxalate and uranyl acetate. The grids are finally floated on drops of 1% tylose in 0.5% uranyl acetate before examination.

Because two-step indirect cytochemical procedures are used for pre-embedding and post-embedding cytochemistry, the amount of antigenicity retained in the cells, and the specificities and affinities of the antibodies used, are critical in obtaining adequate labeling. The cells have to be fixed in a way that aims to retain maximum antigenicity, and "mild" fixation in formaldehyde is usually recommended for this purpose. The retention of antigenicity in cryosections is generally superior to that of other techniques because fewer tissue processing steps are involved; although the cells are fixed before freezing, there is no subsequent exposure to alcohols or embedding resins.

C. Antibodies against Lipoproteins and Lipoprotein Receptors

A central requirement for the intracellular localization of lipoproteins and lipoprotein receptors is the availability of suitable high affinity antibodies. Although a large number of polyclonal and monoclonal antibodies directed against all the lipoproteins and their apolipoprotein constituents has become available over the years, successful electron microscopical localization has been beset with difficulty. The reason for this relates to the preparation procedures used for electron microscopy. Even "mild" formaldehyde fixation may adversely influence the antigenicity of lipoproteins, resulting in a much reduced labeling intensity. Formaldehyde is known to react with the amino groups of proteins and to cross-link polypeptide chains, and so may impair or even abolish immunological reactions.[72,73] As for most procedures some form of chemical fixation is required to preserve cell structure, such fixation-induced changes of antigenicity cannot be avoided. We therefore approached the problem of how to improve lipoprotein recognition in fixed cells from the opposite angle — by raising polyclonal antibodies to lipoproteins or apolipoproteins that had been chemically modified by formaldehyde fixation *before* rabbit immunization.[74] In this way, we hoped to obtain antibodies that were at least in part directed against the antigenic determinants that are normally altered by fixation. Antibodies to prefixed apolipoprotein A-I (anti-fix.apo A-I) and LDL (anti-fix. LDL) were raised in this way and characterized with respect to properties relevant to immunocytochemistry. Our results demonstrated that antibodies against formaldehyde-fixed lipoproteins and apolipoproteins can indeed be more effective than the standard antibodies against their unmodified counterparts in intracellular immunolocalization studies, and this will be illustrated in Section IV.

For lipoprotein receptors (as opposed to lipoproteins), effective antibodies have so far only been raised against one type of lipoprotein receptor, the LDL receptor.[75] A few papers have reported the use of monoclonal anti-LDL receptor antibodies for the ultrastructural localization of LDL receptors in the plasma membrane of cultured fibroblasts,[56-58,75] but so far there are only two papers on the intracellular distribution of LDL receptors studied by pre-embedding and post-embedding immunolabeling techniques.[76,77]

III. MECHANISMS OF CELLULAR CHOLESTEROL INFLUX

A. Receptor-Mediated Endocytosis of Lipoproteins

1. Display and Dynamics of Lipoprotein Receptors in the Plasma Membrane

Cells organize and display their various classes of receptor in different ways, according to the functions they are required to perform. Here, we will discuss a selection of our recent findings on the LDL receptor, binding sites for β-VLDL and lipoprotein (a), and the macrophage scavenger receptor. HDL receptors will be considered later, in the context of cholesterol efflux (Section V. A. 1.). A summary of the distribution patterns of the various classes of lipoprotein receptor is given in Table 2.

LDL receptors were initially thought to be displayed in a dispersed fashion on the plasma membrane, with clustering, involving long-distance lateral migration, occurring after LDL binding.[78] Aggregation of receptor-LDL complexes into clusters was envisaged as the prelude to the classical events of receptor-mediated endocytosis, which we will examine in more detail in the next section. The original view that LDL receptors start their sojourn at the cell surface as individual widely dispersed units came from experiments using ferritin-LDL conjugates and thin-section electron microscopy.[78] Later work with ferritin-anti-receptor antibody labeling, also examined by thin sectioning, did not lead to revision of this view.[75] However, the true surface distributions as viewed from above the cell are difficult to deduce from the limited perpendicular views provided by thin sections. Our experiments, exploiting the technical advantages afforded by gold labeling and surface replicas, suggested a rather different picture of LDL receptor display.

When fibroblasts, maintained in lipoprotein-deficient medium, are incubated with LDL-gold complexes at or just below 4°C for 1 h, standard surface replica and label-fracture preparations consistently reveal the gold label distributed as aggregates on the cell surfaces with virtually no solitary gold particles in the intervening areas. Typically, the individual gold particles are loosely organized within the aggregates, separated by distances of the order of 20 to 100 nm. By using thin sectioning, and the technique of filipin treatment after LDL-gold exposure in label-fracture preparations, it is clear that the aggregates of gold label occur both in indented, clathrin-coated domains and in flat, coated and noncoated domains of the plasma membrane, and they tend to be more closely packed in the former than in the latter. Figure 1e–h illustrates these typical patterns of LDL receptor distribution as seen in the label-fracture preparations. The purpose of using the sterol-binding agent filipin in these experiments was not to detect cholesterol[54] but to distinguish between clathrin-coated and noncoated regions of the membrane. As with other protein-rich membrane domains, clathrin-coated areas are not perturbed by filipin, irrespective of sterol content,[54,79] thus allowing their presence beneath the membrane to be detected as domains free of filipin-induced pits in the en face replica views of the membrane.

TABLE 2
Distribution Patterns of Lipoprotein Receptors in Cultured Cells

Receptor	Lipoprotein[a]	Cell type	Pattern	
			4°C	37°C
LDL	LDL β-VLDL (cholesterol-fed rabbits)	Human skin Fibroblast HepG2	Exclusively in clusters	Exclusively in clusters
β-VLDL	β-VLDL (Type III hyper-cholesterolemia)	Mouse peritoneal macrophages	Randomly dispersed all over the cell surface	Progressive clustering of ligand/receptor complexes
LRP	β-VLDL (cholesterol-fed rabbits)	HepG2	Randomly dispersed and in clusters	—
Scavenger	AcLDL	Mouse peritoneal macrophages	Randomly dispersed in characteristic zone at the plasma membrane surface	Progressive clustering of ligand/receptor complexes
HDL	HDL3 HDL-Fast	Mouse peritoneal macrophages	Randomly dispersed and in clusters all over the cell surface	Progressive clustering of ligand/receptor complexes
	HDL-Slow	ditto	Randomly dispersed	—

[a] The lipoproteins listed here are those used in the gold labeling experiments described, and are not intended as a comprehensive list of all the lipoproteins that bind to the receptors listed in column 1.

These results, and those obtained on fibroblasts manipulated experimentally to remove pre-existing receptors and stimulate insertion of recycled or new receptors,[56] suggest that LDL receptors are inserted and displayed in clusters, at least under the experimental conditions we have investigated. However, this interpretation was disputed, and the original hypothesis of a dispersed distribution continued to be promoted.[80-82] We therefore designed further experiments to discriminate between the competing views.[58] To avoid using LDL-gold complexes, which were suggested as a possible cause of artifactual clustering, we used alternative detection systems. These included labeling via anti-apolipoprotein B-100 antibodies, labeling via anti-LDL receptor antibodies, and direct visualization of native (unlabeled) LDL molecules bound to the cell surface. The results from all these procedures pointed to the same conclusion — LDL receptors are indeed normally displayed in loose aggregates and not in individually dispersed form, just as suggested by our original LDL-gold experiments.[58] A recent study, in which LDL receptor distribution and clathrin lattices were simultaneously visualized on membranes torn from the upper surface of cultured cells, similarly reveals the label predominantly in aggregates and not in dispersed form.[83] This study, however, shows the receptor aggregates to be always associated with clathrin lattices, whereas our filipin-label-fracture results suggest a significant proportion of aggregates (presumed to be the more recently inserted ones) are present in nonclathrin-coated domains of the membrane. Further studies are in progress to investigate this apparent discrepancy.

We have also compared the binding patterns of LDL and β-VLDL on the fibroblast and hepatocyte plasma membranes.[58] The β-VLDL used in these experiments was obtained from the serum of cholesterol-fed rabbits. When fibroblasts were incubated with β-VLDL-gold complexes for 1 h at 4°C, aggregates of gold label, identical to those observed in the standard LDL-gold experiments, were consistently observed, reflecting binding of β-VLDL via its apo B and E moieties to the LDL (apo B/E) receptor. Hepatocytes exposed to LDL-gold under the same conditions gave precisely the same results, indicating that the cluster display pattern appears to be a characteristic of the LDL receptor across cell types. However, when hepatocytes were treated with β-VLDL-gold, again under the same conditions, a different labeling pattern was seen. Aggregates of gold particles were as abundant as in the fibroblast and liver/LDL-gold experiments, but in addition, a large number of widely dispersed individual gold particles was present. The clusters are interpreted as the binding of the β-VLDL to the LDL receptor, and one possible explanation for the additional dispersed label is provided by the presence in hepatocytes of the E receptor, LRP.[52,53,58] If this explanation is correct, then it would seem that LRP, in contrast to the LDL receptor, has a widely dispersed distribution in the plasma membrane.

Several studies have examined whether Lp(a) binds to the LDL receptor for subsequent catabolism through the LDL-receptor pathway. Whereas three groups of investigators have reported that Lp(a) can bind specifically to the LDL receptor,[84-86] another group, after comparing the uptake of LDL and Lp(a) in fibroblasts from normal subjects and subjects with homozygous familial hypercholesterolemia, concluded that Lp(a) is not a ligand for the LDL receptor.[87] In the course of our work, we discovered that Lp(a) was heterogenous with regard to its ability to bind to lysine-Sepharose and that two lipoprotein species could be isolated.[88] This affinity column retained 40 to 81% of the total Lp(a) in the density fraction 1.055 to 1.15 g/ml from five individuals. The remaining unretained Lp(a) species with no apparently functional lysine binding site was similar to the retained species in its electrophoretic mobility, lipid, protein, and apolipoprotein composition, and the heterogeneity was not related to apo(a) size polymorphism. The binding of the two species of Lp(a) was studied at the ultrastructural level in human fibroblasts using two cytochemical approaches: direct labeling with gold-labeled Lp(a) and indirect immunocytochemical labeling of bound native Lp(a). In the first approach, Lp(a)-gold complexes were allowed to bind to fibroblasts at 4°C and cell surface replicas prepared. Both Lp(a) species showed only very low levels of labeling in the form of occasional clusters of just a few gold particles. The same cultures exposed to LDL-gold containing a corresponding quantity of apo B-100 revealed the usual abundance of LDL receptors (Figure 1e). In the second approach, unlabeled Lp(a) were initially incubated with fibroblasts at 4°C, and the antigen apo(a) was then localized by incubation with anti-Lp(a) antibody followed by gold-labeled protein A. The results from these experiments confirmed those obtained with Lp(a)-gold; few binding sites for Lp(a) were detectable on the cells. These results show that Lp(a) does not bind to the LDL receptor to any significant extent.

In the macrophage, only limited numbers of LDL receptors of the classical "fibroblast-type" are normally expressed. However, LDL that has been modified chemically by acetylation loses the ability to bind to the LDL receptor; it becomes recognizable by a different high-affinity receptor characteristic of the macrophage, the AcLDL receptor or scavenger receptor.[50,51] We have studied the surface distribution and dynamics of this receptor in surface replicas of cultured mouse peritoneal macrophages. After exposure to AcLDL-gold complexes for 1 h at 4°C, these cells show diffusely distributed gold label, largely confined to a circular band between the central (nuclear) region and the peripheral zone of the plasma membrane. Only by warming macrophages that are first exposed to AcLDL-gold conjugates at 4°C, does a time-dependent redistribution of receptors into aggregates occur. By 4 min at 37°C, most of the gold particles have become aggregated, by 8 min compact clustering in coated pits is observed, and by 15 min virtually all the label has disappeared from the cell surface. Thus, warming leads to progressive coalescence and ultimate endocytosis of the initially diffusely distributed AcLDL-receptor complexes.

The following conclusions can be drawn from these results: 1) AcLDL receptors are displayed in diffuse fashion over the "intermediate zone" of the macrophage plasma membrane and 2) the binding of AcLDL to its receptors induces long-range clustering of the occupied receptors into coated pits. Labeled surface replicas reveal that β-VLDL binding sites, like AcLDL receptors, are initially diffusely distributed in the membrane, becoming clustered upon warming. The distribution and dynamics of AcLDL and β-VLDL receptors thus differ markedly from those of the LDL receptor. LDL receptors, in contrast to AcLDL and β-VLDL receptors, are always confined to presumptive coated pit regions even prior to ligand binding, and they do not migrate appreciably in the plane of the membrane.

2. *Uptake and Intracellular Fate of Lipoproteins*

The cytochemical and ultrastructural techniques outlined in section II have led to major advances in our ability to follow the sequences of internalization and track the subsequent pathways of macromolecules through cellular compartments. Cells internalize diverse receptor-ligand complexes in astonishingly

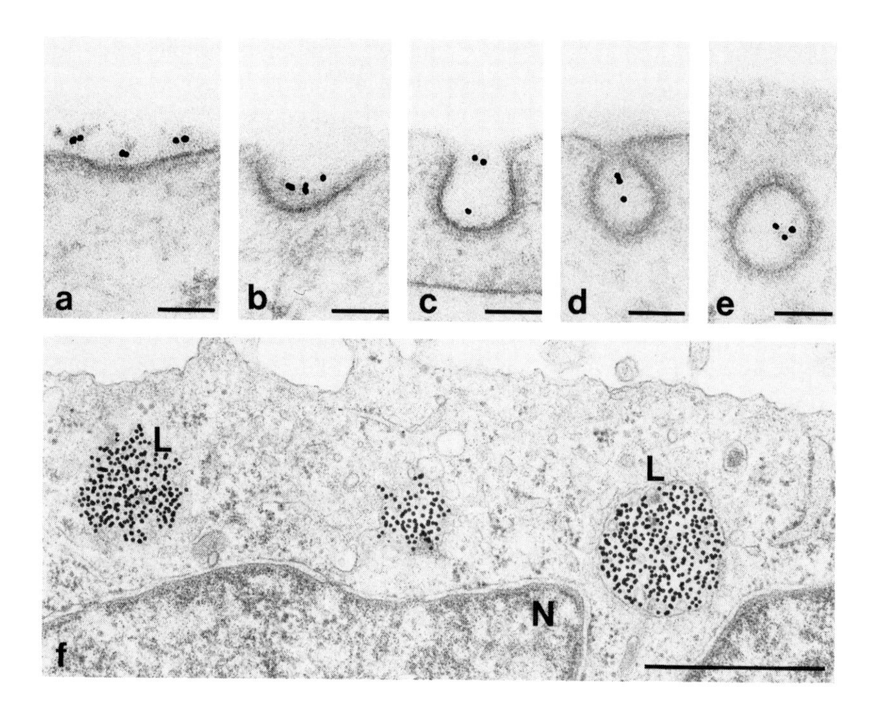

FIGURE 2. Sequence of thin section electron micrographs illustrating the classical pathway of receptor-mediated endocytosis of LDL: (a–d) depict the progressive formation of coated invaginations in which LDL-gold conjugates are concentrated, leading to their internalization in a coated vesicle (e), and ultimately to degradation of LDL in lysosomes (f). L = lysosomes, N = nucleus. (Bars: (a–e) 0.1 μm; (f) 0.5 μm).

similar fashion, but as we have seen, there are striking differences in receptor display patterns, and so too are there in the intracellular fates of different ligands and receptors (for review see Goldstein et al.[3]).

The receptor-mediated endocytosis of LDL is representative of a system in which ligands are catabolized in lysosomes and receptors reutilized. The intracellular pathway of LDL and the LDL receptor have been extensively characterized and described in several excellent reviews,[2,3] and they are illustrated at the ultrastructural level with our labeling techniques in Figures 1e–h and Figure 2. LDL, via apo B-100, binds to LDL receptors. Clathrin coats (Figure 3) form at the undersurface of membrane domains containing receptor-LDL aggregates. The domains invaginate to form coated pits, and eventually become detached within the cell as coated vesicles (Figure 2a-e). The coat is removed within seconds and the uncoated vesicles fuse with the endosomal compartment of the cell. Within endosomes, the LDL dissociate from the receptors, and the receptors are recycled back to the cell surface. The LDL are transported to lysosomes where they are degraded (Figure 2f), liberating amino acids from apo-B and freeing cholesterol from the cholesterol esters present in the lipoprotein.

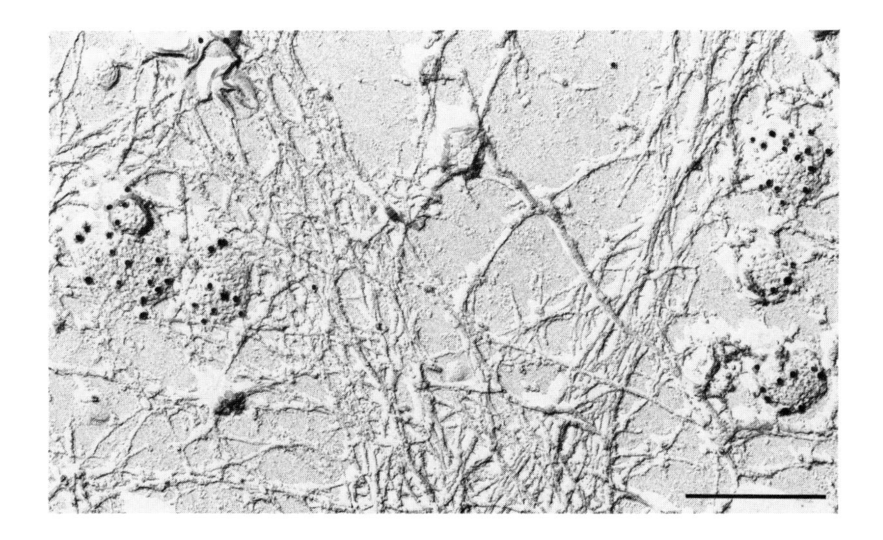

FIGURE 3. Immunocytochemical localization of clathrin on the protoplasmic surface of the fibroblast plasma membrane after lysis squirting. Regions with the "honeycomb pattern" characteristic of clathrin are specifically labeled, using antibodies against clathrin followed by protein A-gold (17 nm). Gold label appears attached to extended, rather flat clathrin sheets found between microfilaments and also to raised coated pits or coated vesicles. (Bar: 0.5 μm).

The cholesterol released upon lysosomal hydrolysis of the LDL suppresses both 3-hydroxy-3-methylglutaryl CoA reductase, the rate-limiting step in cholesterol biosynthesis, and synthesis and expression of the receptor. Excess cholesterol not needed for membrane or steroid hormone synthesis is reesterified by acyl CoA:cholesterol acyl transferase (ACAT), which is activated by LDL uptake, and stored as cholesteryl oleate in the cytoplasm. It is through these mechanisms that the cells are protected from excessive accumulation of cholesterol or cholesteryl esters, and do not become foam cells.

The processes that lead to foam cell formation include the β-VLDL[40] and the AcLDL receptor pathways in macrophages.[26,27] Receptor-mediated endocytosis of these lipoproteins operates via clathrin-coated pits and vesicles, as with classical LDL receptor-mediated endocytosis, but in contrast to the LDL system, the process is not switched off in the presence of high concentrations of the lipoprotein. Cultured mouse peritoneal macrophages exposed to AcLDL or human β-VLDL (E2/E2) for 12 to 24 h at 37°C develop abundant large multivesicular organelles (Figure 4). To determine whether these foamy organelles are forms of lysosome, we carried out enzyme cytochemistry to establish whether the lysosomal enzymes trimetaphosphatase and acid phosphatase were present. In addition, we used immunogold cytochemistry to localize the lysosomal membrane protein Lpg 110 and the DAMP method to detect acidic compartments. DAMP (3-(2,4-dinitroanilino)-3′-amino-N-methyldipropylamine) accumulates in acidic compartments when incubated

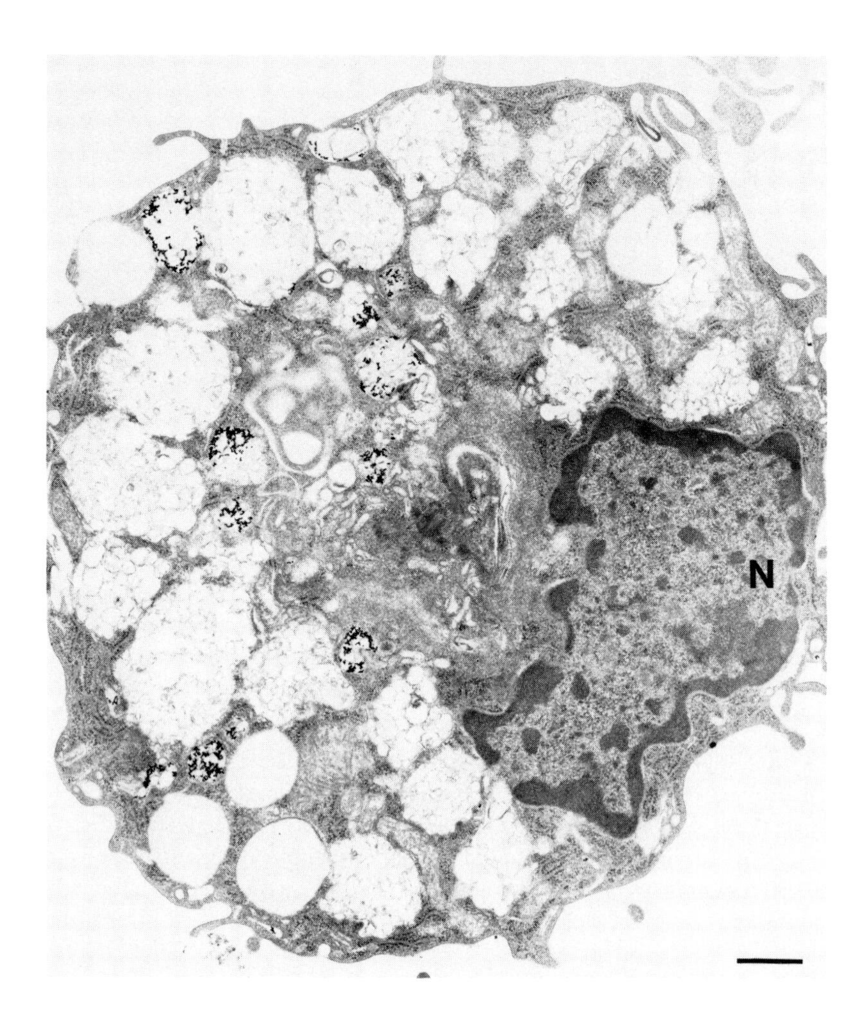

FIGURE 4. Thin-section electron micrograph of a mouse peritoneal macrophage cultivated under normal culture conditions for 24 h at 37°C, and then in the presence of 70 μg/ml AcLDL for 12 h at 37°C followed by AcLDL-gold for 15 min. The cytoplasm of the macrophages is filled with numerous large multivesicular organelles, so-called "foamy organelles". Gold label has accumulated in some of these "foamy organelles" indicating transport of AcLDL-gold to them. N = nucleus. (Bar: 1 μm).

with viable cells; detection of DAMP in these compartments is done using anti-DNP antibodies. As shown in Figure 5, all these procedures demonstrated the lysosomal nature of the foamy organelles.

When macrophages that have been fed AcLDL or β-VLDL for 12 to 24 h at 37°C are then incubated for a further 6 to 12 h in a lipoprotein-free medium at the same temperature, the foamy organelles disappear, and instead numerous lipid droplets appear (Figure 6). What happens at the biochemical level is that

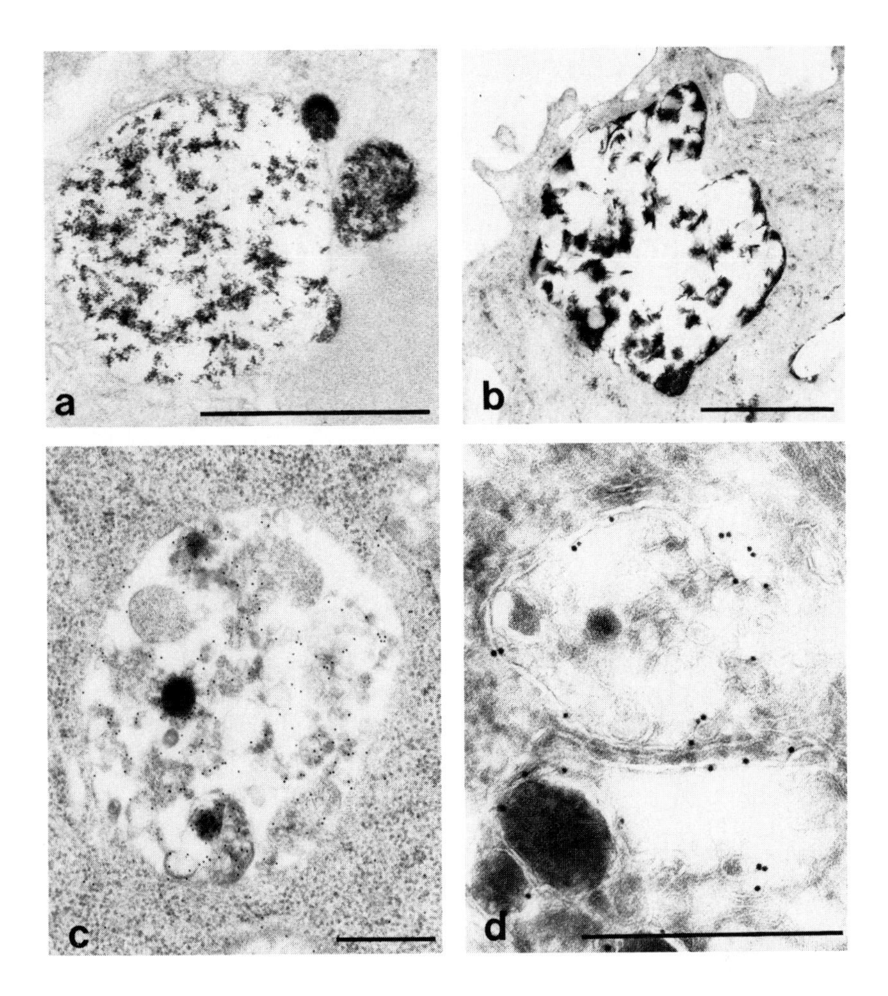

FIGURE 5. Cytochemistry and immunocytochemistry of foamy organelles. (a) Trimetaphosphatase reaction products, (b) Acid phosphatase reaction products, (c) DAMP localization and (d) Lpg 110 localization in foamy organelles, identifying them as lysosomes. (Bars: 0.5 μm).

cholesterol ester from the internalized lipoproteins is hydrolyzed in the lysosomes. Cholesterol is then released into the cytoplasm, and reesterified to form cholesteryl ester lipid droplets. A continuous cycle of rehydrolysis and esterification then ensues.

The responses of macrophages to feeding with AcLDL differs in detail according to the source of the cells. Macrophages from the plasma of normal human subjects, for example, differ from mouse peritoneal macrophages in the speed of lipid droplet accumulation. Of particular interest are the responses of macrophages from patients with Tangier disease, a rare autosomal recessive disorder characterized by deficiency of HDL and deposition of cholesteryl esters in various organs and tissues.

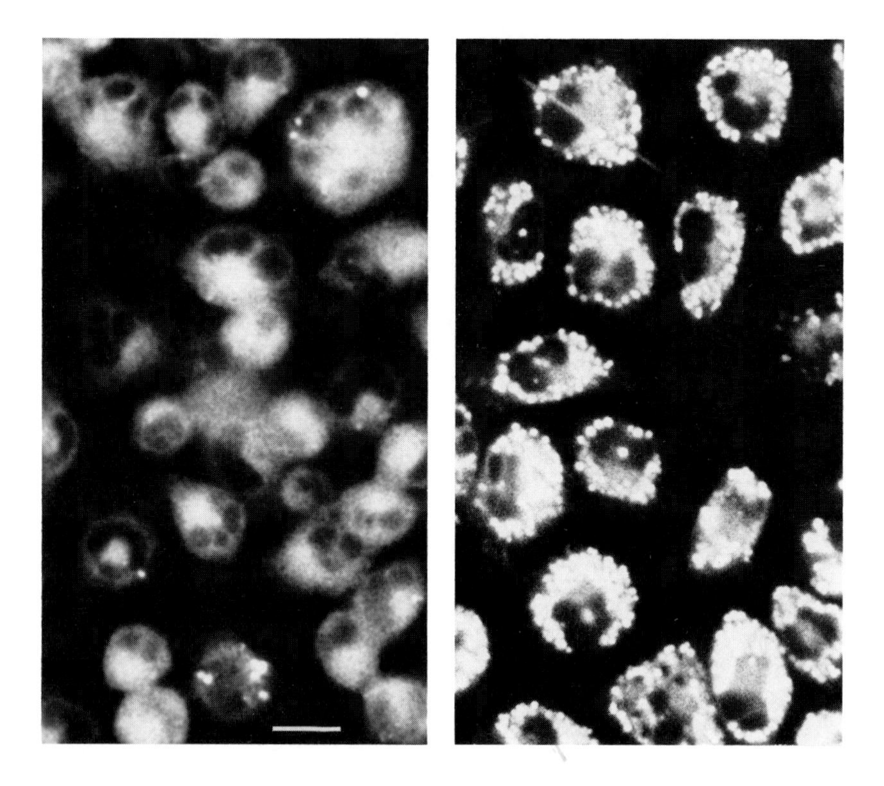

FIGURE 6. Nile red fluorescence of lipid in control macrophages (a) and macrophages after incubation in AcLDL (b) observed by light microscopy. For the cells in (b), 70 μg/ml AcLDL was given for 12 h at 37°C and then the cells were incubated further in the absence of AcLDL for 12 h at 37°C. The cytoplasm of the cells in (b) is filled with numerous lipid droplets, giving the typical foam cell-like morphology. (Bar: 10 μm).

Under standard culture conditions, the ultrastructure of human control and Tangier macrophages appears broadly similar. The cells contain numerous mitochondria and cisternae of rough endoplasmic reticulum, abundant ribosomes and several vacuoles of variable size. Only occasional single multivesicular bodies, lysosomes, and lipid droplets are observed. The Golgi apparatus is well developed, appearing more extensive in the Tangier-macrophage than in the control cells. Exposure of cells from both sources to AcLDL for 24 h at 37°C reveals spectacular ultrastructural changes.[89] Control macrophages accumulate abundant small membrane-free lipid droplets, which fill much of the cytoplasm. In Tangier macrophages, fewer cytoplasmic lipid droplets appear. Instead, the cytoplasm becomes packed with distinctive vacuoles, not observed in control macrophages. Two types of vacuole, referred to as Type I and Type II vacuoles, are observed. Type I vacuoles are irregular spheroid or ovoid electron-lucent structures, filled with fine flocculent or fibrillar material, distributed as aggregates in the cell. Type II vacuoles are larger, contain scrolls and lamellae of

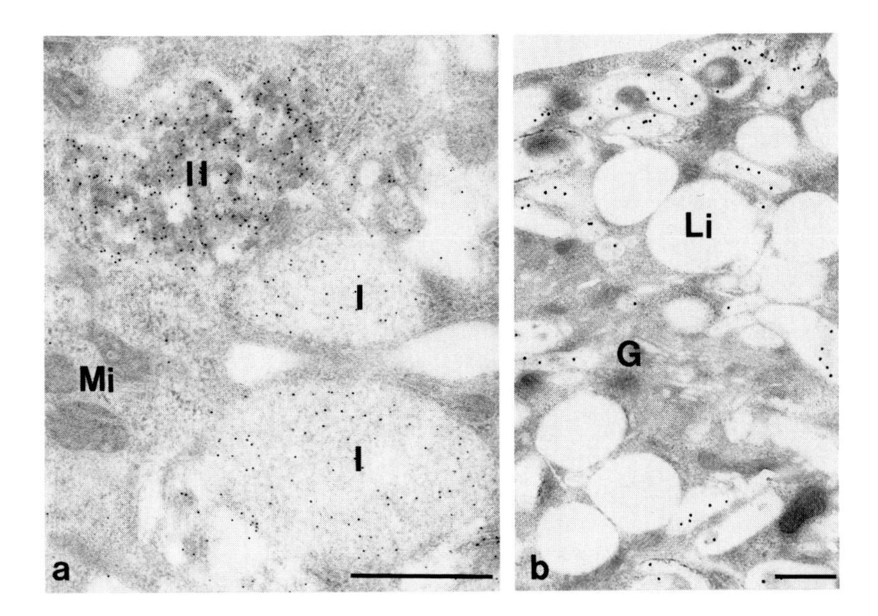

FIGURE 7. Transmission electron micrographs showing localization of cathepsin D in Tangier (a) and control macrophages (b) cultivated in the presence of 70 μg/ml acetylated LDL for 24 h at 37°C. Cathepsin D was revealed by post-embedding immunoelectron microscopy of Lowicryl sections with monospecific antibodies to cathepsin D and the protein A-gold technique. (a) In Tangier macrophages both type I (I) and type II (II) vacuoles are found to be intensely and specifically labeled. Although gold particles (8 nm) are abundant over both types of vacuoles, they appear to be most concentrated over the more electron-dense regions of type II vacuoles. Gold particles are rarely seen over the other organelles such as the nuclei, mitochondria (Mi), cisternae of the endoplasmic reticulum, Golgi apparatus, or the cytoplasm. (b) In control macrophages small pale lysosomes are distributed among the abundant lipid droplets (Li). Gold label (12 nm) within these lysosomes is mainly confined to electron-dense fibrillar material within their interiors. The lipid droplets and Golgi apparatus (G) are completely free of label, as are other organelles and the cytoplasm. (Bars: 0.5 μm).

variable electron opacity, and are filled with more electron-dense material. To determine whether the two vacuole types in Tangier macrophages are forms of lysosome, immunogold cytochemistry was carried out to localize the lysosomal marker enzyme cathepsin D. This was done by post-embedding immunoelectron microscopy using an affinity-purified polyclonal rabbit anti-human cathepsin D antibody. Both Type I and Type II vacuoles were intensely and specifically labeled, indicating a lysosomal identity for both forms of vacuoles (Figure 7a). In control macrophages, the size of the lysosomal compartment was much smaller than that of Tangier macrophages, and the individual lysosomes were less intensely stained with the cathepsin D antibody (Figure 7b).

Whether the two distinct morphological forms of vacuole reflect different stages in maturation of a single type of lysosome or are separate types of lysosomes engaged in different degradative activities is uncertain. Earlier observations suggested that Tangier macrophages are characterized by an abnormal cellular processing of HDL precursors. This results in their diversion to lysosomes for degradation instead of resecretion, coupled with a dysregulation of cellular lipid metabolism.[25] The presence of a marked accumulation of lysosomes in Tangier macrophages is consistent with a block in the normal mechanism of cholesterol transfer from the lysosomal compartment to lipid droplet storage. Accompanying this lysosome accumulation, there is a marked hyperplasia of the Golgi apparatus in Tangier macrophages. The Golgi apparatus becomes widely distributed throughout the cell and develops markedly dilated *trans*cisternae. These changes, not present in control human macrophages exposed to AcLDL, may arise from extra demand for lysosomal enzyme production by the Golgi apparatus under the conditions of blocked cholesterol transfer, or they may reflect overproduction of sphingomyelin and phospholipids.[24]

Tangier disease is thus associated with abnormalities of macrophage intracellular organelles and cellular lipid metabolism which suggest a translocation disorder affecting lysosomal processing and the Golgi apparatus. Interestingly, fibroblasts from Tangier patients exposed to AcLDL show similar changes in the Golgi apparatus but not in the lysosomal compartment. This situation may stem from the fundamental difference in the way the two cell types metabolize cholesterol. That a parallel alteration in the Golgi apparatus does occur in the two cell types reinforces the view that this is a fundamental abnormal feature of the disease, and opens the possibility of using fibroblast cell cultures as a model system for investigating the cellular basis of Tangier disease. The advantage of such a system would be the ease of continuous cultivation and the standardization of experimental conditions on a single, defined cell line.

IV. *IN SITU* LOCALIZATION OF APOLIPOPROTEINS AND LIPOPROTEINS IN TISSUE

In contrast to the cultured cell systems, few cytochemical studies have previously been done on lipoproteins in tissue, largely because of methodological problems that such studies pose. Recently, however, we have achieved a measure of success in two investigations of this type. These involve the subcellular localization of apo A-1 in human liver, and the localization of a range of lipoproteins in the human atherosclerotic plaque.

A. Localization of Apo A-I in Human Liver

Apo A-I is the major protein constituent of plasma HDL, representing about 60% of protein mass. In humans, apo A-I is synthesized by intestinal and liver cells, and is secreted as a basic precursor protein that is subsequently processed to more acidic isoproteins found in plasma HDL.[90] A number of studies suggest that apo A-I functions as a ligand for HDL binding to cells[91-96] and both HDL and apo A-I levels are inversely-related to risk of coronary artery disease.[97] HDL and Apo A-I metabolism are thus of considerable importance in atherosclerosis, but attempts to localize apo A-I at the ultrastructural level by immunoelectron microscopy with conventional antibodies to untreated apo A-I have previously been unsuccessful.

As outlined earlier (section II.C), to overcome the problem of reduced antigenicity due to fixation and embedding, we raised polyclonal antibodies to formaldehyde-fixed apo A-I (anti-fix.apo A-I). The aim was to obtain antibodies that were at least in part directed against antigenic determinants altered by fixation.[74] Purified apo A-I was fixed with 4% formaldehyde and used to immunize rabbits. The antiserum was purified by protein A-Sepharose followed by affinity chromatography with the fixed antigen coupled to vinylsulfone-activated agarose. The specificity of the antibodies was ascertained by enzyme-linked immunosorbent assay (ELISA) and Western blot analysis against different fixed and unfixed lipoproteins. The antibodies reacted specifically with apo A-I and recognized the fixed as well as the unfixed protein. In ELISA, the reaction of the antibodies with the fixed antigen was markedly enhanced compared with that to the unfixed antigen, confirming that the antibodies were indeed directed against epitopes modified by fixation, as had been intended.

Before conducting experiments on liver tissue, we tested the efficacy of the antibodies for immunocytochemical labeling using HepG2 cells. Standard light microscope immunofluorescence techniques revealed distinct patches of intracellular staining. By applying the pre-embedding immunoperoxidase technique, the feasibility of immunolocalizing synthesized apo A-I within cells was demonstrated at the ultrastructural level (Figures 8, 9). Apo A-I was localized in various cellular compartments, especially the cisternae of the rough endoplasmic reticulum (Figure 8), the Golgi complexes (Figure 9), and in Golgi-associated vesicles. No positive staining occurred in these organelles in control sections. Other subcellular compartments such as nuclei, mitochondria, and lysosomes were consistently negative.

To localize apo A-I in human liver tissue, we used the anti-fix.apo A-I antibodies and protein A-gold for immunocytochemistry of ultrathin cryosections (Figure 10). As with the HepG2 cells, immunostaining of the Golgi area was prominent. Positively stained vacuolar structures and secretion granules were common in the *trans* Golgi complex and in the pericanalicular region. Positive staining was also consistently found in the space of Dissé. These findings indicate that the synthetic and secretory pathway of apo A-1 follows the classical endoplasmic reticulum-Golgi apparatus-secretory granule route, release presumably being mediated by exocytosis of secretory granules. The success of this study suggested that the approach of using antibodies to fixed antigens had wider potential for localization of apolipoproteins in tissue and in cells, and this we have exploited further in the study of lipoproteins *in situ* in the atherosclerotic plaque.

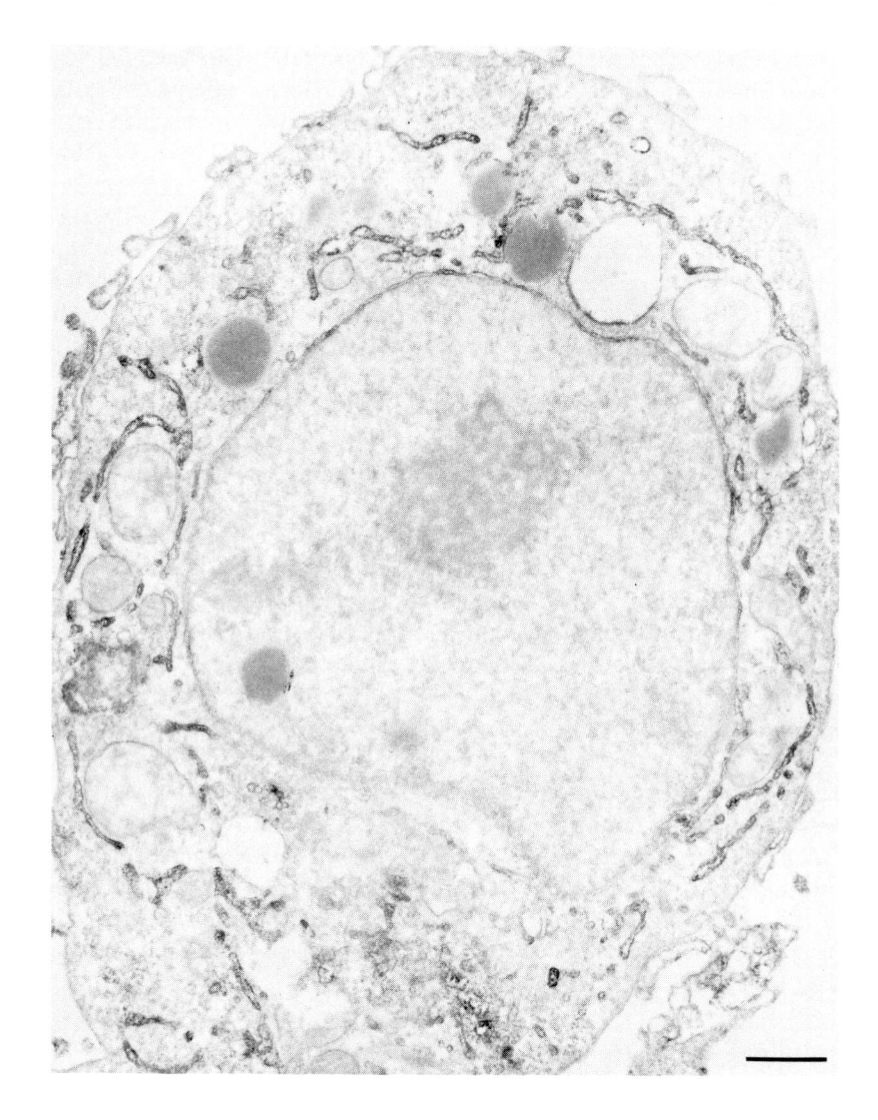

FIGURE 8. Thin-section electron micrograph of a HepG2 cell stained by the indirect pre-embedding immunoperoxidase method for apolipoprotein A-I. Electron dense immunoperoxidase reaction products that indicate the antigenic sites for apolipoprotein A-I are present in the cisternae of the rough endoplasmic reticulum. (Bar: 1 μm).

B. Localization of Lipoproteins in Human Atherosclerotic Plaque

How hypercholesterolemia, the most clearly recognized risk factor in atherosclerosis,[98-102] exerts its effect on the genesis and progression of lesions is currently one of the most actively investigated areas in atherosclerosis research.

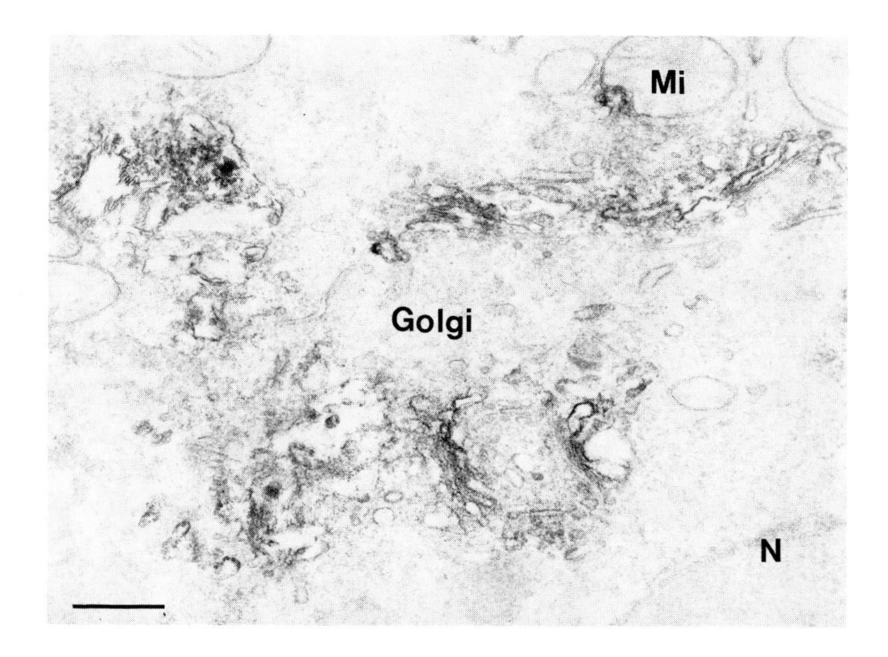

FIGURE 9. Transmission electron micrograph of a HepG2 cell stained by the indirect pre-embedding immunoperoxidase method for apolipoprotein A-I. Electron-dense immunoperoxidase reaction products that indicate the antigenic sites for apolipoprotein A-I are present in the Golgi complex (Golgi), cisternae of the rough endoplasmic reticulum and secretion granules. Mi = mitochondria, N = nucleus. (Bar: 0.5 μm).

It is postulated that an increased transport of plasma lipoproteins across the arterial wall arises from endothelial injury; this leads to lipoprotein uptake by macrophages and consequent formation of foam cells. Though macrophages are the principal source of foam cells, smooth muscle cells also have an ability for lipid accumulation.[103] Much of our understanding of the role of lipoproteins in the pathogenesis of atherosclerosis has come from *in vitro* experiments with cell culture systems. As we have seen in the preceding discussion, both chemically modified LDL[26-33,35,46,84-86,104-106] and β-VLDL[37,38,107-112] in these systems promote foam cell formation, and it is largely from such studies that the postulated role of HDL in "reverse cholesterol transport" has been deduced.[41-43,95,113-120] Data on the *in situ* distribution of lipoproteins within the atherosclerotic plaque itself have remained limited, however, and those available are in many respects contradictory. Histochemical studies have been carried out at the light microscopic level, but the resolution is insufficient to reveal how different lipoproteins are organized in the various cellular compartments of the plaque.[121,122] More precise localization of these molecules within the lesion would further our knowledge of the composition of the plaque and the role of lipoproteins in progression and regression.

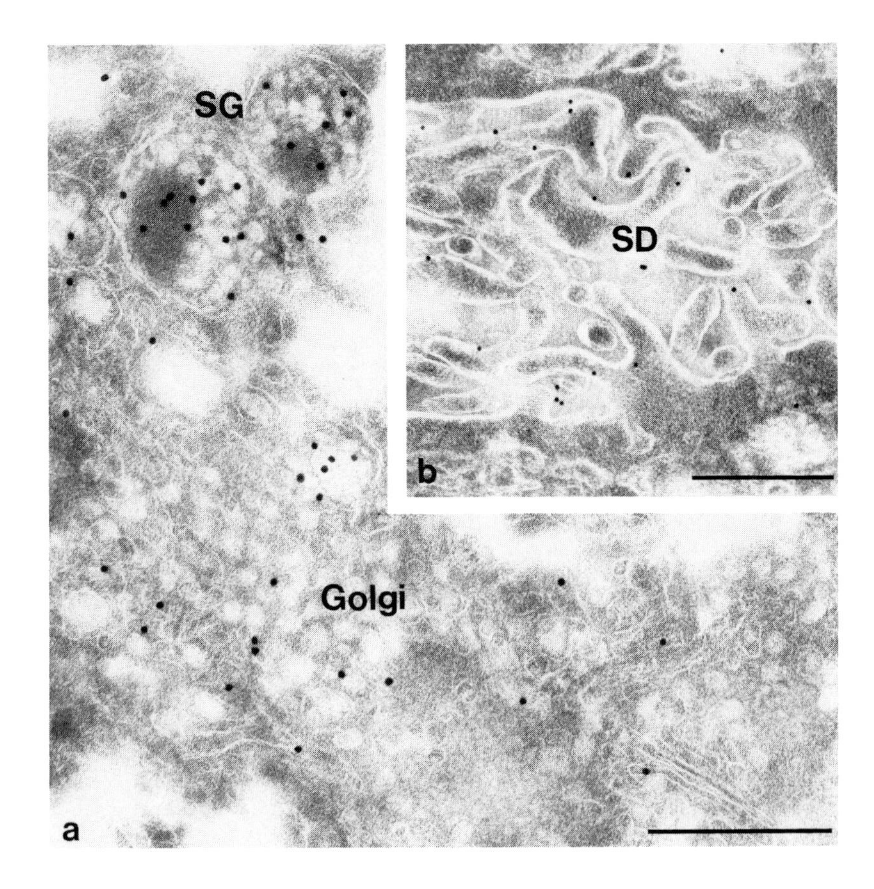

FIGURE 10. Transmission electron micrographs of hepatocytes of human liver stained for apolipoprotein A-I by post-embedding immunocytochemistry on ultrathin cryosections, using the protein A-gold method: (a) the cisternae of the Golgi complex (Golgi) and the secretion granules (SG) at the trans side of the Golgi complex are strongly labeled indicating the sites of apolipoprotein A-I; (b) Immunogold label is also consistently present over the space of Dissè (SD). (Bars: 0.5 μm).

We have therefore used the higher resolution afforded by electron microscopy combined with polyclonal antibodies for immunogold labeling to detect various classes of lipoprotein in the plaque. Antibodies against the following were used: LDL, apo B-100, HDL, formaldehyde-modified apo A-I, and formaldehyde-modified LDL (two different antibodies). For simultaneous characterization of the cellular constituents and compartments of lesions, we used a polyclonal antibody against alpha-actin to identify smooth muscle cells and an affinity-purified polyclonal rabbit anti-human cathepsin D antibody to identify lysosomes. In single labeling experiments to detect lipoproteins we used the protein A-gold technique, and in double-labeling experiments for simultaneous identification of other cellular constituents we used protein A-gold followed by protein G-gold.

The investigation was carried out on atherosclerotic plaques obtained from human femoral artery. The plaques, characterized as advanced lesions by light microscopy, contained abundant foam cells of macrophage and smooth muscle cell origin embedded in extensive extracellular matrix that contained deposits of crystalline cholesterol. We succeeded in localizing HDL, LDL, and the apolipoproteins A-I and B-100 in various subcellular compartments of macrophages and smooth muscle cells and in the extracellular matrix. Though there were some differences in labeling intensity, the antibodies to the various epitopes of LDL gave similar results to one another, as did those against HDL. Figure 11 illustrates the distribution of LDL labeling. LDL was widely distributed over cellular compartments and extracellular matrix. HDL labeling, by contrast, occurred predominantly over cellular compartments. Both lipoproteins were abundant in the cytoplasm of macrophages and smooth muscle cells. They were most concentrated in macrophages over electron-dense lamellar organelles, which were identified as lysosomes using antibodies against cathepsin D. Double-labeling with antibodies against LDL and cathepsin D confirmed colocalization of LDL and cathepsin D within these structures.

Bearing in mind the involvement of lysosomes in the endocytotic pathway (section III. A. 2.), we interpret the intense labeling of LDL and HDL in these organelles as a reflection of active lipoprotein internalization from the extracellular matrix. Foam cells within the plaque appear to take up LDL and HDL and direct them to a lysosomal compartment rich in cathepsin D for hydrolysis. This indicates that a proportion of internalized HDL is not successfully resecreted to remove cholesterol. The demonstration of LDL within a cathepsin-D-rich compartment accords with *in vitro* biochemical studies[123,124] reporting that intralysosomal degradation of lipoproteins, especially apo B, involves an initial limited endoproteolytic attack by cathepsin D.

V. MECHANISMS OF CELLULAR CHOLESTEROL EFFLUX

A. Reverse Cholesterol Transport

The concept of reverse cholesterol transport involves transport of cholesterol from the peripheral tissues (e.g., arterial wall) to the liver. From there it can be eliminated in the bile either via synthesis of bile acids or secretion of free cholesterol.[41] HDL play a central role in this process,[41-43,95,113-120] acting as the vehicles for cholesterol uptake from peripheral tissues and subsequent transport. Cholesterol-enriched HDL may be taken up directly by liver cells, or the cholesterol may be transferred via physicochemical exchange mechanisms to other lipoproteins such as VLDL. The VLDL are converted to VLDL-remnants and LDL, which then, in turn, can be removed from the plasma by receptor-mediated uptake in liver cells. Our work has suggested that further mechanisms, independent of HDL, may contribute to reverse cholesterol transport. We have discovered that macrophages can release excess cholesterol by direct secretion of lysosome-derived cholesterol and phospholipid-containing "lamellar bodies", and that lipid-laden macrophages can, with appropriate stimulation, be

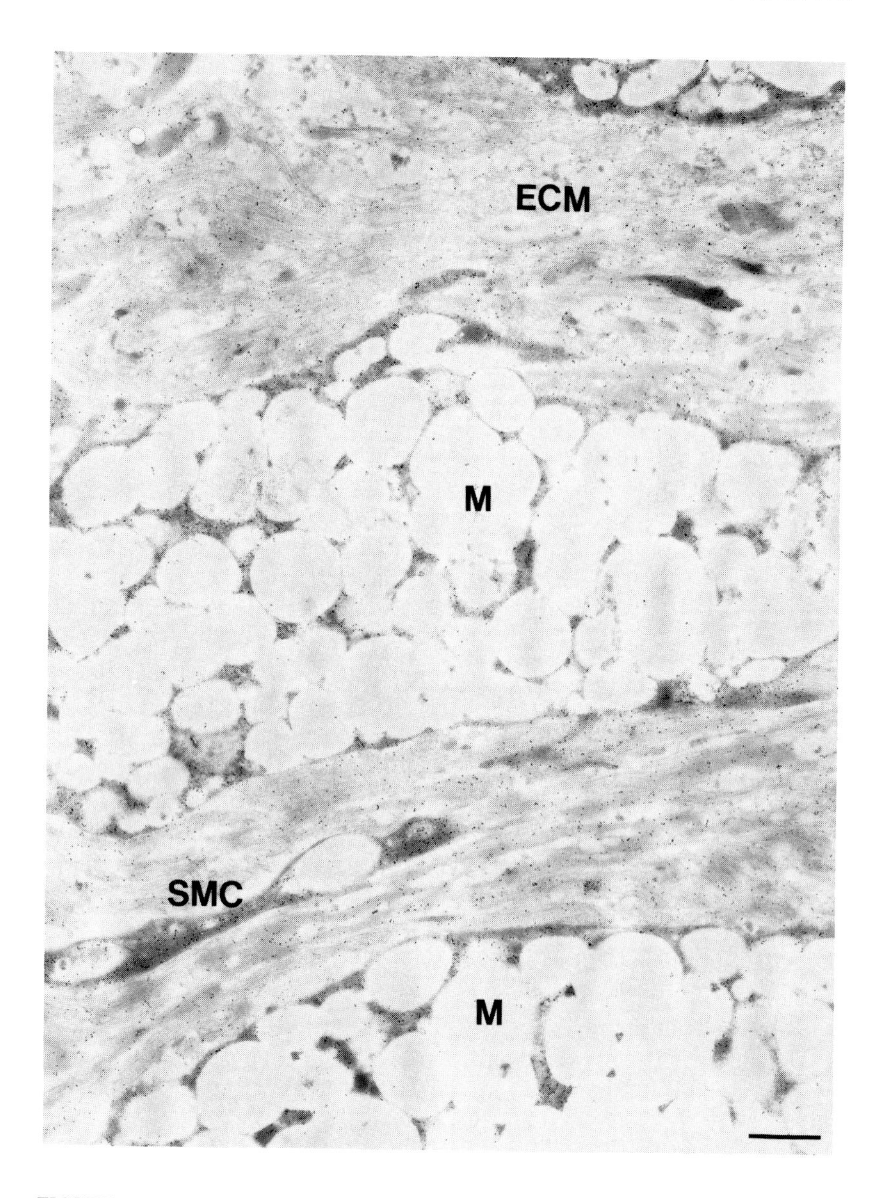

FIGURE 11. Transmission electron micrograph of a Lowicryl K4M-embedded atherosclerotic plaque from a human femoral artery stained for LDL by post-embedding immunocytochemistry with anti-fix. LDL-antibodies followed by protein A-gold. Gold particles are abundant over the cytoplasm of foam cells derived from macrophages (M) and smooth muscle cells (SMC) as well as in the surrounding extracellular matrix (ECM). Gold particles appear to be most concentrated over lysosomes. The lipid droplets and the other organelles are poorly labeled. (Bar: 1 μm).

induced to migrate. This reinforces the view that foam cells are not inevitably destined to perish in the plaque, but may participate in cholesterol efflux to a variable degree that is determined by external factors.

1. HDL Receptor-Dependent Release of Cholesterol

Macrophages depend on cholesterol accepters to prevent their transformation into foam cells during times of increased cholesterol intake. HDL are among the most effective cholesterol accepter molecules known. However, the precise nature of the cellular interaction of HDL and HDL subclasses with macrophages and the mechanisms of cholesterol secretion from macrophages are not well understood. It has been hypothesized that the HDL subclass without apolipoprotein E (low cholesterol content) may absorb unesterified cholesterol by direct interaction with the plasma membrane. The physicochemical mechanisms of cholesterol transfer would be facilitated by a low cholesterol to phosphatidylcholine ratio in the outer surface of the HDL particle compared with that of the plasma membrane. This would lead to an imbalance between cytoplasmic cholesterol and the cholesterol content of the plasma membrane; with re-equilibration, cholesterol from the cytoplasmic pool replaces that lost from the plasma membrane.

Cholesterol accumulation in macrophages stimulates synthesis of large amounts of apo E and secretion of apo E/phospholipid discs, a process independent of HDL.[125] The current concept of reverse cholesterol transport encompasses the possibility that HDL, during the cytoplasmic cholesterol esterification process mediated by the enzyme lecithin: cholesterol acyltransferase (LCAT), not only absorb unesterified cholesterol from the cell surfaces but also incorporate apo E from extracellular apo E/phospholipid discs. Cholesteryl esters and apo A-I could also be transferred from HDL onto the apo E/phospholipid discs. The net result would be formation of cholesteryl ester-rich and apo E-rich HDL1 that, upon reaching the liver, would be recognized and internalized via the hepatic apo E receptor (LRP).

We have examined in detail the cellular interaction of HDL and HDL subclasses with cultured mouse peritoneal macrophages. That HDL receptors are expressed by macrophages and other cells has been demonstrated in a number of studies,[126-128] and using biochemical and morphological methods we have also confirmed that apo A-I-containing apo E-free HDL3 bind to specific macrophage surface receptors. These HDL receptors can be visualized electron microscopically on the cell surface using gold-labeled HDL and surface replicas. When the experiments are carried out at 4°C, the gold-labeled HDL particles bind in a combined preclustered and diffusely distributed display pattern. Raising the incubation temperature to 37°C results in rapid concentration of the gold label into aggregates, followed by uptake via coated pits and coated vesicles. The internalization process has been further documented by staining thin sections of macrophages (prefed native HDL) with anti-apo A-I immunoperoxidase, and by following the fate of fluorescence-labeled HDL by light microscopy and quantitative flow cytometry. Binding studies with ^{125}I-labeled HDL define the HDL binding site as a high affinity receptor; the ligand for binding to the receptor is thought to be Apo A-I. After internalization, the HDL particles are transported to endosomes, but in contrast to LDL or AcLDL, they are not normally degraded to any significant extent in lysosomes. On their transcellular route, HDL come into contact with cytoplasmic lipid droplets where they take up cholesterol and apo E. They are then resecreted as intact cholesterol and apo E-rich HDL particles which are ultimately catabolized by the liver.[126]

The co-ordinate action of lipid-modifying enzymes and physicochemical exchange events results in a series of HDL subclasses differing in density, particle size, charge, and protein/lipid composition. With the aid of preparative free flow isotachophoresis, Schmitz and co-workers have separated three HDL subpopulations with fast, intermediate, and slow electrophoretic mobilities.[91,129] The fast migrating HDL particles are rich in apo A-I and phosphatidylcholine. The subpopulation with intermediate mobility is rich in apo A-II, apo E, apo C, cholesteryl esters and sphingomyelin. The slow migrating HDL subpopulation consists of particles rich in apo A-I and apo A-IV and is associated with high LCAT activity.

The fast migrating HDL particles bind with high affinity to HDL receptors on macrophages, whereas slow migrating HDL particles exhibit high nonspecific binding with only a small amount of specific interaction.[91] Using the gold-label surface replication technique, we compared the labeling patterns of HDL3 and fast and slow migrating HDL particles on the surface of macrophages. Gold-labeled HDL3 and fast migrating HDL particles both showed a preclustered distribution with dispersed label in the intervening areas, as described above. The labeling intensity with fast migrating HDL particles was lower than that with HDL3 (121 ± 30 s.e.m and 200 ± 75 gold particles μm^{-2}, respectively). The slow migrating HDL particles showed lower binding still (32 ± 13 gold particles μm^{-2}) and were randomly distributed. These morphological results fit the binding characteristics determined by biochemical methods. Since both fast and slow migrating HDL subpopulations are good promoters of cholesterol efflux, it may be that the HDL subpopulations rich in apo A-I promote cholesterol efflux via specific interaction with HDL receptors, whereas apo A-IV rich HDL extract cholesterol from the plasma membrane via the action of LCAT during a predominantly nonspecific interaction with the cell surface.

2. *HDL Receptor-Independent Release of Cholesterol*

In addition to the above HDL receptor-dependent mechanisms, cholesterol efflux from macrophages may also be facilitated by an HDL receptor-independent secretion of lamellar bodies[19,20] (Figure 12). When mouse peritoneal mac-

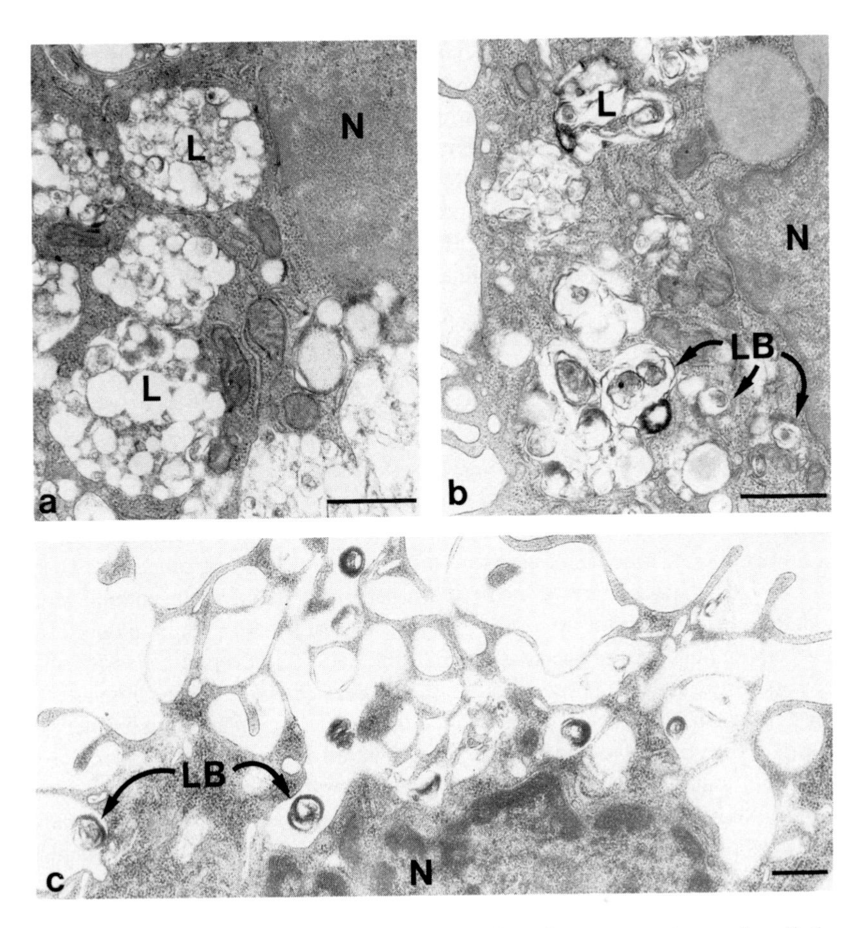

FIGURE 12. Transmission electron micrographs of peritoneal mouse macrophages cultured in the presence of 70 μg/ml acetylated LDL for 12 h at 37°C (a) and sequentially in the presence of 70 μg/ml acetylated LDL and 2 μ mol/l nifedipine for 8 h at 37°C (b,c). (a) The cytoplasm of macrophages cultured in the presence of acetylated LDL is filled with numerous large multivesicular organelles, so-called foamy organelles, with a diameter of about 0.5 to 1.5 μm. They represent the lysosomal compartments of the cell (L). (b) Macrophages cultured sequentially in the presence of 70 μg/ml acetylated LDL and 2 μ mol/l nifedipine develop foamy organelles (L) containing a large number of membrane-bound vacuoles, which exhibit multiple layers of membrane arranged in concentric whorls. With longer exposure time these lamellar bodies (LB) become increasingly condensed. The foamy organelles exhibit a progressive fragmentation into single lamellar bodies that move towards the cell periphery. (c) The lamellar bodies (LB) are released by the cells into the extracellular space and are either in intimate contact with the cell membrane or are free in the extracellular environment. The lamellar bodies released by the cells have a diameter of about 100 to 300 nm. N = nucleus. (Bars: 0.5 μm).

rophages are cholesterol-loaded, treatment with the Ca^{++}-antagonist nifedipine inhibits formation of lipid droplets, resulting instead in the production of lamellar bodies within multivesicular foamy organelles of lysosomal origin (Figure 12a,b). Prolonged incubation leads to the detachment of membrane-surrounded lamellar bodies from the lysosomes, their movement to the cell periphery, and their secretion into the surrounding medium (Figure 12b,c). This process can occur in the absence of HDL. The lamellar bodies measure 100 to 300 nm in diameter, and are rich in unesterified cholesterol (68%) and phospholipids (21%), phosphatidylcholine accounting for 70 to 80% of the phospholipid. Precisely how the lamellar bodies are released into the extracellular environment is not yet understood. Evidence from intracellular and extracellular lipid analyses and our morphological studies on control and experimentally treated cultures suggest that secretion of the cholesterol-containing lamellar bodies is a naturally occurring process in the macrophage that is enhanced by nifedipine.

A different type of lamellar body, originating from lipid droplets rather than lysosomes, is induced by treatment of macrophages with the ACAT-inhibitor octimibate.[19,20] These lamellar bodies are also bounded by membranes, newly synthesized at the margins of the lipid droplets by elements of the endoplasmic reticulum with which they interact. In contrast to the lamellar bodies induced by Ca^{++}-antagonist treatment, those originating from lipid droplets after ACAT-inhibitor treatment are not secreted. In the absence of HDL, they accumulate in the cytoplasm, but when HDL are added to the medium, these lamellar bodies interact with endosomes containing internalized HDL particles and rapidly decrease in number parallel with HDL-mediated cholesterol efflux.

From biochemical data on cellular cholesterol metabolism, characterization of the secreted lamellar bodies and other morphological data, we conclude that macrophages release cholesterol by two major pathways: (1) an HDL-receptor-mediated release of unesterified cholesterol that is enhanced with ACAT inhibition, and (2) an HDL-receptor-independent secretion of cholesterol which can be amplified by Ca^{++}-antagonists. Both the HDL-receptor-dependent and the HDL-receptor-independent pathways appear to protect macrophages and possibly other cholesterol-loaded cells from overaccumulation of cholesterol and hence foam cell transformation.

B. Role of Macrophages in the Regression of Atherosclerotic Plaques

From microscopical observations on fatty streaks and atherosclerotic lesions, it appears that foam cells eventually stop moving within the intima and accumulate, instead of migrating back to the blood stream with their lipid load. There is, however, experimental evidence that fatty streaks and more advanced lesions may undergo regression and that foam cell mobilization is involved in this process.[130-136] In animal models involving lipid feeding, several authors have reported that a significant decrease in the number of lipid-laden macrophages in intimal lesions occurs after a short period on a regression diet;[137-139] Gerrity, in particular, has presented morphological evidence for the departure of foam cells from lesions.[138,139] These studies suggest the existence of a monocyte-macrophage lipid clearance mechanism in which the numbers of macrophage/foam cells in a lesion are determined by the balance between monocyte entry into the arterial wall and departure of foam cells from it. In theory then, this balance will be affected by the number of circulating monocytes recruited to lesions and by whether monocyte-derived foam cells are activated to migrate out of the lesion or remain immobilized within it. However, the extent to which foam cells of established human atherosclerotic plaques are, in practice, able to migrate and the conditions under which this might occur, remain unclear.

To try and elucidate some of the mechanisms involved in these processes, we have investigated *in vitro* the capacity of foam cells to undergo migration, and examined how lipid uptake and cholesterol accumulation affect this process. The system we use involves lipid-loading mouse peritoneal macrophages with oxidized LDL, and then measuring their mobility in an *in vitro* migration assay in the presence or absence of the chemotactic agent, zymosan activated mouse serum (ZAMS). The migration assay is illustrated in Figure 13.

The assay was carried out as follows. Agar medium was placed in 35-mm diameter disposable Petri dishes. Seven holes each, 2.1 mm in diameter, were made in each agar plate using a sterile needle connected to a vacuum pump, and 10^5 macrophages were transferred to each hole. After attachment of the cells, 800 μl of control medium with 10% fetal calf serum was added to one plate as a control (see no. 1, Figure 9). The same quantity of control medium supplemented with 50 μg/ml oxidized LDL was poured onto the remaining plates. One set of these plates was left for 48 h (no. 2, Figure 14). Two further sets of plates were incubated with oxidized LDL alone for 24 h, and then either mouse serum (no. 3, Figure 14) or a 10 × dilution of ZAMS (no. 4, Figure 14) was added to the oxidized LDL-containing medium. The cells were incubated for a further 24 h. The agar plates were then fixed with glutaraldehyde. After removal of the gel, the cells adherent to the plastic dish were stained with Wright's Giemsa solution for 60 min. The individual areas of migration were magnified through a photomagnifier and measured by planimetry.

Using this system, we confirmed that foam cells produced by incubating macrophages with oxidized LDL are significantly less mobile than control macrophages (Figure 13, 14). The experiments with ZAMS were designed to establish whether this reduced ability to migrate was a fixed feature of the foam cell or whether, in the presence of chemotactic agents, migration ability could be restored. We found that ZAMS causes a marked enhancement of mobility in lipid-laden macrophages to levels matching those of controls (Figure 14). Immunofluorescence microscopy using antibodies against actin revealed that ZAMS-activated resumption of motility is associated with striking changes in organization of the actin cytoskeleton. Instead of prominent stress fibers, ZAMS-treated cells feature a more diffuse and irregular distribution of actin. The identities of the active factors, and whether they originate from ZAMS itself or are produced by the macrophages in response to treatment, remain to be

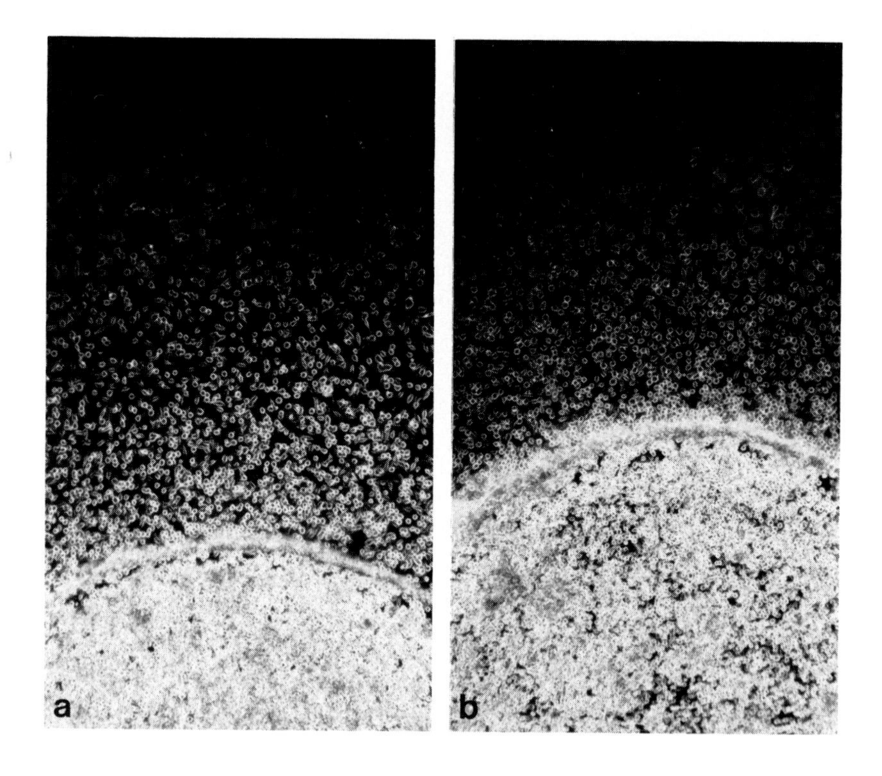

FIGURE 13. Migration assay using macrophages cultivated in normal culture medium (a) and macrophages cultivated in the presence of 50 μg/ml oxidized LDL for 24 h at 37°C (b). The cells in (b) have lost the capacity to migrate shown by the cells in (a). When cells in (b) are treated with the chemotactic agent ZAMS, their ability to migrate is restored to the same level as that shown by the cells in (a).

determined. What is important, however, is the demonstration that the immobilization of foam cells resulting from uptake and accumulation of oxidized LDL is reversible. If foam cells in lesions *in vivo* can similarly be stimulated to migrate, then approaches to lesion regression based on this strategy may ultimately become possible.

VI. CONCLUDING REMARKS

In conclusion, modern ultrastructural cytochemical techniques, in combination with data from biochemistry and molecular biology, have enabled us to build up a comprehensive picture of the mechanisms underlying the diverse interactions of lipoproteins with cells. To date, most of the studies have been directed at elucidating fundamental cellular mechanisms, but more practical applications, addressing such problems as the cellular basis for lesion regression, have now been initiated. Studies on this topic may be expected increasingly to be the focus of future research, and may ultimately help guide approaches to therapy.

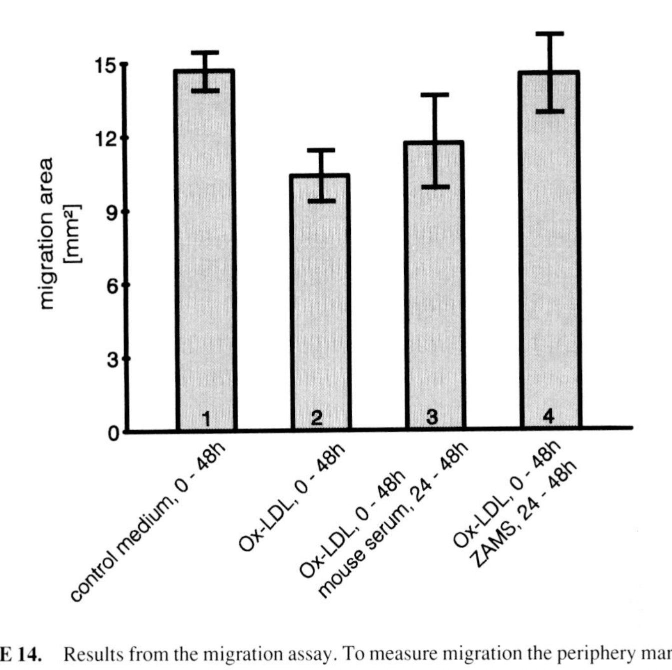

FIGURE 14. Results from the migration assay. To measure migration the periphery marking the furthest point of migration from the control starting well was traced and the total area determined (mm^2). Samples incubated as follows: (1) Control medium (Dulbecco's minimal essential medium) containing 10% fetal calf serum. Total incubation period 48 h. (2) Medium containing oxidized LDL (50 μg/ml), 48 h. (3) Medium containing oxidized LDL (50 μg/ml), 48 h. After the first 24 h, 10% mouse serum added. (4) Medium containing oxidized LDL (50 μg/ml), 48 h. After the first 24 h, 10% ZAMS added. (OxLDL [2] significantly reduces mobility compared with control medium [1], $P < 0.01$; in the presence of ZAMS, migration is restored to control values [4]).

ACKNOWLEDGMENTS

We acknowledge the many contributions of our colleagues, Drs. A. Hesz, V.W. Armstrong, A.K. Walli, G. Schmitz, H. Dieplinger, A. Roessner, and E. Vollmer. The cited experimental data from our laboratories include the efforts of our coworkers Drs. B. Harrach, R. Semich, M. Pataki, F. Robenek, and Dipl.-Biol. B. Kaesberg. For excellent technical assistance we thank S. Otter, K. Schlattmann, and M. Opalka.

We thank M. Rohe for outstanding secretarial support in manuscript preparation.

H. Robenek was supported by the Deutsche Forschungsgemeinschaft (SFB 223, SFB 310). N.J. Severs acknowledges support from the British Heart Foundation. Both authors are supported by a NATO Collaborative Research Grant (CRG.910122).

REFERENCES

1. **Goldstein, J.L. and Brown, M.S.,** The low-density lipoprotein pathway and its relation to atherosclerosis, *Ann. Rev. Biochem.*, 46, 897, 1977.
2. **Mahley, R.W. and Innerarity, T.L.,** Lipoprotein receptors and cholesterol homeostasis, *Biochim. Biophys. Acta*, 737, 197, 1983.
3. **Goldstein, J.L., Brown, M.S., Anderson, R.G.W., Russel, D.W., and Schneider, W.J.,** Receptor-mediated endocytosis: concepts merging from the LDL receptor system, *Ann. Rev. Cell Biol.*, 1, 1, 1985.
4. **Bradley, W.A. and Gianturco, S.H.,** Lipoprotein receptors in cholesterol metabolism, in *Biology of Cholesterol*, Yeagle, P.L., Ed., CRC Press, Boca Raton, FL, 1988, 95.
5. **Hobbs, H.H., Russel, D.W., Brown, M.S., and Goldstein, J.L.,** The LDL receptor locus in familial hypercholesterolemia: mutational analysis of a membrane protein, *Ann. Rev. Genet.*, 24, 133, 1990.
6. **Goldstein, J.L. and Brown, M.S.,** Atherosclerosis: the low-density lipoprotein receptor hypothesis, *Metabolism*, 26, 1257, 1977.
7. **Goldstein, J.L., Anderson, R.G.W., and Brown, M.S.,** Coated pits, coated vesicles and receptor-mediated endocytosis, *Nature*, 279, 679, 1979.
8. **Schneider, W.J., Beisiegel, U., Goldstein, J.L., and Brown, M.S.,** Purification of the low density lipoprotein receptor, an acidic glycoprotein of 164,000 molecular weight, *J. Biol. Chem.*, 257, 2664, 1982.
9. **Ross, R.,** The pathogenesis of atherosclerosis — an update, *N. Engl. J. Med.*, 314, 488, 1986.
10. **Ross, R., Masuda, J., Raines, E.W., Gown, A.M., Katsuda, S., Sasahara, M., Malden, L.T., Masuko, H., and Sato, H.,** Localization of PDGF-β protein in macrophages in all phases of atherogenesis, *Science*, 248, 1009, 1990.
11. **Brown, M.S. and Goldstein, J.L.,** Lipoprotein metabolism in the macrophage: implications for cholesterol deposition in atherosclerosis, *Ann. Rev. Biochem.*, 52, 223, 1983.
12. **Gerrity, R.G.,** The role of the monocyte in atherogenesis. I. Transition of blood-borne monocytes into foam cells in fatty lesions, *Am. J. Pathol.*, 103, 181, 1981.
13. **Joris, I., Zand, T., Nunnari, J.J., Krolikowski, F.J., and Majno, G.,** Studies on the pathogenesis of atherosclerosis. I. Adhesion and emigration of mononuclear cells in the aorta of hypercholesterolemic rats, *Am. J. Pathol.*, 113, 341, 1983.
14. **Klurfeld, D.M.,** Identification of foam cells in human atherosclerotic lesions as macrophages using monoclonal antibodies, *Arch. Pathol. Lab. Med.*, 109, 445, 1985.
15. **Stary, H.C.,** Macrophages, macrophage foam cells, and eccentric intimal thickening in the coronary arteries of young children, *Atherosclerosis*, 64, 91, 1987.
16. **Aquel, N.M., Bell, R.Y., Walmann, H., and Mitchinson, M.J.,** Identification of macrophages and smooth muscle cells in human atherosclerosis using monoclonal antibodies, *J. Pathol.*, 146, 197, 1985.
17. **Watanabe, T., Hirata, M., Yoshikawa, Y., Nagafuchi, Y., and Toyoshima, H.,** Role of macrophages in atherosclerosis, *Lab. Invest.*, 53, 80, 1985.
18. **Robenek, H., Schmitz, G., and Assmann, G.,** Topography and dynamics of receptors for acetylated and malondialdehyde-modified low-density lipoprotein in the plasma membrane of mouse peritoneal macrophages as visualized by colloidal gold in connjunction with surface replicas, *J. Histochem. Cytochem.*, 32, 1017, 1984.
19. **Schmitz, G., Robenek, H., Beuck, M., Krause, R., Schurek, A., and Niemann, R.,** Ca^{++} antagonists and ACAT inhibitors promote cholesterol efflux from macrophages by different mechanisms. I. Characterization of intracellular lipid metabolism, *Arteriosclerosis*, 8, 46, 1988.
20. **Robenek, H. and Schmitz, G.,** Ca^{++} antagonists and ACAT inhibitors promote cholesterol efflux from macrophages by different mechanisms. II. Characterization of intracellular morphological changes, *Arteriosclerosis*, 8, 57, 1988.

21. **Schmitz, G., Robenek, H., Lohmann, U., and Assmann, G.,** Interaction of high density lipoproteins with cholesteryl ester-laden macrophages: biochemical and morphological characterization of cell surface receptor binding, endocytosis and resecretion of high density lipoproteins by macrophages, *EMBO J.*, 4, 613, 1985.
22. **Robenek, H. and Schmitz, G.,** Receptor domains in the plasma membrane of cultured mouse peritoneal macrophages, *Eur. J. Cell Biol.*, 39, 77, 1985.
23. **Schmitz, G., Beuck, M., Fischer, H., and Robenek, H.,** Regulation of phospholipid biosynthesis during cholesterol influx and high density lipoprotein-mediated cholesterol efflux in macrophages, *J. Lipid Res.*, 31, 1741, 1990.
24. **Schmitz, G., Fischer, H., Beuck, M., Hoecker, K.-P., and Robenek, H.,** Dysregulation of phospholipid synthesis in Tangier monocyte-derived macrophages, *Arteriosclerosis*, 10, 1010, 1990.
25. **Schmitz, G., Assmann, G., Robenek, H., and Brennhausen, B.,** Tangier disease: a disorder of intracellular membrane traffic, *Proc. Natl. Acad. Sci. U.S.A.*, 82, 6305, 1985.
26. **Goldstein, J.L., Ho, Y.K., Basu, S.K., and Brown, M.S.,** Binding site on macrophages that mediates uptake and degradation of acetylated low density lipoprotein, producing massive cholesterol deposition, *Proc. Natl. Acad. Sci. U.S.A.*, 76, 333, 1979.
27. **Brown, M.S., Goldstein, J.L., Krieger, M., Ho, Y.K., and Anderson, R.G.W.,** Reversible accumulation of cholesteryl esters in macrophages incubated with acetylated lipoproteins, *J. Cell Biol.*, 82, 597, 1979.
28. **Parthasarathy, S.,** Oxidation of low-density lipoprotein by thiol compounds leads to its recognition by the acetyl LDL receptor, *Biochim. Biophys. Acta*, 917, 337, 1987.
29. **Kita, T., Ishii, K., Yokode, M., Kume, N., Nagano, Y., Arai, H., and Kawai, C.,** The role of oxidized low density lipoprotein in the pathogenesis of atherosclerosis, *Eur. Heart J.*, 11, 122, 1990.
30. **Parthasarathy, S., Printz, D.J., Boyd, D., Joy, L., and Steinberg, D.,** Macrophage oxidation of low density lipoprotein generates a modified form recognized by the scavenger receptor, *Arteriosclerosis*, 6, 505, 1986.
31. **Steinberg, D., Parthasarathy, S., Carew, T.E., Khoo, J.C., and Witztum, J.C.,** Beyond cholesterol: modifications of low-density lipoprotein that increase its atherogenicity, *N. Engl. J. Med.*, 320, 915, 1989.
32. **Sparrow, C.P., Parthasarathy, S., and Steinberg, D.,** A macrophage receptor that recognizes oxidized low density lipoprotein but not acetylated low density lipoprotein, *J. Biol. Chem.*, 264, 2599, 1989.
33. **Goldstein, J.L., Ho, Y.K., Brown, M.S., Innerarity, T.L., and Mahley, R.W.,** Cholesteryl-ester accumulation in macrophages resulting from receptor-mediated uptake and degradation of hypercholesterolemic canine β-very low density lipoprotein, *J. Biol. Chem.*, 258, 1839, 1980.
34. **Polacek, D., Byrne, R.E., and Scanu, A.M.,** Modification of low density lipoproteins by polymorphonuclear cell elastase leads to enhanced uptake by human monocyte-derived macrophages via the low density lipoprotein receptor pathway, *J. Lipid Res.*, 29, 797, 1989.
35. **Fogelman, A.M., Shechter, I., Seager, J., Hokom, M., Child, J.S., and Edwards, P.A.,** Malondialdehyde alteration of low density lipoproteins leads to cholesteryl ester accumulation in human monocyte-macrophages, *Proc. Natl. Acad. Sci. U.S.A.*, 77, 2214, 1980.
36. **Ellsworth, J.L., Fong, L.G., Kraemer, F.B., and Cooper, A.D.,** Differences in the processing of chylomicron remnants and β-VLDL by macrophages, *J. Lipid Res.*, 31, 1399, 1990.
37. **Henson, D.A., St. Clair, R.W., and Lewis, J.C.,** β-VLDL and acetylated-LDL binding to pigeon monocyte macrophages, *Atherosclerosis*, 78, 47, 1989.
38. **Jaakkola, O., Ylä-Herttuala, S., Särkioja, T., and Nikkari, T.,** Macrophage foam cells from human aortic fatty streaks take up β-VLDL and acetylated LDL in primary culture, *Atherosclerosis*, 79, 173, 1989.

39. **Innerarity, T.L., Arnold, K.S., Weisgraber, K.H., and Mahley, R.W.,** Apolipoprotein E is the determinant that mediates the receptor uptake of beta-very low density lipoproteins by mouse macrophages, *Arteriosclerosis*, 6, 114, 1986.
40. **Robenek, H., Schmitz, G., and Greven, H.,** Cell surface distribution and intracellular fate of human β-very low density lipoprotein in cultured peritoneal mouse macrophages: a cytochemical and immunocytochemical study, *Eur. J. Cell Biol.*, 43, 110, 1987.
41. **Glomset, J.A.,** The plasma lecithin:cholesterol acyltransferase reaction, *J. Lipid Res.*, 9, 155, 1968.
42. **Badimore, J.J., Badimore, L., Galvez, A., Dische, R., and Fuster, V.,** High density lipoprotein plasma fractions inhibit aortic fatty streaks in cholesterol-fed rabbits, *Lab. Invest.*, 60, 455, 1989.
43. **Ho, Y.K., Brown, M.S., and Goldstein, J.L.,** Hydrolysis and excretion of cytoplasmic cholesteryl esters by macrophages: stimulation by high density lipoproteins and other agents, *J. Lipid Res.*, 21, 391, 1980.
44. **Assmann, G.,** *Lipid Metabolism and Atherosclerosis*, Schattauer Verlag, Stuttgart, Germany, 1982.
45. **Rifai, N.,** Lipoproteins and apolipoproteins, *Arch. Pathol. Lab. Med.*, 110, 694, 1986.
46. **Palinski, W., Rosenfeld, M.E., Ylä-Herttuala, S., Gurtner, G.C., Socher, S.S., Butler, S.W., Parthasarathy, S., Carew, T.E., Steinberg, D., and Witztum, J.L.,** Low density lipoprotein undergoes oxidative modification *in vivo*, *Proc. Natl. Acad. Sci. U.S.A.*, 86, 1372, 1989.
47. **Hurt, E. and Camejo, G.,** Effect of arterial proteoglycans on the interaction of LDL with human monocyte-derived macrophages, *Atherosclerosis*, 67, 115, 1987.
48. **Vijayagopol, P., Srinivasan, S.R., Radhakrishnamurthy, B., and Berenson, G.S.,** Factors regulating the metabolism of low-density lipoprotein proteoglycan complex in macrophages, *Biochim. Biophys. Acta*, 1042, 204, 1990.
49. **Goldstein, J.L. and Brown, M.S.,** Binding and degradation of low density lipoproteins by cultured human fibroblasts. Comparison of cells from a normal subject and from a patient with homozygous familial hypercholesterolemia, *J. Biol. Chem.*, 249, 5153, 1974.
50. **Kodama, T., Freeman, M., Rohrer, L., Zabrecky, J., Matsudaira, P., and Krieger, M.,** Type I macrophage scavenger receptor contains α-helical and collagen-like coiled coils, *Nature*, 343, 531, 1990.
51. **Rohrer, L., Freeman, M., Kodama, T., Penman, M., and Krieger, M.,** Coiled-coil fibrous domains mediate ligand binding by macrophage scavenger receptor type II, *Nature*, 343, 570, 1990.
52. **Herz, J., Kowal, R.C., Goldstein, J.L., and Brown, M.S.,** Proteolytic processing of the 600 kd low density lipoprotein receptor-related protein (LRP) occurs in a trans-Golgi compartment, *EMBO J.*, 9, 1769, 1990.
53. **Herz, J., Hamann, U., Rogne, S., Myklebost, O., Gausepohl, H., and Stanley, K.K.,** Surface location and high affinity for calcium of a 500 kD liver membrane protein closely related to the LDL-receptor suggest a physiological role as lipoprotein receptor, *EMBO J.*, 7, 4119, 1988.
54. **Severs, N.J. and Robenek, H.,** Detection of microdomains in biomembranes: an appraisal of recent developments in freeze-fracture cytochemistry, *Biochim. Biophys. Acta*, 737, 373, 1983.
55. **Robenek, H., Rassat, J., Hesz, A., and Grünwald, J.,** A correlative study on the topographical distribution of the receptors for low density lipoprotein (LDL) conjugated to colloidal gold in cultured human skin fibroblasts employing thin section, freeze-fracture, deep-etching, and surface replication techniques, *Eur. J. Cell Biol.*, 27, 242, 1982.
56. **Robenek, H.,** Distribution and mobility of receptors in the plasma membrane, in *Freeze-Fracture Studies of Membranes*, Hui, S.W., Ed., CRC Press, Boca Raton, FL, 1989, 61.

57. **Robenek, H.,** Topography and internalization of cell surface receptors as analyzed by affinity — and immunolabeling combined with surface replication and ultrathin sectioning techniques, in *Electron Microscopy of Subcellular Dynamics*, Plattner, H., Ed., CRC Press, Boca Raton, FL, 1989, 141.
58. **Robenek, H., Harrach, B., and Severs, N.J.,** Display of low density lipoprotein receptors is clustered, not dispersed, in fibroblast and hepatocyte plasma membranes, *Arterioscler. Thromb.*, 11, 261, 1991.
59. **Nicol, A., Nermut, M.V., Doeinck, A., Robenek, H., Wiegand, C., and Jockusch, B.M.,** Visualization of cytoskeletal elements of the ventral plasma membrane of fibroblasts by gold immunolabeling, *J. Histochem. Cytochem.*, 35, 499, 1987.
60. **Semich, R., Gerke, V., Robenek, H., and Weber, K.,** The p36 substrate of pp60 src kinase is located at the cytoplasmic surface of the plasma membrane of fibroblasts; an immunoelectron microscopic analysis, *Eur. J. Cell Biol.*, 50, 313, 1989.
61. **Semich, R. and Robenek, H.,** Organization of the cytoskeleton and the focal contacts of bovine aortic endothelial cells cultured on type I and III collagen, *J. Histochem. Cytochem.*, 38, 59, 1990.
62. **Nermut, M.V.,** Strategy and tactics in electron microscopy of cell surfaces, *Electron Micros. Rev.*, 2, 171, 1989.
63. **Pinto da Silva, P. and Kan, F.W.K.,** Label-fracture: a method for high resolution labeling of cell surfaces, *J. Cell Biol.*, 99, 1156, 1984.
64. **Severs, N. J.,** Freeze-fracture cytochemistry: review of methods, *J. Electron Microsc.Tech.*, 13, 175, 1989.
65. **Brown, D.,** Polyvinyl coating — an improvement of the freeze-fracture technique, *J. Microsc.*, 121, 283, 1981.
66. **McLean, I.W. and Nakane, P.K.,** Periodate-lysine-paraformaldehyde fixative. A new fixative for immunoelectron microscopy, *J. Histochem. Cytochem.*, 22, 1077, 1974.
67. **Brown, D.,** Low-temperature embedding and cryosectioning in the immunocytochemical study of membrane recycling, in *Electron Microscopy of Subcellular Dynamics*, Plattner, H., Ed., CRC Press, Boca Raton, FL, 1989, 179.
68. **Roth, J.,** The protein A-gold (pAg) technique — a qualitative and quantitative approach for antigen localization on thin sections, in *Techniques in Immunocytochemistry*, Vol. 1, Bullock, G.P. and Petrusz, P., Eds., Academic Press, London, 1982, 107.
69. **Bendayan, M.,** Protein A-gold electron microscopic immunocytochemistry: methods, applications and limitations, *J. Electron Microsc. Tech.*, 1, 243, 1984.
70. **Tokuyasu, K.T.,** Immunocytochemistry on ultrathin frozen sections, *Histochem. J.*, 12, 381, 1980.
71. **Völker, W., Frick, B., and Robenek, H.,** A simple device for low temperature polymerization of Lowicryl K4M resin, *J. Microsc.*, 138, 91, 1985.
72. **Holund, B., Clausen, P.P., and Clemmensen, I.,** The influence of fixation and tissue preparation on the immunohistochemical demonstration of fibronectin in human tissues, *Histochemistry*, 72, 291, 1981.
73. **Meloan, S.N., Barton, B.P., Puchtler, H., Waldrop, E.S., and Hobbs, J.L.,** Effects of formaldehyde and methacarn fixation on prekeratin, *Ga. J. Sci.*, 42, 31, 1984.
74. **Harrach, B. and Robenek, H.,** Polyclonal antibodies against formaldehyde-modified apolipoprotein A-I. An approach to circumventing fixation-induced loss of antigenicity in immunocytochemistry, *Arteriosclerosis*, 10, 564, 1990.
75. **Anderson, R.G.W., Brown, M.S., Beisiegel, U., and Goldstein, J.L.,** Surface distribution and recycling of the low density lipoprotein receptor as visualized with antireceptor antibodies, *J. Cell Biol.*, 93, 523, 1982.
76. **Pathak, R.K., Merkle, R.K., Cummings, R.D., Goldstein, J.L., Brown, M.S., and Anderson, R.G.W.,** Immunocytochemical localization of mutant low density lipoprotein receptors that fail to reach the Golgi complex, *J. Cell Biol.*, 106, 1831, 1988.

77. **Pathak, R.K., Yokode, M., Hammer, R.E., Hofmann, S.L., Brown, M.S., Goldstein, J.L., and Anderson, R.G.W.,** Tissue-specific sorting of the human LDL receptor in polarized epithelia of transgenic mice, *J. Cell Biol.*, 111, 347, 1990.
78. **Anderson, R.G.W., Brown, M.S., and Goldstein, J.L.,** Role of coated endocytic vesicle in the uptake of receptor-bound low density lipoprotein in human fibroblasts, *Cell*, 10, 351, 1977.
79. **Steer, C.J., Bisher, M., Blumenthal, R., and Steven, A.C.,** Detection of membrane cholesterol by filipin in isolated rat liver coated vesicles is dependent upon removal of the clathrin coat, *J. Cell Biol.*, 99, 315, 1984.
80. **Wofsy, C., Echavarria-Heras, H., and Goldstein, B.,** Effect of preferential insertion of LDL receptors near coated pits, *Cell Biophys.*, 7, 197, 1985.
81. **Sanan, D.A., Van der Westhuyzen, D.R., Gevers, W., and Coetzee, G.A.,** The surface distribution of low density lipoprotein receptors on cultured fibroblasts and endothelial cells, *Histochemistry*, 86, 517, 1987.
82. **Sanan, D.A., Van der Westhuyzen, D.R., Gevers, W., and Coetzee, G.A.,** Early appearance of dispersed low density lipoprotein receptors on the fibroblast surface during recycling, *Eur. J. Cell Biol.*, 48, 327, 1989.
83. **Sanan, D.A. and Anderson, R.G.W.,** Simultaneous visualization of LDL receptor distribution and clathrin lattices on membranes torn from the upper surface of cultured cells, *J. Histochem. Cytochem.*, 39, 1017, 1991.
84. **Krempler, F., Kostner, G.M., Roscher, A., Haslauer, F., Bolzano, K., and Sandhofer, F.,** Studies on the role of specific cell surface receptors in the removal of lipoprotein (a) in man, *J. Clin. Invest.*, 71, 1431, 1983.
85. **Havekes, L., Vermeer, B.J., Brugman, T., and Emeis, J.,** Binding of Lp(a) to the low density lipoprotein receptor of human fibroblasts, *FEBS Lett.*, 132, 169, 1981.
86. **Floren, C.-H., Albers, J.J., and Biermann, E.L.,** Uptake of Lp(a) lipoprotein by cultured fibroblasts, *Biochem. Biophys. Res. Commun.*, 102, 636, 1981.
87. **Martmann-Moe, K. and Berg, K.,** Lp(a) lipoprotein enters cultured fibroblasts independently of the plasma membrane low density lipoprotein receptor, *Clin. Genet.*, 20, 352, 1981.
88. **Armstrong, V.W., Harrach, B., Robenek, H., Helmhold, M., Walli, A.K., and Seidel, D.,** Heterogeneity of human lipoprotein LP(a): cytochemical and biochemical studies on the interaction of two Lp(a) species with the LDL receptor, *J. Lipid Res.*, 31, 429, 1990.
89. **Robenek, H. and Schmitz, G.,** Abnormal processing of Golgi elements and lysosomes in Tangier disease, *Arterioscler. Thromb.*, 11, 1007, 1991.
90. **Zannis, V.J., Kurnit, D.M., and Breslow, J.L.,** Hepatic apo-A-I and intestinal apo-A-I are synthesized in precursor isoprotein forms by organ cultures of human fetal tissues, *J. Biol. Chem.*, 257, 536, 1982.
91. **Schmitz, G., Brüning, E., Williamson, E., and Nowicka, G.,** The role of HDL in reverse cholesterol transport and its disturbances in Tangier disease and HDL deficiency with xanthomas, *Eur. Heart J.*, 11 (Suppl. E), 197, 1990.
92. **Schmitz, G., Niemann, R., Brennhausen, B., Krause, R., and Assmann, G.,** Regulation of high density lipoprotein receptors in cultured macrophages: role of acyl CoA: cholesterol acyltransferase, *EMBO J.*, 4, 2773, 1985.
93. **Rifici, V.A. and Eder, H.A.,** A hepatocyte receptor for high density lipoproteins specific for apolipoprotein A-I, *J. Biol. Chem.*, 259, 13,814, 1984.
94. **Hwang, J. and Menon, K.M.J.,** Binding of apolipoprotein A-I and A-II after recombination with phospholipid vesicles to the high density lipoprotein receptor of luteinized rat ovary, *J. Biol. Chem.*, 260, 5660, 1985.
95. **Fidge, N.H. and Nestel, P.J.,** Identification of apolipoproteins involved in the interaction of human high density lipoproteins$_3$ with receptors on cultured cells, *J. Biol. Chem.*, 260, 3570, 1985.
96. **Fong, B.S., Salter, A.M., Jimenez, J., and Angel, A.,** The role of apolipoprotein A-I and apolipoprotein A-II in high-density lipoprotein binding to human adipocyte plasma membrane, *Biochim. Biophys. Acta*, 920, 105, 1987.

97. **Maciejko, J.J., Holmes, D.R., Kottke, B.A., Zinsmeister, A.R., Dihn, D.M., and Mao, S.J.T.,** Apolipoprotein A-I as a marker of angiographically assessed coronary artery disease, *N. Engl. J. Med.*, 309, 385, 1983.
98. **Munro, J.M. and Cotran, R.W.,** Biology of Disease. The pathogenesis of atherosclerosis: atherogenesis and inflammation, *Lab. Invest.*, 58, 249, 1988.
99. **Goldstein, J.L. and Brown, M.S.,** Familial hypercholesterolemia, in *The Metabolic Basis of Inherited Disease*, Stanbury, J.B., Wyngarden, J.B., and Fredrickson, D.S., Eds., McGraw-Hill, New York, 1983, 672.
100. **Thompson, G.R., Soutar, A.K., Spengel, F.A., Jodhav, A., Gavigan, S.J., and Myant, N.B.,** Defects of receptor-mediated low density lipoprotein catabolism in homozygous familial hypercholesterolemia and hypothyroidism *in vivo*, *Proc. Natl. Acad. Sci. U.S.A.*, 78, 2591, 1981.
101. **Brown, M.S. and Goldstein, J.L.,** How LDL receptors influence cholesterol and atherosclerosis, *Sci. Am.*, 251 (5), 58, 1984.
102. **Goldstein, J.L. and Brown, M.S.,** The LDL receptor locus and the genetics of familial hypercholesterolemia, *Ann. Rev. Genet.*, 13, 259, 1979.
103. **Wolfbauer, G., Glick, J.M., Minor, L.K., and Rothblat, G.H.,** Development of the smooth muscle foam cell: uptake of macrophage lipid inclusions, *Proc. Natl. Acad. Sci. U.S.A.*, 83, 7760, 1986.
104. **Pitas, R.E., Innerarity, T.L., and Mahley, R.W.,** Foam cells in explants of atherosclerotic rabbit aortas have receptors for beta-very low density lipoproteins and modified low density lipoproteins, *Arteriosclerosis*, 3, 2, 1983.
105. **Haberland, M.E., Fogelman, A.M., and Edwards, P.A.,** Specificity of receptor-mediated recognition of malondialdehyde-modified low density lipoproteins, *Proc. Natl. Acad. Sci. U.S.A.*, 79, 1712, 1982.
106. **Baker, D.P., Van Lenten, B.J., Fogelman, A.M., Edwards, P.A., Kenn, C., and Berliner, J.A.,** LDL, scavenger, and β-VLDL receptors on aortic endothelial cells, *Arteriosclerosis*, 4, 248, 1984.
107. **Koo, C., Wernette-Hammond, L., and Innerarity, T.L.,** Uptake of canine β-VLDL by mouse peritoneal macrophages is mediated by a LDL receptor, *J. Biol. Chem.*, 261, 11,194, 1986.
108. **Gianturco, S.H., Brown, S.A., Via, D.P., and Bradley, W.A.,** The beta-VLDL receptor pathway in murine P388d cells, *J. Lipid Res.*, 27, 412, 1986.
109. **Hui, D.Y., Innerarity, T.L., Milne, R.W., Marcel, Y.L., and Mahley, R.W.,** Binding of chylomicron remnants and β-very low density lipoproteins to hepatic and extrahepatic lipoprotein receptors, *J. Biol. Chem.*, 259, 15,060, 1984.
110. **Van Lenten, B.J., Fogelman, A.M., Hokom, M.M., Benson, L., Haberland, M.E., and Edwards, P.A.,** Regulation of the uptake and degradation of β-very low density lipoprotein in human monocyte macrophages, *J. Biol. Chem.*, 258, 5151, 1983.
111. **Adelman, S.J. and St. Clair, R.W.,** Metabolism of beta-VLDL by peritoneal macrophages from atherosclerosis-susceptible white carneau and resistant show racer pigeons: stimulation of cholesteryl ester accumulation by a beta-VLDL devoid of apo E, *Arteriosclerosis*, 5, 506a, 1985.
112. **Van Lenten, B.J., Fogelman, A.M., Jackson, R.L., Shapiro, S., Haberland, M.E., and Edwards, P.A.,** Receptor-mediated uptake of remnant lipoproteins by cholesterol-loaded human monocyte-macrophages, *J. Biol. Chem.*, 260, 8783, 1985.
113. **Nakai, T., Otto, P.S., Kennedy, D.L., and Whayne, F.F., Jr.,** Rat high density lipoprotein HDL_3 uptake and catabolism by isolated rat liver parenchymal cells, *J. Biol. Chem.*, 251, 4914, 1976.
114. **Soltys, P.A., Portman, O.W., and O'Malley, J.P.,** Binding properties of high density lipoprotein subfractions and low density lipoproteins to rabbit hepatocytes, *Biochim. Biophys. Acta*, 713, 300, 1982.

115. **Bachorik, P.S., Franklin, F.A., Virgil, D.G., and Kwiterovich, P.O.,** High-affinity uptake and degradation of apolipoprotein E free high-density lipoprotein and low-density lipoprotein in cultured porcine hepatocytes, *Biochemistry*, 21, 5675, 1982.
116. **Biesbrock, R., Oram, J.F., Albers, J.J., and Biermann, E.L.,** Specific high-affinity binding of high density lipoproteins to cultured human skin fibroblasts and arterial smooth muscle cells, *J. Clin. Invest.*, 71, 525, 1983.
117. **Eisenberg, S.,** High density lipoprotein metabolism, *J. Lipid Res.*, 25, 1017, 1984.
118. **Hoeg, J.M., Demosky, S.J.,Jr., Edge, S.B., Gregg, R.E., Osborne, J.C., Jr., and Brewer, H.B., Jr.,** Characterization of a human hepatic receptor for high density lipoproteins, *Arteriosclerosis*, 5, 228, 1985.
119. **Schmitz, G., Brennhausen, B., and Robenek, H.,** Regulation of macrophage cholesterol homeostasis, in *Recent Developments in Lipid and Lipoprotein Research. Cholesterol Transport Systems and Their Relation to Atherosclerosis*, Steinmetz, A., Kaffarnik, H., and Schneider, J., Eds., Springer-Verlag, Berlin, 1989, 22.
120. **Schmitz, G. and Robenek, H.,** Significance of the interaction between lipoprotein subfractions and macrophages for reverse cholesterol transport, in *Recent Developments in Lipid and Lipoprotein Research. Lipoprotein Subfractions Omega-3 Fatty Acids*, Klör, H.U., Ed., Springer-Verlag, Berlin, 1989, 82.
121. **Niendorf, A., Rath, M., Wolf, K., Peters, H., Arps, U., Beisiegel, U., and Dietel, M.,** Morphological detection and quantification of lipoprotein (a) deposition in atheromatous lesions of human aorta and coronary arteries, *Virchows Arch.* [A], 417, 105, 1990.
122. **Vollmer, E., Brust, J., Roessner, A., Bosse, A., Burwinkel, F., Kaesberg, B., Harrach, B., Robenek, H., and Böcker, W.,** Distribution patterns of apolipoprotein-A1, apolipoprotein-A2, and apolipoprotein-B in the wall of atherosclerotic vessels, *Virchows Arch.* [A], 419, 79, 1991.
123. **Van Lenten, B.J. and Fogelman, A.M.,** Processing of lipoproteins in human monocyte-macrophages, *J. Lipid Res.*, 31, 1455, 1990.
124. **Van der Westhuyzen, D.R., Gevers, W., and Coetzee, G.A.,** Cathepsin-D-dependent initiation of the hydrolysis by lysosomal enzymes of apoprotein B from low-density lipoproteins, *Eur. J. Biochem.*, 112, 153, 1980.
125. **Basu, S.K., Brown, M.S., Ho, Y.K., Havel, J.R., and Goldstein, J.L.,** Mouse macrophages synthesize and secrete a protein resembling apo E, *Proc. Natl. Acad. Sci. U.S.A.*, 78, 7545, 1981.
126. **Schmitz, G., Robenek, H., and Assmann, G.,** Role of the high density lipoprotein-receptor cycle in macrophage-cholesterol metabolism, in *Atherosclerosis Reviews*, Gotto, A., Ed., Raven Press, New York, 1987, 95.
127. **Murakami, M., Horiuchi, S., Takata, K., and Morino, Y.,** Distinction in the mode of receptor-mediated endocytosis between high density lipoprotein and acetylated high density lipoprotein: evidence for high density lipoprotein receptor-mediated cholesterol transfer, *J. Biochem.*, 101, 729, 1987.
128. **Assmann, G. and Funke, H.,** HDL metabolism and atherosclerosis, *J. Cardiovasc. Pharmacol.*, 16, 15, 1990.
129. **Nowicka, G., Brüning, T., Böttcher, A., Kahl, G., and Schmitz, G.,** The macrophage interaction of HDL subclasses separated by free flow isotachophoresis, *J. Lipid Res.*, 31, 1947, 1990.
130. **Tucker, C.F., Catsulis, C., Strong, J.P., and Eggen, D.A.,** Regression of early cholesterol-induced aortic lesions in rhesus monkeys, *Am. J. Pathol.*, 65, 493, 1971.
131. **Stary, H.C., Strong, J.P., and Eggen, D.A.,** Differences in the degradation rate of intracellular lipid droplets in the intimal smooth muscle cells and macrophages of regressing atherosclerotic lesions of primates, in *Atherosclerosis V*, Gotto, A.M., Smith, L.C., and Allen, B., Eds., Springer-Verlag, Berlin, 1980, 753.

132. **Fritz, K.E., Augustyn, J.M., Jarmolych, J., and Daoud, A.S.,** Sequential study of biochemical changes during regression of swine aortic atherosclerotic lesions, *Arch. Pathol. Lab. Med.*, 105, 240, 1981.
133. **Daoud, A.S., Jarmolych, J., Augustyn, J.M., and Fritz, K.E.,** Sequential morphologic studies of regression of advanced atherosclerosis, *Arch. Pathol. Lab. Med.*, 105, 233, 1981.
134. **Small, D.M.,** Progression and regression of atherosclerotic lesions. Insights from lipid physical biochemistry, *Arteriosclerosis*, 8, 103, 1988.
135. **Dudrick, S.J.,** Regression of atherosclerosis by the intravenous infusion of specific biochemical nutrient substrates in animals and humans, *Ann. Surg.*, 206, 296, 1987.
136. **Wissler, R.W. and Vesselinovitch, D.,** Can atherosclerotic plaques regress? Anatomic and biochemical evidence from nonhuman animal models, *Am. J. Cardiol.*, 65, 33, 1990.
137. **Daoud, A.S., Fritz, K.E., Jarmolych, J., and Frank, A.S.,** Role of macrophages in regression of atherosclerosis, in *Atherosclerosis*, Vol. 454, Lee, K.T., Ed., Annals of the New York Academy of Sciences, 1985, 101.
138. **Gerrity, R.G.,** The role of the monocyte in atherogenesis. II. Migration of foam cells from atherosclerotic lesions, *Am. J. Pathol.*, 103, 191, 1981.
139. **Gerrity, R.G. and Naito, H.K.,** Lipid clearance from fatty streak lesions by foam cell migration, *Artery*, 8, 215, 1980.

INDEX

B

C

D

E

F

G

H

M

N

O

P

Q

R

S

T